Contents

Preface

In June of 2014, New York State adopted a more difficult Algebra I course based on the Common Core curriculum and changed the Regents exam to reflect this new course. On the new Algebra I Regents exam, students are expected not only to solve algebra problems and answer algebra-related questions but also to *explain their reasoning*. This is a major shift in the testing of algebra concepts, so a thorough review is even more critical for success.

This book will introduce you to the concepts covered throughout the new Algebra I Regents course as well as to the different types of problems you will encounter on the Regents exam at the end of the year.

WHO SHOULD USE THIS BOOK?

Students and teachers alike can use this book as a resource for preparing for the Algebra Common Core Regents exam.

Students will find this book to be a great study tool because it contains a review of Algebra I concepts, useful examples, and practice problems of varying difficulty that can be practiced throughout the school year to reinforce what they are learning in class. The most ideal way to prepare for the Algebra Common Core Regents exam is to work through the practice problems in the review sections and then the recently administered Regents exams at the end of the book.

Teachers can use this book as a tool to help structure an Algebra I course that will culminate with the Regents exam. The topics in the book are arranged by priority, so the sections in the beginning of the book are the ones from which more of the questions on the test are drawn. There are 13 sections, dedicated to all topics for Algebra I, each with practice exercises and solutions.

WHY IS THIS BOOK A HELPFUL RESOURCE?

Becoming familiar with the specific types of questions on the Algebra I Regents exam is crucial to performing well on this test. There are questions in which the math may be fairly easy but the way in which the question is asked makes the question seem much more difficult. For example, the question "Find all zeros of the function $f(x) = 2x + 6$" is a fancy way of asking the much simpler sounding "Solve for x if $2x + 6 = 0$." *Knowing exactly what the questions are asking is a big part of success on this test.*

Algebra has been around for thousands of years, and its basic concepts have never changed. So fundamentally, this new algebra curriculum is not very different from the algebra taught in schools two years ago, ten years ago, or twenty years ago. But the exam that follows this new Common Core-based course, with more complicated ways of asking questions and presenting problems, requires a specifically presented study plan that's more important than ever.

Gary Rubinstein

How to Use This Book

This book is designed to help you get the most out of your review for the new Regents exam in Algebra I (Common Core). Use this book to improve your understanding of the Algebra I topics and improve your grade.

TEST-TAKING TIPS

The first section in this book contains test-taking tips and strategies to help prepare you for the Algebra I Regents exam. This information is valuable, so be sure to read it carefully and refer to it during your study time. Remember: no single problem-solving strategy works for all problems—you should have a toolbox of strategies to pick from as you're facing unfamiliar or difficult problems on the test.

PRACTICE WITH KEY ALGEBRA I FACTS AND SKILLS

The second section in this book provides you with key Algebra I facts, useful skills, and practice problems with solutions. It provides you with a quick and easy way to refresh the skills you learned in class.

REGENTS EXAMS AND ANSWERS

The final section of the book contains actual Algebra I Regents exams that were administered in June 2014, August 2014, and June 2015. These exams and thorough answer explanations are probably the most useful tool for your review, as they let you know what's most important. By answering the questions on these exams, you will be able to identify your strengths and weaknesses and then concentrate on the areas in which you may need more study.

Remember, the answer explanations in this book are more than just simple solutions to the problems—they contain facts and explanations that are crucial to success in the Algebra I course and on the Regents exam. Careful review of these answers will increase your chances of doing well.

SELF-ANALYSIS CHARTS

Each of the Algebra I Regents exams ends with a Self-Analysis Chart. This chart will further help you identify weaknesses and direct your study efforts where needed. In addition, the chart classifies the questions on each exam into an organized set of topic groups. This feature will also help you to locate other questions on the same topic in the other Algebra I exam.

IMPORTANT TERMS TO KNOW

The terms that are listed in the glossary are the ones that have appeared most frequently on past Integrated Algebra and the most recently Algebra I (Common Core) exams. All terms and their definitions are conveniently organized for a quick reference.

Test-Taking Tips and Strategies

Knowing the material is only part of the battle in acing the new Algebra I Regents exam. Things like improper management of time, careless errors, and struggling with the calculator can cost valuable points. This section contains some test-taking strategies to help you perform your best on test day.

TIP 1

Time Management

SUGGESTIONS

- *Don't rush.* The Algebra I Regents exam is three hours long. While you are officially allowed to leave after 90 minutes, you really should stay until the end of the exam. Just as it wouldn't be wise to come to the test an hour late, it is almost as bad to leave a test an hour early.
- *Do the test twice.* The best way to protect against careless errors is to do the entire test twice and compare the answers you got the first time to the answers you got the second time. For any answers that don't agree, do a "tie breaker" third attempt. Redoing the test and comparing answers is much more effec-

tive than simply looking over your work. Students tend to miss careless errors when looking over their work. By redoing the questions, you are less likely to make the same mistake.

- *Bring a watch.* What will happen if the clock is broken? Without knowing how much time is left, you might rush and make careless errors. Yes, the proctor will probably write the time elapsed on the board and update it every so often, but its better safe than sorry.

The TI–84 graphing calculator has a built in clock. Press the [MODE] to see it. If the time is not right, go to SET CLOCK and set it correctly. The TI–Nspire does not have a built-in clock.

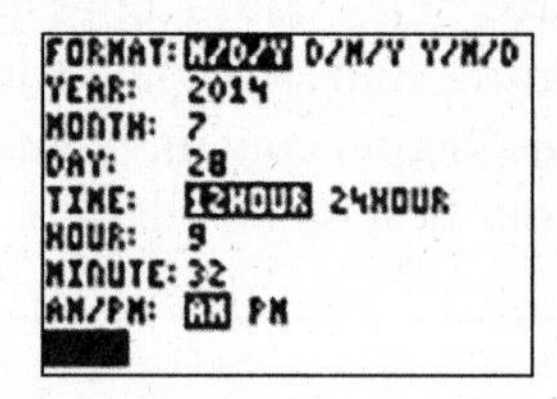

TIP 2
Know How to Get Partial Credit

SUGGESTIONS

- *Know the structure of the exam.* The Algebra Regents exam has 37 questions. The first 24 of those questions are multiple-choice worth two points each. There is no partial credit if you make a mistake on one of those questions. Even the smallest careless error, like missing a negative sign, will result in no credit for that question. Parts Two, Three, and Four are free-response questions with no multiple-choice. Besides giving a numerical answer, you may be asked to explain your reasoning. Part Two has eight free-response questions worth two points each. The smallest careless error will cause you to lose one point, which is half the value of the question. Part Three has four free-response questions worth four points each. These questions generally have multiple parts. Part Four has one free-response question worth six points. This question will have multiple parts.

- *Explain your reasoning.* When a free-response question asks to "Justify your answer," "Explain your answer," or "Explain how you determined your answer," the grader is expecting a few clearly written sentences. For these, you don't want to write too little since the grader needs to see that you understand why you did the different steps you did to solve the equation. You also don't want to write too much because if anything you write is not accurate, points can be deducted.

Here is an example followed by two solutions. The first would not get full credit, but the second would.

Example 1

Use algebra to solve for x in the equation $\frac{2}{3}x + 1 = 11$. Justify your steps.

Solution 1 (part credit):

$\frac{2}{3}x+1=11$ $-1=-1$ $\frac{2}{3}x=10$ $x=15$	I used algebra to get the x by itself. The answer was $x = 15$.

Solution 2 (full credit):

$\frac{2}{3}x+1=11$ $-1=-1$ $\frac{2}{3}x=10$ $\frac{3}{2}\cdot\frac{2}{3}x=\frac{3}{2}\cdot 10$ $1x=15$ $x=15$	I used the subtraction property of equality to eliminate the +1 from the left-hand side. Then to make it so the x had a 1 in front of it, I used the multiplication property of equality and multiplied both sides of the equation by the reciprocal of $\frac{2}{3}$, which is $\frac{3}{2}$. Then since $1 \cdot x = x$, the left-hand side of the equation just became x and the right-hand side became 15.

- Computational errors vs. conceptual errors

  In the Part Three and Part Four questions, the graders are instructed to take off one point for a "computational error" but half credit for a "conceptual error." This is the difference between these two types of errors.

  If a four point question was $x - 1 = 2$ and a student did it like this,

$$
\begin{aligned}
x - 1 &= 2 \\
+1 &= +1 \\
x &= 4
\end{aligned}
$$

  the student would lose one point out of 4 because there was one computational error since $2 + 1 = 3$ and not 4.

  Had the student done it like this,

$$
\begin{aligned}
x - 1 &= 2 \\
-1 &= -1 \\
x &= 1
\end{aligned}
$$

  the student would lose half credit, or 2 points, since this error was conceptual. The student thought that to eliminate the –1, he should subtract 1 from both sides of the equation.

  Either error might just be careless, but the conceptual error is the one that gets the harsher deduction.

TIP 3

Know Your Calculator

SUGGESTIONS

- Which calculator should you use? The two calculators used for this book are the TI-84 and the TI-Nspire. Both are very powerful. The TI-84 is somewhat easier to use for the functions needed for this test. The TI-Nspire has more features for courses in the future. The choice is up to you. This author prefers the TI-84 for the Algebra Regents. Graphing calculators come with manuals that are as thick as the book you are holding. There are also plenty of video tutorials online for learning how to use advanced features of the calculator. To become an expert user, watch the online tutorials or read the manual.
- Clearing the memory. You may be asked at the beginning of the test to clear the memory of your calculator. When practicing for the test, you should clear the memory too so you are practicing under test-taking conditions.

  This is how you clear the memory.

For the TI-84:

Press [2ND] and then [+] to get to the MEMORY menu. Then press [7] for Reset.

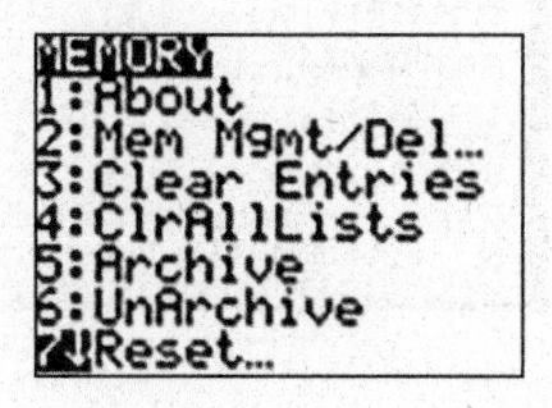

Use the arrows to go to [ALL] for All Memory. Then press [1].

Press [2] for Reset.

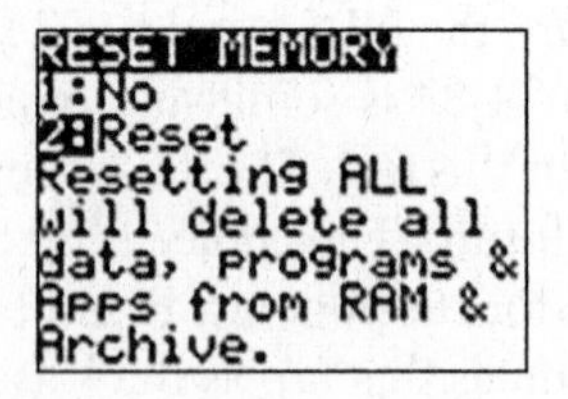

The calculator will be reset as if in brand new condition. The one setting that you may need to change is to turn the diagnostics on if you need to calculate the correlation coefficient.

For the TI-Nspire:

The TI-Nspire must be set to Press-To-Test mode when taking the Algebra Regents. Turn the calculator off by pressing [ctrl] and [home]. Press and hold [esc] and then press [home].

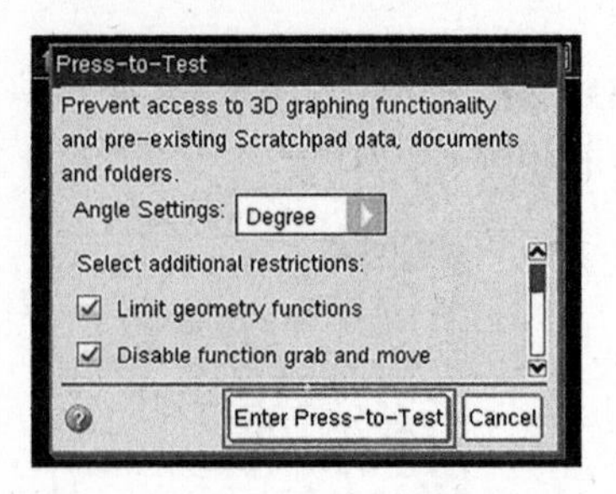

While in Press-to-Test mode, certain features will be deactivated. A small green light will blink on the calculator so a proctor can verify the calculator is in Press-to-Test mode.

To exit Press-to-Test mode, use a USB cable to connect the calculator to another TI-Nspire. Then from the home screen on the calculator in Press-to-Test mode, press [doc], [9] and select Exit Press-to-Test.

- Use parentheses
  The calculator always uses the order of operations where multiplication and division happen before addition and subtraction. Sometimes, though, you may want the calculator to do the operations in a different order.

  Suppose at the end of a quadratic equation, you have to round $x = \frac{-1+\sqrt{5}}{2}$ to the nearest hundredth. If you enter (–) (1) (+) (2ND) (x^2) (5) (/) (2), it displays

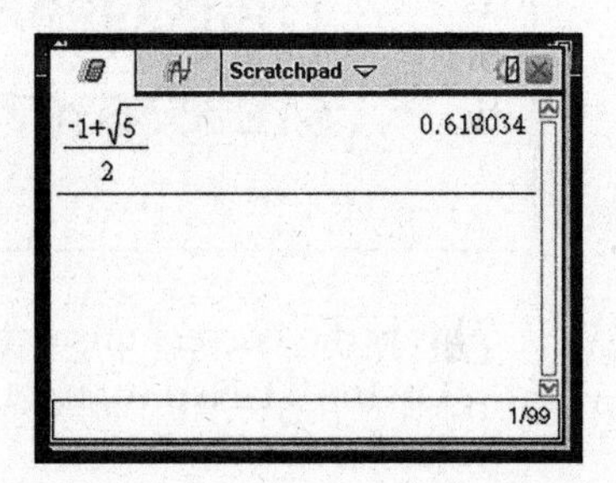

which is not the correct answer.

One reason is that for the TI-84 there needs to be a closing parentheses (or on the TI-Nspire, press [right arrow] to move out from under the radical sign) after the 5 in the square root symbol. Without it, it calculated $-1+\sqrt{\frac{5}{2}}$. More needs to be done, though, since

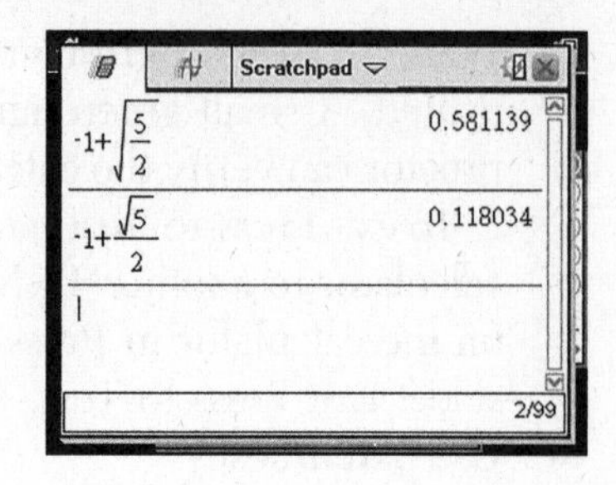

still is not correct. This is the solution to $-1+\sqrt{\dfrac{5}{2}}$.

To get this correct, there also needs to be parentheses around the entire numerator, $-1+\sqrt{5}$.

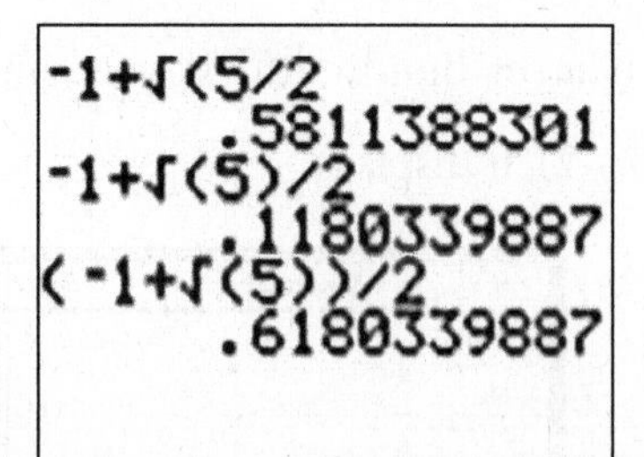

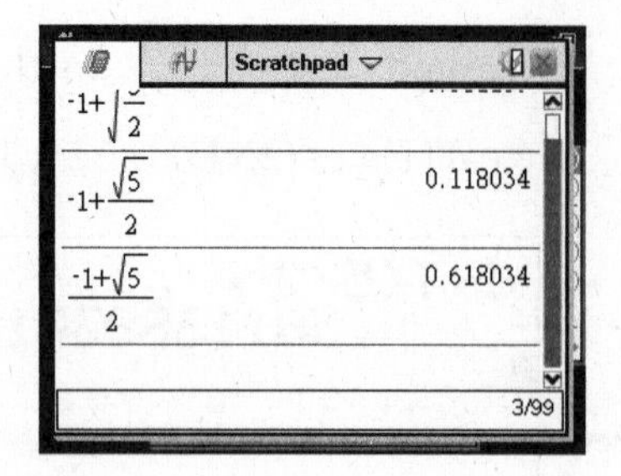

This is the correct answer.

On the TI-Nspire, fractions this can also be done with [templates].

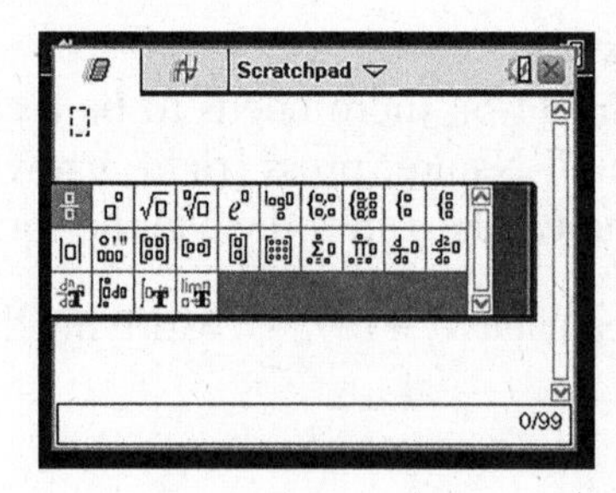

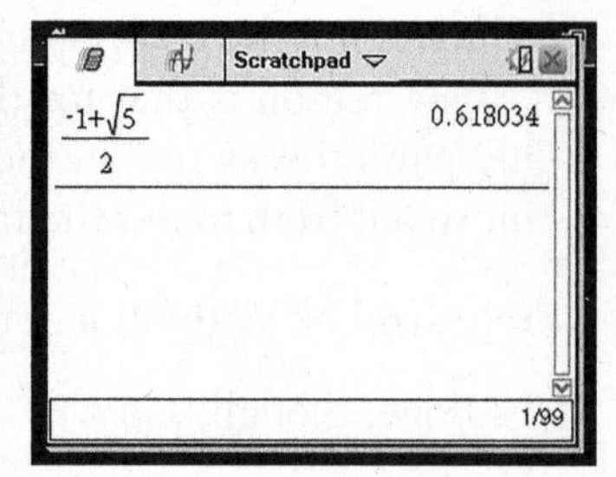

- Using the ANS feature
  The last number calculated with the calculator is stored in something called the ANS variable. This ANS variable will appear if you start an expression with a +, –, ×, or ÷. When an answer has a lot of digits in it, this saves time and is also more accurate.

  If for some step in a problem you need to calculate the decimal equivalent of $\frac{1}{7}$, it will look like this on the TI-84:

For the TI-Nspire, if you try the same thing, it leaves the answer as $\frac{1}{7}$. To get the decimal approximation, press [ctrl] and [enter] instead of just [enter].

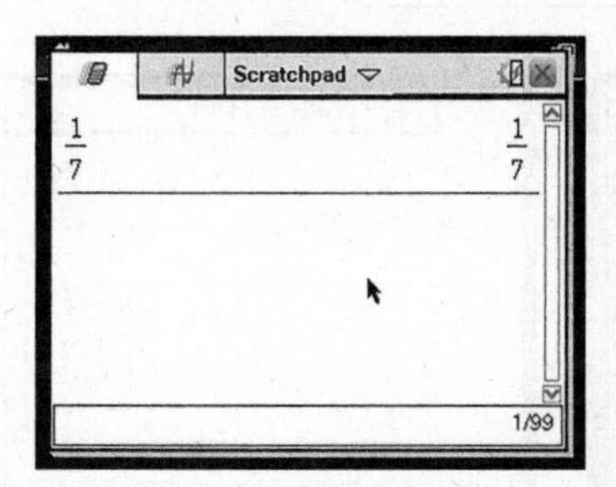

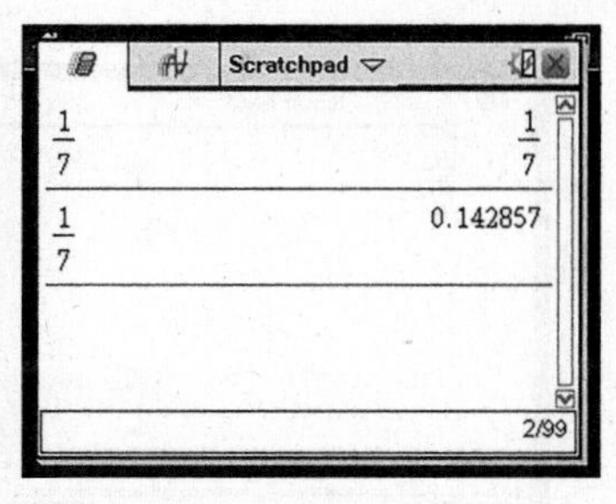

Now if you want to multiply this by 3, just press [×], and the calculator will display "Ans*"; press [3] and [enter].

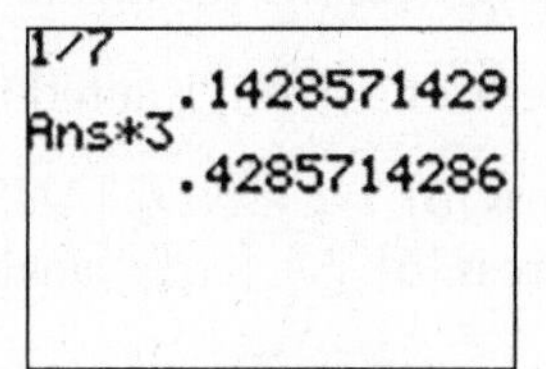

Scratchpad
$\frac{1}{7}$ $\frac{1}{7}$
$\frac{1}{7}$ 0.142857
Ans·3
2/99

Scratchpad
$\frac{1}{7}$ $\frac{1}{7}$
$\frac{1}{7}$ 0.142857
0.14285714285714·3 0.428571
3/99

The ANS variable can also help you do calculations in stages. To calculate $x = \frac{-1+\sqrt{5}}{2}$ without using so many parentheses as before, it can be done by first calculating $-1 + \sqrt{5}$ and then pressing [÷] and [2] and Ans will appear automatically.

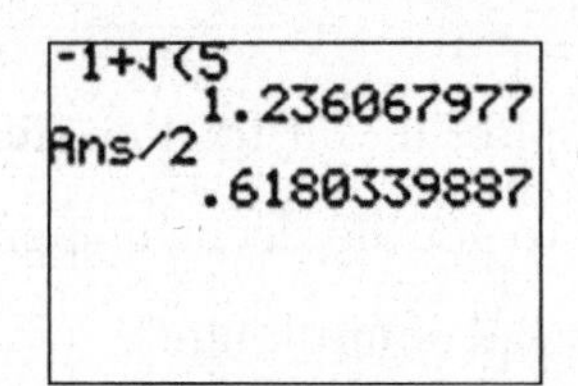

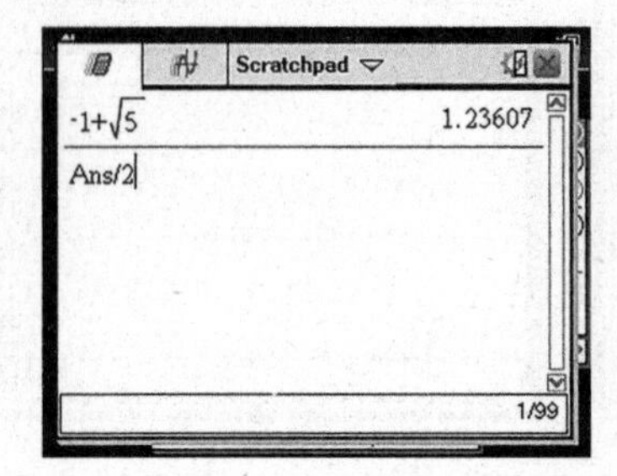

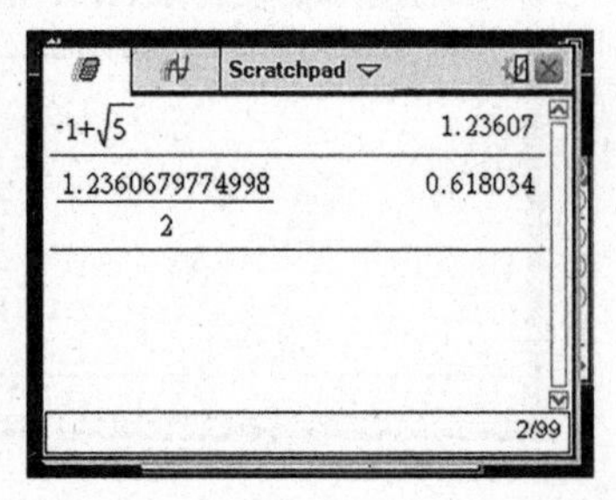

The ANS variable can also be accessed by pressing [2ND] and [–] at the bottom right of the calculator. If after calculating the decimal equivalent of 1/7 you wanted to subtract $\frac{1}{7}$ from 5, for the TI-84 press [5], [–], [2ND], [ANS], and [ENTER]. For the TI-Nspire press [5], [–], [ctrl], [ans], and [enter].

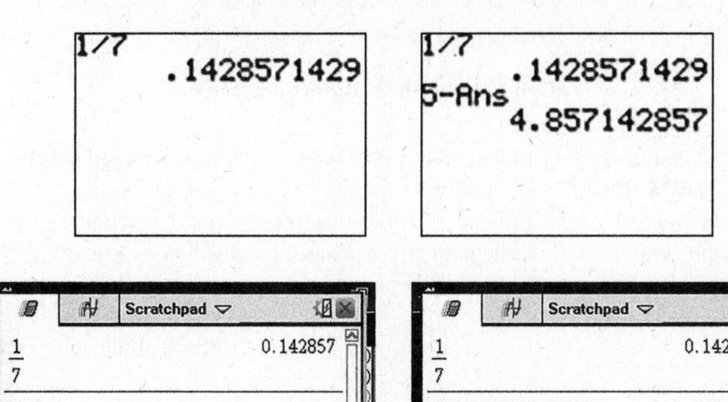

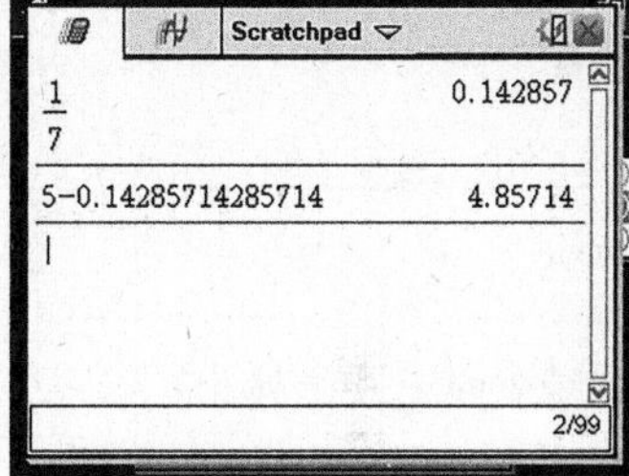

TIP 4

Use the Reference Sheet

SUGGESTIONS

- In the back of the Algebra Regents booklet is a reference sheet that contains 17 conversion facts, such as inches to centimeters and quarts to pints, and also 17 formulas. Many of these conversion facts and formulas will not be needed for an individual test, but the quadratic formula and the arithmetic sequence formula are the two that will come in the handiest.

High School Math Reference Sheet

1 inch = 2.54 centimeters
1 meter = 39.37 inches
1 mile = 5280 feet
1 mile = 1760 yards
1 mile = 1.609 kilometers

1 kilometer = 0.62 mile
1 pound = 16 ounces
1 pound = 0.454 kilogram
1 kilogram = 2.2 pounds
1 ton = 2000 pounds

1 cup = 8 fluid ounces
1 pint = 2 cups
1 quart = 2 pints
1 gallon = 4 quarts
1 gallon = 3.785 liters
1 liter = 0.264 gallon
1 liter = 1000 cubic centimeters

Triangle	$A = \frac{1}{2}bh$
Parallelogram	$A = bh$
Circle	$A = \pi r^2$
Circle	$C = \pi d$ or $C = 2\pi r$
General Prisms	$V = Bh$
Cylinder	$V = \pi r^2 h$
Sphere	$V = \frac{4}{3}\pi r^3$
Cone	$V = \frac{1}{3}\pi r^2 h$
Pyramid	$V = \frac{1}{3}Bh$

Pythagorean Theorem	$a^2 + b^2 = c^2$
Quadratic Formula	$x = \frac{-b \pm \sqrt{b^2 - 4ac}}{2a}$
Arithmetic Sequence	$a_n = a_1 + (n - 1)d$
Geometric Sequence	$a_n = a_1 r^{n-1}$
Geometric Series	$S_n = \frac{a_1 - a_1 r^n}{1 - r}$ where $r \neq 1$
Radians	1 radian = $\frac{180}{\pi}$ degrees
Degrees	1 degree = $\frac{\pi}{180}$ radians
Exponential Growth/Decay	$A = A_0 e^{k(t - t_0)} + B_0$

- How Many Points Do You Need to Pass?
  The Algebra Regents exam is scored out of a possible 86 points. Unlike most tests given in the year by your teacher, the score is not then turned into a percent out of 86. Instead each test has a conversion sheet that varies from year to year. For the June 2014 test, the conversion sheet looked like this.

Raw Score	Scale Score	Raw Score	Scale Score	Raw Score	Scale Score
86	100	57	75	28	64
85	99	56	74	27	63
84	97	55	74	26	62
83	96	54	74	25	61
82	95	53	73	24	60
81	94	52	73	23	59
80	92	51	73	22	58
79	91	50	72	21	56
78	90	49	72	20	55
77	89	48	72	19	54
76	88	47	72	18	52
75	87	46	71	17	50
74	86	45	71	16	49
73	85	44	71	15	47
72	84	43	70	14	45
71	83	42	70	13	42
70	82	41	70	12	40
69	82	40	70	11	38
68	81	39	69	10	35
67	80	38	69	9	32
66	79	37	69	8	30
65	79	36	68	7	26
64	78	35	68	6	23
63	78	34	67	5	20
62	77	33	67	4	16
61	77	32	66	3	12
60	76	31	66	2	9
59	76	30	65	1	4
58	75	29	64	0	0

On this test, 30 points became a 65, 57 points became a 75, and 73 points became an 85. This means that for this examination a student who got 30 out of 86, which is just 35% of the possible points, would get a 65 on this exam. 57 out of 86 is 66%, but this scaled to a 75. 73 out of 86, however, is actually 85% and became an 85. So in the past there has been a curve on the exam for lower scores, though the scaling is not released until after the exam.

Some Key Algebra I Facts and Skills

1. PROPERTIES OF ALGEBRA AND SOLVING LINEAR EQUATIONS WITH ALGEBRA

1.1 ONE-STEP ALGEBRA EQUATIONS

An algebra problem, like $x - 5 = 2$, is one that can be solved by changing both sides of the equation until the variable x is isolated. There are four main properties that can be used in solving algebra problems.

- The **addition property** of equality
  The equation $x - 5 = 2$ is solved by adding 5 to both sides of the equation. When you add to both sides of an equation, you are using the addition property of equality.

$x - 5 = 2$	the given equation
$+5 = +5$	addition property of equality
$x = 7$	x is isolated. The solution is 7.

- The **subtraction property** of equality
  The equation $x + 2 = 7$ is solved by subtracting 2 from both sides of the equation. When you subtract from both sides of an equation, you are using the subtraction property of equality.

$x + 2 = 7$	the given equation
$-2 = -2$	subtraction property of equality
$x = 5$	x is isolated. The solution is 5.

- The **division** property of equality
  The equation $2x = 10$ is solved by dividing both sides of the equation by 2. When you divide both sides of the equation, you are using the division property of equality.

$\frac{2x}{2} = \frac{10}{2}$ the given equation

division property of equality

$x = 5$ x is isolated. The solution is 5.

- The **multiplication** property of equality
  The equations $\frac{x}{5} = 3$ and $\frac{2}{3}x = 8$ can be solved by multiplying both sides of the equation by the same number.

$\frac{x}{5} = 3$ the given equation

$5 \cdot \frac{x}{5} = 5 \cdot 3$ multiplication property of equality

$x = 15$ x is isolated. The solution is 15.

$\frac{2}{3}x = 8$ the given equation

$\frac{3}{2} \cdot \frac{2}{3}x = \frac{3}{2} \cdot 8$ multiplication property of equality

$x = 12$ x is isolated. The solution is 12.

1.2 TWO-STEP ALGEBRA PROBLEMS

When an equation has the form $mx + b = y$, it takes two steps to solve for x. The first step is to eliminate the b, which is called the **constant**. The second step is to eliminate the m, which is called the **coefficient**. The b is eliminated with either the addition or the subtraction property of equality. The m is eliminated with either the division or the multiplication property of equality.

- For the equation $3x - 7 = 11$

$3x - 7 = 11$	the given equation
$+7 = +7$	addition property of equality
$\frac{3x}{3} = \frac{18}{3}$	division property of equality
$x = 6$	x is isolated. The solution is 6.
$\frac{2}{3}x + 5 = 11$	the given equation
$-5 = -5$	subtraction property of equality
$\frac{2}{3}x = 6$	
$\frac{3}{2} \cdot \frac{2}{3}x = \frac{3}{2} \cdot 6$	multiplication property of equality
$x = 9$	x is isolated. The solution is 9.

1.3 COMBINING LIKE TERMS BEFORE SOLVING

If the equation has multiple x terms or multiple constants, the equation should first be simplified by combining like terms. After all like terms have been combined, the question will usually be a two-step algebra problem and can be solved with the methods from Section 1.2.

$2x + 5 + 3x - 2 = 23$	the given equation
$2x + 3x + 5 - 2 = 23$	terms are rearranged so the x terms are together and the constants are together. This step is optional.
$5x + 3 = 23$	Like terms have been combined.
$-3 = -3$	subtraction property of equality
$\frac{5x}{5} = \frac{20}{5}$	division property of equality
$x = 4$	x is isolated. The solution is 4.

1.4 VARIABLES ON BOTH SIDES OF THE EQUATION

When there are x terms on both sides of the equation, the addition property of equality or the subtraction property of equality can be used to change the equation into one where the x terms are all on the same side of the equation.

$5x - 3 = 12 + 2x$	the given equation
$-2x = -2x$	subtraction property of equality eliminates the x term from the right-hand side of the equation
$3x - 3 = 12$	
$+ 3 = + 3$	addition property of equality
$\frac{3x}{3} = \frac{15}{3}$	division property of equality
$x = 5$	x is isolated. The solution is 5.

1.5 EQUATIONS WITH MORE THAN ONE VARIABLE

An equation with more than one variable can be solved the same way as an equation with just one variable. The **solution** will not be a number in these problems, but an **expression** with variables and numbers in it.

$ax + 2 = c$	the given equation
$-2 = -2$	subtraction property of equality
$ax = c - 2$	the c and the 2 cannot be combined since they are unlike terms
$\frac{ax}{a} = \frac{c-2}{a}$	division property of equality
$x = \frac{c-2}{a}$	x is isolated. The solution for x is not a number but an expression in terms of c and a. The answer is $\frac{c-2}{a}$.

Practice Exercises

1. Antonio started the question $2x + 1 = 11$ by writing $2x = 10$. Which property justifies this step?
(1) Commutative property of addition
(2) Distributive property of multiplication over addition
(3) Addition property of equality
(4) Subtraction property of equality

2. Mila used the multiplication property to justify the first step in solving an equation. The original equation was $\frac{x}{2} + 4 = 10$. What could the equation have been transformed into after this step?

(1) $\frac{x}{2} = 6$
(2) $\frac{x}{2} + 6 + 12$
(3) $x + 8 = 20$
(4) $x + 2 = 5$

3. What is the solution set for the equation $x - 6 = 7$?
(1) {1}
(2) {6}
(3) {11}
(4) {13}

4. What value of x makes the equation $3x + 7 = 22$ true?
(1) 1
(2) 3
(3) 5
(4) 7

5. Find the solution set for the equation $5(x + 4) = 35$.

(1) {1} (3) {3}

(2) {2} (4) {4}

6. Solve for d in terms of c, e, and f.

$cd - e = f$.

(1) $\frac{f-e}{c}$ (3) $\frac{f}{c}+e$

(2) $\frac{f+e}{c}$ (4) $\frac{f}{c}-e$

7. Solve for m in terms of a, b, and c.

$b - ma = c$.

(1) $\frac{b-c}{-a}$ (3) $\frac{c-b}{-a}$

(2) $(c - b) - a$ (4) $(c + b) - a$

8. Solve for r in terms of c and π.

$c = 2\pi r$

(1) $\frac{2c}{\pi}$ (3) $\frac{c}{2\pi}$

(2) $\frac{2\pi}{c}$ (4) $\frac{2}{c\pi}$

Solutions

1. The first step of the process is to subtract 1 from each side of the equation. This is called the subtraction property of equality. The correct choice is **(4)**.

2. Though it is more common to begin this question by subtracting 4 from both sides of the equation, in this case she does it by multiplying both sides of the equation by 2. The left-hand side becomes $x + 8$ and the right-hand side becomes 20. The correct choice is **(3)**.

3. Isolate the x by adding 6 to both sides of the equation. The equation then becomes $x = 13$. The correct choice is **(4)**.

4. Subtract 7 from both sides of the equation to get $3x = 15$. Divide both sides of the equation by 3 to get $x = 5$. The correct choice is **(3)**.

5. One way to solve this equation is to first distribute the 5 through the left-hand side to get the equation $5x + 20 = 35$, then subtract 20 from both sides of the equation to get $5x = 15$, and finally to divide both sides of the equation by 5 to get $x = 3$. Another way is to first divide both sides of the equation by 5 to get $x + 4 = 7$, and then subtract 4 from both sides of the equation to get $x = 3$. The correct choice is **(3)**.

6. First add e to both sides of the equation to get $cd = f + e$. Then divide both sides of the equation by c to get $d = (f + e)/c$. The correct choice is **(2)**.

7. First subtract b from both sides of the equation to get $-ma = c - b$. Then divide both sides by $-a$ to get $m = \frac{c-b}{-a}$. The correct choice is **(3)**.

8. Divide both sides of the equation by 2π to get $\frac{c}{2\pi} = r$. The correct choice is **(3)**.

2. POLYNOMIAL ARITHMETIC

2.1 CLASSIFYING POLYNOMIALS

A **polynomial** is an expression like $2x + 5$ or $3x^2 - 5x + 3$. The **terms** of a polynomial are separated by + or – signs. The polynomial $2x + 5$ has two terms. The polynomial $3x^2 - 5x + 3$ has three terms. The terms of a polynomial have a **coefficient** and a **variable part**. The term $3x^2$ has a coefficient of 3 and a variable part of x^2. A term with no variable part is called a **constant**.

- A polynomial with three terms is called a **trinomial**.
- A polynomial with two terms is called a **binomial**.
- A polynomial with one term is called a **monomial**.

2.2 MULTIPLYING AND DIVIDING MONOMIALS

- To *multiply* one monomial by another, multiply the coefficients and multiply the variable parts by adding the exponents on the same variables.

$$8x^3 \cdot 2x$$

Multiply the coefficients $8 \cdot 2 = 16$.
Multiply the variable parts by adding the exponents $x^3 \cdot x^1 = x^4$.
The solution is $16x^4$.

- To *divide* one monomial by another, divide the coefficients and divide the variable parts by subtracting the exponents on the same variables.

$$8x^3 \div 2x$$

Divide the coefficients. $8 \div 2 = 4$.
Divide the variable parts by subtracting the exponents

$$x^3 \div x^1 = x^2$$

The solution is $4x^2$.

This question can also be expressed as $\frac{8x^3}{2x} = 4x^2$

2.3 COMBINING LIKE TERMS

Like terms are terms that have the same variable part. For example, $3x^2$ and $2x^2$ are like terms because the variable part for both is x^2. Like terms can be added or subtracted by adding or subtracting the coefficients and by not changing the variable part.

$$3x^2 + 2x^2 = 5x^2$$
$$3x^2 - x^2 = 3x^2 - 1x^2 = 2x^2$$

- To simplify $2x + 3 + 4x - 5$, combine the like terms with the variable part of x. $2x + 4x = 6x$. Also combine the constants $3 - 5 = -2$. This expression simplifies to $6x - 2$.

If terms are not like terms, they cannot be combined by adding or subtracting. $3x^2 + 5x^3$ cannot be combined because the exponents are different, so they are not like terms.

2.4 MULTIPLYING MONOMIALS AND POLYNOMIALS

- To *multiply* a polynomial by a monomial, use the **distributive property**.

$$2\,(3x + 5) = 2 \cdot 3x + 2 \cdot 5 = 6x + 10$$

This works for more complex monomials and polynomials also.

$$2x^2\,(5x^2 - 7x + 3) = 2x^2 \cdot 5x^2 + 2x^2 \cdot -7x + 2x^2 \cdot 3$$
$$= 10x^4 - 14x^3 + 6x^2$$

2.5 ADDING AND SUBTRACTING POLYNOMIALS

- To *add* two polynomials, remove the parentheses from both and combine like terms.

$$(5x + 2) + (3x - 4) = 5x + 2 + 3x - 4 = 8x - 2$$

- To **subtract** two polynomials, remove the parentheses of the polynomial on the left, then negate all the terms of the polynomial on the right, and remove the parentheses before combining like terms.

$$(5x + 2) - (3x - 4) = 5x + 2 - 3x + 4 = 2x + 6$$

2.6 MULTIPLYING BINOMIALS

- To *multiply* binomials, use the FOIL process.

$$(2x + 3)(5x - 2)$$

The F stands for firsts. Multiply $2x \cdot 5x$, the first term in each of the parentheses. $10x^2$.

The O stands for outers. Multiply $2x \cdot -2$, the terms on the far left and on the far right. $-4x$.

The I stands for inners. Multiply $3 \cdot 5x$, the two terms in the middle. $15x$.

The L stands for lasts. Multiply $3 \cdot -2$, the second term in each of the parentheses. -6

These four answers become $10x^2 - 4x + 15x - 6$. Combine like terms to get $10x^2 + 11x - 6$ in simplified form.

2.7 FACTORING POLYNOMIALS

Factoring a number, like 15, is when two numbers are found that can be multiplied to become that number, $15 = 3 \cdot 5$. Factoring polynomials is more involved and there are certain patterns to be aware of.

- **Greatest Common Factor Factoring**

The terms of some polynomials have a greatest common factor. This can be factored out like a reverse use of the distributive property.

In $6x^2 + 8x$, the terms have a common factor of $2x$. Write $2x$ outside the parentheses and divide each term by $2x$ to determine what goes inside the parentheses.

$$2x(3x + 4)$$

- **Difference of Perfect Squares Factoring**

The expression $a^2 - b^2$ can be factored into $(a - b)(a + b)$. This works anytime both terms of a binomial are perfect squares and there is a minus sign between the two terms.

$$x^2 - 9 = x^2 - 3^2 = (x - 3)(x + 3)$$

- **Reverse FOIL**

A **trinomial** like $x^2 + 8x + 15$ can be factored if there are two numbers that have a sum equal to the coefficient of the x term, 8, and a product equal to the constant 15. Since $3 + 5 = 8$ and $3 \cdot 5 = 15$,

$$x^2 + 8x + 15 = (x + 3)(x + 5).$$

For $x^2 + 3x - 10$, the numbers that have a sum of 3 and a product of -10 are -2 and 5.

$$x^2 + 3x - 10 = (x - 2)(x + 5)$$

2.8 MORE COMPLICATED FACTORING

Sometimes none of the factoring patterns seems to match the polynomial that needs to be factored. When this happens, see if it is possible to rewrite it in a way that resembles the pattern better.

- The polynomial $x^4 - 9$ can be rewritten as $(x^2)^2 - 3^2$, which now has the difference of perfect squares pattern.

$$x^4 - 9 = (x^2)^2 - 3^2 = (x^2 - 3)(x^2 + 3)$$

- The polynomial $x^4 + 8x^2 + 15$ can be rewritten as

$$(x^2)^2 + 8(x^2) + 15.$$

$$x^4 + 8x^2 + 15 = (x^2)^2 + 8(x^2) + 15 = (x^2 + 3)(x^2 + 5)$$

Practice Exercises

1. Classify this polynomial $5x^2 + 3$.
(1) Monomial
(2) Binomial
(3) Trinomial
(4) None of the above

2. Classify this polynomial $7x^2 - 3x + 2$.
(1) Monomial
(2) Binomial
(3) Trinomial
(4) None of the above

3. Multiply $3x^3 \cdot 4x^5$.
(1) $7x^8$
(2) $7x^{15}$
(3) $12x^8$
(4) $12x^{15}$

4. Which expression is equivalent to $2x^2 + 5x^2$?
(1) $10x^2$
(2) $7x^2$
(3) $7x^4$
(4) The expression cannot be simplified any further.

5. Simplify $6x(2x + 3)$.
(1) $8x^2 + 18x$
(2) $12x^2 + 18x$
(3) $12x^2 + 3$
(4) $20x$

6. Simplify $3x(5x^2 - 2x + 3)$.
(1) $15x^3 - 2x + 3$
(2) $15x^3 - 6x^2 + 9x$
(3) $5x^2 + x + 3$
(4) $15x^3 + 6x^2 - 9x$

7. Simplify $(3x - 4) - (5x - 3)$.
(1) $-2x - 7$
(2) $-2x - 1$
(3) $2x - 1$
(4) $2x - 7$

8. $(2x + 3)(3x - 1) =$
(1) $6x^2 - 3$
(2) $6x^2 + 11x - 3$
(3) $6x^2 + 7x - 3$
(4) $6x^2 - 7x - 3$

9. Factor $x^2 - 2x - 15$.
(1) $(x - 3)(x - 5)$
(2) $(x + 3)(x - 5)$
(3) $(x - 15)(x + 1)$
(4) $(x + 15)(x - 1)$

10. Factor $x^4 + 7x^2 + 12$.
(1) $(x^2 + 6)(x^2 + 2)$
(2) $(x^2 - 3)(x^2 - 4)$
(3) $(x^2 + 3)(x^2 + 4)$
(4) This cannot be factored.

Solutions

1. There are two terms, $5x^2$ and 3 separated by a + sign. A polynomial with two terms is called a binomial. The correct choice is **(2)**.

2. There are three terms, $7x^2$, $3x$, and 2 separated by + and – signs. A polynomial with three terms is called a trinomial. The correct choice is **(3)**.

3. To multiply two monomials, first multiply the coefficients, $3 \cdot 4 = 12$. Then multiply the variable parts. Remember that when multiplying variables, you add the exponents. $x^3 \cdot x^5 = x^{(3+5)} = x^8$. The solution is $12x^8$. The correct choice is **(3)**.

4. Since these are like terms with variable part x^2, they can be combined. The solution will also have a variable part of x^2 with a coefficient equal to the sum of the two coefficients. Since $2 + 5 = 7$, $2x^2 + 5x^2 = 7x^2$. The correct choice is **(2)**.

5. Using the distributive property it becomes $6x \cdot 2x + 6x \cdot 3 = 12x^2 + 18x$. The correct choice is **(2)**.

6. Using the distributive property it becomes $3x \cdot 5x^2 + 3x(-2x) + 3x(3) = 15x^3 - 6x^2 + 9x$. The correct choice is **(2)**.

7. Distribute the – sign through the parentheses on the right. The expression becomes $3x - 4 - 5x + 3$. Combine like terms to get $-2x - 1$. The correct choice is **(2)**.

8. Use the FOIL process. The firsts are $2x \cdot 3x = 6x^2$. The outers are $2x \cdot (-1) = -2x$. The inners are $3 \cdot 3x = 9x$. The lasts are $3 \cdot (-1) = -3$. Combine these four terms to get $6x^2 - 2x + 9x - 3$. Combine like terms to get $6x^2 + 7x - 3$. The correct choice is **(3)**.

9. To factor this quadratic trinomial, find two numbers that have a product of –15 and a sum of –2. The numbers are –5 and +3. The factors, then, are $(x - 5)(x + 3)$. The correct choice is **(2)**.

10. This trinomial can be written as $(x^2)^2 + 7(x^2) + 12$. It has the same structure, then, as a quadratic trinomial and can be factored by finding two numbers that have a product of 12 and a sum of 7. The two numbers are +3 and +4. The factors are $(x^2 + 3)$ and $(x^2 + 4)$. The correct choice is **(3)**.

3. QUADRATIC EQUATIONS

3.1 METHODS OF SOLVING QUADRATIC EQUATIONS

A **quadratic equation** is an equation that can be written in the form $ax^2 + bx + c = 0$. For example, $x^2 + 4x - 5 = 0$ is a quadratic equation. There are three ways to solve a quadratic equation.

- 1. Solve by factoring. If possible, factor the left-hand side of the equation.

$$x^2 + 4x - 5 = 0$$
$$(x + 5)(x - 1) = 0$$

Since the only way that two things can have a product of zero is if at least one of them is zero, this means that either $(x + 5)$ or $(x - 1)$ must equal zero.

$$x + 5 = 0 \text{ or } x - 1 = 0$$
$$-5 = -5 \qquad +1 = +1$$
$$x = -5 \text{ or } \quad x = 1$$

$x = -5$ or $x = 1$ are the solutions to this quadratic equation.

- 2. Solve by completing the square. First eliminate the constant from the left-hand side by adding or subtracting.

$$x^2 + 4x - 5 = 0$$
$$+5 = +5$$
$$x^2 + 4x = 5$$

Next, *divide* the coefficient of the x by 2, square that answer, and add that number to both sides of the equation.

$$\left(\frac{4}{2}\right)^2 = 2^2 = 4$$
$$x^2 + 4x + 4 = 5 + 4$$
$$x^2 + 4x + 4 = 9$$

The left-hand side of the equation will factor.

$$(x + 2)(x + 2) = 9$$
$$(x + 2)^2 = 9$$

- Take the square root of both sides of the equation, putting a ± in front of the square root of the right-hand side.

$$\sqrt{(x+2)^2} = \pm\sqrt{9}$$
$$x + 2 = \pm 3$$
$$-2 = -2$$
$$x = -2 \pm 3$$
$$x = -2 + 3 \text{ or } x = -2 - 3$$
$$x = 1 \text{ or } x = -5$$

- 3. Solve with the quadratic formula. Any quadratic equation of the form $ax^2 + bx + c = 0$ can be solved with the equation $x = \dfrac{-b \pm \sqrt{b^2 - 4ac}}{2a}$.

For $x^2 + 4x - 5 = 0$, $a = 1$, $b = 4$, $c = -5$.

$$x = \frac{-4 \pm \sqrt{4^2 - 4 \cdot 1 \cdot (-5)}}{2 \cdot 1} = \frac{-4 \pm \sqrt{16+20}}{2} = \frac{-4 \pm \sqrt{36}}{2} = \frac{-4 \pm 6}{2}$$

$$x = \frac{-4+6}{2} = \frac{2}{2} = 1 \text{ or } x = \frac{-4-6}{2} = -\frac{10}{2} = -5$$

3.2 THE RELATIONSHIP BETWEEN FACTORS AND ZEROS

If a quadratic equation has factors $(x - p)$ and $(x - q)$ then the roots of the equation are p and q. If the equation has roots (or zeros) p and q, the factors are $(x - p)$ and $(x - q)$.

For example, the factors of the equation $x^2 - 7x + 10$ are $(x - 5)$ and $(x - 2)$. Therefore, the zeros of the equation are 5 and 2.

If the roots of a quadratic equation are −2 and 8, then the factors are $(x - (-2))$ and $(x - 8)$. The $(x - (-2))$ can be expressed as $(x + 2)$.

3.3 WORD PROBLEMS INVOLVING QUADRATIC EQUATIONS

Some real-world scenarios can be modeled with quadratic equations.

- **Area Problems**

The width of a rectangle is 3 units more than the length. If the area of the rectangle is 70 square units, what are the length and width of the rectangle?

A = 70 $l + 3$

l

Since area is length times width, for this scenario

$$
\begin{aligned}
70 &= l \cdot w \\
70 &= l \cdot (l + 3) \\
70 &= l^2 + 3l \\
0 &= l^2 + 3l - 70
\end{aligned}
$$

Any of the methods can be used to solve for the answers $l = -10$ and $l = 7$. Since the length must be positive, the answer is 7 units.

- **Projectile Problems**

The height of a projectile after t seconds can be modeled with a quadratic equation. If the equation for the height of a baseball is $h = -16t^2 + 48t + 64$, when will the baseball land on the ground?

When the ball lands on the ground, the height will be 0.

$0 = -16t^2 + 48t + 64$ is the equation.

- This can be solved by any of the methods for the answers $t = -1$ and $t = 4$. Since the amount of time must be positive, the solution is 4 seconds.

Practice Exercises

1. Find all solutions to $(x + 2)^2 = 64$.

(1) $\sqrt{62}, -\sqrt{62}$
(2) 6, –10
(3) 6
(4) –10

2. Use completing the square to find both solutions for x in the equation $x^2 + 8x + 16 = 9$.

(1) –1, –7
(2) –1, –2
(3) –2, –3
(4) –3, –4

3. Find all solutions to $x^2 + 6x = 0$.

(1) 0
(2) –6
(3) 0, –6
(4) 0, 6

4. Solve $x^2 + 10x + 24 = 0$ by factoring.

(1) 4, –6
(2) –4, –6
(3) –4, 6
(4) 4, 6

5. If the roots of an equation are 3 and –6, what could the equation be?

(1) $(x - 3)(x - 6) = 0$
(2) $(x - 3)(x + 6) = 0$
(3) $(x + 3)(x - 6) = 0$
(4) $(x + 3)(x + 6) = 0$

6. If the roots of a polynomial are 1 and –8, what could be the factors?

(1) $(x - 1)$ and $(x + 8)$
(2) $(x - 1)$ and $(x - 8)$
(3) $(x + 1)$ and $(x + 8)$
(4) $(x + 1)$ and $(x - 8)$

7. Solve $x^2 + 4x - 7 = 0$ with the quadratic formula.

(1) 1.3
(2) $2 \pm \sqrt{11}$
(3) $2 \pm \sqrt{12}$
(4) $-2 \pm \sqrt{11}$

8. The width of a rectangle is 10 inches longer than its length. If the area of the rectangle is 56 square inches, which equation could be used to determine its length (l)?

(1) $l(l + 10) = 56$
(2) $l(l - 10) = 56$
(3) $2l + 2(l + 10) = 56$
(4) $2l - 2(l + 10) = 56$

9. The height of a projectile in feet at time t is determined by the equation $h = -16t^2 + 128t + 320$. At what time will the projectile be 560 feet high?

(1) 4 seconds
(2) 5 seconds
(3) 6 seconds
(4) 7 seconds

10. The height of a projectile in feet at time t is determined by the equation $h = -16t^2 + 112t + 128$. At what time will the projectile be 0 feet high?

(1) 5 seconds
(2) 6 seconds
(3) 7 seconds
(4) 8 seconds

Solutions

1. Take the square root of both sides to get $x + 2 = \pm 8$. The equation $x + 2 = 8$ has solution $x = 6$. The equation $x + 2 = -8$ has solution $x = -10$. The correct choice is **(2)**.

2. Since the constant, 16, is already equal to the square of half the coefficient $\left(\frac{8}{2}\right)^2$, the left-hand side of the equation is already a perfect square trinomial. It can be factored as $(x + 4)^2 = 9$. Then take the square root of both sides to get $x + 4 = \pm 3$. The equation $x + 4 = 3$ has solution $x = -1$. The equation $x + 4 = -3$ has solution $x = -7$. The correct choice is **(1)**.

3. Factor out an x to get the equation $x(x + 6) = 0$. This equation is true when either $x = 0$ or when $x + 6 = 0$, which becomes $x = -6$. The correct choice is **(3)**.

4. The two numbers that have a product of 24 and a sum of 10 are +4 and +6. The factors are $(x + 4)(x + 6)$. The solutions to the equation $(x + 4)(x + 6) = 0$ are when $x + 4 = 0$, which becomes $x = -4$ and also when $x + 6 = 0$, which becomes $x = -6$. The correct choice is **(2)**.

5. When a is a root of an equation, $(x - a)$ is a factor. If the roots are 3 and -6, the factors can be $(x - 3)$ and $(x - (-6)) = (x + 6)$. The correct choice is **(2)**.

6. If 1 is a root, $(x - 1)$ is a factor. If -8 is a root $(x - (-8)) = (x + 8)$ is a factor. The correct choice is **(1)**.

7. $a = 1, b = 4, c = -7$.

$$x = \frac{-4 \pm \sqrt{4^2 - 4(1)(-7)}}{2} = \frac{-4 \pm \sqrt{16+28}}{2} = \frac{-4 \pm \sqrt{44}}{2}$$

$$= \frac{-4 \pm 2\sqrt{11}}{2} = -2 \pm \sqrt{11}.$$

The correct choice is **(4)**.

8. The area of a rectangle is $l \cdot w$. If l is the length and $l + 10$ is the width, the area is $l(l + 10)$. If the area is known to be 56, the equation that can be used to solve for l is $l(l + 10) = 56$. The correct choice is **(1)**.

9. When 560 is substituted for h, the equation becomes $560 = -16t^2 + 128t + 320$. Subtract 560 from both sides of the equation to get $0 = -16t^2 + 128t - 240$. Divide both sides by -16 to get $0 = t^2 - 8t + 15$. The right-hand side factors and the equation becomes $0 = (t - 3)(t - 5)$ with solutions $t = 3$ and $t = 5$. Of the choices listed, only 5 seconds is correct. The correct choice is **(2)**.

10. Substituting $h = 0$ into the equation, it becomes $0 = -16t^2 + 112t + 128$. Divide both sides by -16 to get $0 = t^2 - 7t - 8$. The right side factors so the equation becomes $0 = (t - 8)(t + 1)$. The solutions are $t = 8$ and $t = -1$. Since the amount of time must be positive, the -1 is rejected. The correct choice is **(4)**.

4. SYSTEMS OF LINEAR EQUATIONS

4.1 WHAT IS A SYSTEM OF LINEAR EQUATIONS?

A **system of equations** is a set of two equations that each have two variables. The system is *linear* if there are no exponents greater than one on any of the variables.

$$2x + 3y = 21$$
$$5x - 2y = 5$$

is a system of linear equations.

The solution to a system of equations is the set of **ordered pairs** that satisfy both equations at the same time. For the system above, the solution is the ordered pair (3,5) since

$$2(3) + 3(5) = 6 + 15 = 21$$
$$5(3) - 2(5) = 15 - 10 = 5$$

There are two main techniques for solving systems of equations.

4.2 SOLVING A SYSTEM OF EQUATIONS WITH THE SUBSTITUTION METHOD

If one of the two equations is in the form $y = mx + b$, use **the substitution method**.

$$y = 2x + 3$$
$$2x + 4y = 42$$

- Substitute the expression $2x + 3$ for the y in the second equation.

$$2x + 4(2x + 3) = 42$$

Solve for x.

$$2x + 8x + 12 = 42$$
$$10x + 12 = 42$$
$$-12 = -12$$
$$\frac{10x}{10} = \frac{30}{10}$$
$$x = 3$$

- Substitute 3 for x into either of the equations and solve for y.

$$y = 2(3) + 3$$
$$y = 6 + 3$$
$$y = 9$$

The solution is the ordered pair (3, 9).

4.3 SOLVING A SYSTEM OF EQUATIONS WITH THE ELIMINATION METHOD

When both equations are in the form $ax + by = c$, use the **elimination method**. The elimination method is when the two equations are combined in a way that one of the variables is eliminated.

In the system of equations below, the y term will be eliminated if the two equations are added. When the coefficient of one of the variables in one equation is the same number with the opposite sign of the same variable in the other equation, add the equations to eliminate that variable. In this case, the $+2y$ has the opposite coefficient as the $-2y$.

$$\begin{array}{r} 3x + 2y = 19 \\ + \;\; 4x - 2y = 16 \\ \hline \end{array}$$
$$\frac{7x}{7} = \frac{35}{7}$$
$$x = 5$$

- Substitute 5 for x into either of the equations

$$3(5) + 2y = 19$$
$$15 + 2y = 19$$
$$-15 = -15$$
$$\frac{2y}{2} = \frac{4}{2}$$
$$y = 2$$

The solution is the ordered pair (5, 2).

For some systems of equations, one or both of the equations need to be changed so the elimination will happen.

In the system

$$\begin{aligned} 2x - 4y &= 4 \\ 3x + 2y &= 14 \end{aligned}$$

Since 4 is a multiple of 2, multiply both sides of the bottom equation by 2 and the coefficients of the y variables will be the same number with opposite signs.

$$\begin{aligned} 2x - 4y &= 4 \\ 2(3x + 2y) &= 2(14) \end{aligned}$$

$$\begin{aligned} 2x - 4y &= 4 \\ + \quad 6x + 4y &= 28 \\ \hline \frac{8x}{8} &= \frac{32}{8} \\ x &= 4 \end{aligned}$$

$$\begin{aligned} 2(4) - 4y &= 4 \\ 8 - 4y &= 4 \\ -8 &= -8 \\ \frac{-4y}{-4} &= \frac{-4}{-4} \\ y &= 1 \end{aligned}$$

The solution is the ordered pair (4, 1).

In the system

$$\begin{aligned} 3x - 5y &= 17 \\ 2x + 4y &= 4 \end{aligned}$$

- To eliminate the y, determine the least common multiple of 4 and 5, which is 20. Multiply both sides of both equations so that one of the coefficients on the y is –20 and the other is +20. To do this, multiply both sides of the top equation by 4 and multiply both sides of the bottom equation by 5.

$$\begin{aligned} 4(3x - 5y) &= 4(17) \\ 5(2x + 4y) &= 5(4) \end{aligned}$$

$$\begin{aligned} 12x - 20y &= 68 \\ +\ 10x + 20y &= 20 \\ \hline \frac{22x}{22} &= \frac{88}{22} \\ x &= 4 \end{aligned}$$

- Substitute 4 for x in either of the original equations and solve for y.

$$\begin{aligned} 3(4) - 5y &= 17 \\ 12 - 5y &= 17 \\ -12 &= -12 \\ \frac{-5y}{-5} &= \frac{5}{-5} \\ y &= -1 \end{aligned}$$

The solution to the system of equations is the ordered pair (4, –1).

4.4 WORD PROBLEMS INVOLVING SYSTEMS OF EQUATIONS

Some real-world scenarios can be modeled with a system of linear equations. Here is a typical example.

If five slices of pizza and three drinks cost $21 and two slices of pizza and five drinks cost $16, how much is it for just one slice of pizza?

- Let x be the cost of a slice of pizza and y be the cost of a drink.

The system of equations is

$$\begin{aligned} 5x + 3y &= 21 \\ 2x + 5y &= 16 \end{aligned}$$

- To eliminate the y, make the $3y$ and the $5y$ into $15y$ and $-15y$, respectively, by multiplying both sides of the top equation by 5 and both sides of the bottom equation by -3. Then add the two equations, and solve for x.

$$\begin{aligned} 5(5x + 3y) &= 5\,(21) \\ -3(2x + 5y) &= -3\,(16) \end{aligned}$$

$$\begin{array}{r} 25x + 15y = 105 \\ +\ \ -6x - 15y = -48 \\ \hline \end{array}$$

$$\frac{19x}{19} = \frac{57}{19}$$

$$x = 3$$

Since the question just asked for the price of a slice of pizza, it is not necessary to also find the value of y. A slice of pizza costs $3.

Practice Exercises

1. (2, 5) is a solution to which equation?

(1) $x + 2y = 9$
(2) $2x + y = 9$
(3) $8x - y = 9$
(4) $3x + 4y = 25$

2. Which equation has the same solution set as the equation $2x + 3y = 5$?

(1) $8x + 12y = 20$
(2) $8x + 12y = 15$
(3) $6x + 9y = 12$
(4) $4x + 6y = 8$

3. Solve the system of equations

$$3x + 2y = 17$$
$$4x - 2y = 4$$

(1) (4, 3)
(2) (5, 1)
(3) (3, 4)
(4) (1, 6)

4. Solve the system of equations

$$y = 5x + 3$$
$$2x + 6y = 50$$

(1) (1, 8)
(2) (8, 1)
(3) (–1, –8)
(4) (–8, –1)

5. Solve the system of equations

$$
\begin{aligned}
2x + 3y &= -1 \\
-2x + 5y &= -23
\end{aligned}
$$

(1) (–4, 3) (3) (4, –3)
(2) (4, 3) (4) (–4, –3)

6. Solve the system of equations

$$
\begin{aligned}
y &= 3x - 2 \\
4x - 2y &= -4
\end{aligned}
$$

(1) (10, 4) (3) (–10, –4)
(2) (4, 10) (4) (–4, –10)

7. In order to eliminate the x from this system of equations,

$$
\begin{aligned}
12x - 3y &= 21 \\
-2x + 6y &= 2
\end{aligned}
$$

you could
(1) Multiply both sides of the first equation by 2.
(2) Multiply both sides of the second equation by 6.
(3) Multiply both sides of the first equation by –2.
(4) Multiply both sides of the second equation by 1/2.

8. Solve the system of equations

$$
\begin{aligned}
8x - 2y &= 28 \\
4x + 3y &= 6
\end{aligned}
$$

(1) (3, 2) (3) (–3, –2)
(2) (–3, 2) (4) (3, –2)

9. Which system of equations can be used to model the following scenario?

There are 50 animals. Some of the animals have two legs and the rest of them have four legs. In total there are 172 legs.

(1) $x + y = 172$
$2x + 4y = 50$

(2) $x + 50 = y$
$2x + 172 = 4y$

(3) $y + 50 = x$
$4y + 172 = 2x$

(4) $x + y = 50$
$2x + 4y = 172$

10. A pet store has 30 animals. Some are cats and the rest are dogs. The cats cost \$50 each. The dogs cost \$100 each. If the total cost for all 30 animals is \$1,900, how many cats are there?

(1) 8
(2) 20
(3) 22
(4) 24

Solutions

1. Substitute 2 for x and 5 for y into each of the choices. Choice (1) becomes $2 + 2(5) = 9$, which is not true. Choice (2) becomes $2(2) + 5 = 9$, which is true. The correct choice is **(2)**.

2. If both sides of an equation are multiplied by the same number, the new equation has the same solution set as the original equation. If both sides of the equation $2x + 3y = 5$ are multiplied by 4, it becomes $8x + 12y = 20$, which is choice (1). The other choices could be obtained by multiplying the right-hand side of the original equation by one number and the left-hand side of the original equation by another number, which will not produce an equation with the same solution set as the original. The correct choice is **(1)**.

3. Since the top equation has a $+2y$ and the bottom equation has a $-2y$, the equations can be added together, and the y terms will drop out leading to the equation $7x = 21$ or $x = 3$. To solve for y, substitute 3 for x into either of the original equations, like $3(3) + 2y = 17$, $9 + 2y = 17$, $2y = 8$, which leads to $y = 4$. The correct choice is **(3)**.

4. Since the y is isolated in the top equation, substitute $5x + 3$ for y in the bottom equation. It becomes $2x + 6(5x + 3) = 50$, $2x + 30x + 18 = 50$, $32x + 18 = 50$, $32x = 32$, $x = 1$. Only one choice has an x-coordinate of 1. The correct choice is **(1)**.

5. Since the top equation has a $+2x$ and the bottom equation has a $-2x$, the equations can be added together to get $8y = -24$, $y = -3$. Substitute -3 for y into either equation to solve for x. $2x + 3(-3) = -1$, $2x - 9 = -1$, $2x = 8$, $x = 4$. The correct choice is **(3)**.

6. Since the y is isolated in the top equation, substitute $3x - 2$ for y in the bottom equation. $4x - 2(3x - 2) = -4$, $4x - 6x + 4 = -4$, $-2x + 4 = -4$, $-2x = -8$, $x = 4$. To solve for y, substitute 4 for x into either equation. $y = 3(4) - 2 = 12 - 2 = 10$. The correct choice is **(2)**.

7. The x will be eliminated after combining two equations when the coefficient of the x in one of the equations is the opposite of the coefficient of the x in the other equation. For choice (1), if both sides of the top equation are multiplied by 2, it would become $24x - 6y = 42$. $+24$ is not the opposite of -2. For choice (2) if both sides of the second equation are multiplied by 6, it becomes $-12x + 36y = 12$. Since -12 is the opposite of $+12$, this is the best answer. The correct choice is **(2)**.

8. Multiply both sides of the bottom equation by -2 to get $-8x - 6y = -12$. Add this to the top equation to eliminate the x and get $-8y = 16$. Divide both sides by -8 to get $y = -2$. Substitute -2 for y into one of the original equations. $8x - 2(-2) = 28$, $8x + 4 = 28$, $8x = 24$, $x = 3$. The correct choice is **(4)**.

9. If x is the number of two-legged animals and y is the number of four-legged animals, the number of animals is $x + y$ and the number of legs is $2x + 4y$. The system, then, is $x + y = 50$ and $2x + 4y = 172$. The correct choice is **(4)**.

10. If x is the number of dogs and y is the number of cats, the system of equations is

$$x + y = 30$$
$$100x + 50y = 1{,}900$$

To eliminate the x, multiply both sides of the top equation by -100 to get $-100x - 100y = -3{,}000$. Add this to the bottom equation to get $-50y = -1{,}100$ or $y = 22$. The correct choice is **(3)**.

5. GRAPHS OF LINEAR EQUATIONS

5.1 GRAPHING THE SOLUTION SET OF A LINEAR EQUATION BY MAKING A TABLE OF VALUES

The equation $x + y = 10$ has an infinite number of ordered pairs that satisfy it. One way to organize the information before creating a graph is to make a table of values.

This is a chart with three ordered pairs satisfying the equation $x + y = 10$. For a linear equation, only two ordered pairs are needed, but it is wise to do an extra ordered pair in case one of your first two is incorrect.

x	y
2	8
3	7
9	1

Plot the ordered pair (2, 8) on the coordinate plane by locating the point that is two units to the right of the y-axis and 8 units above the x-axis. One way to do this is to start at the origin point where the two axes intersect and move to the right two units from there and then up 8 units.

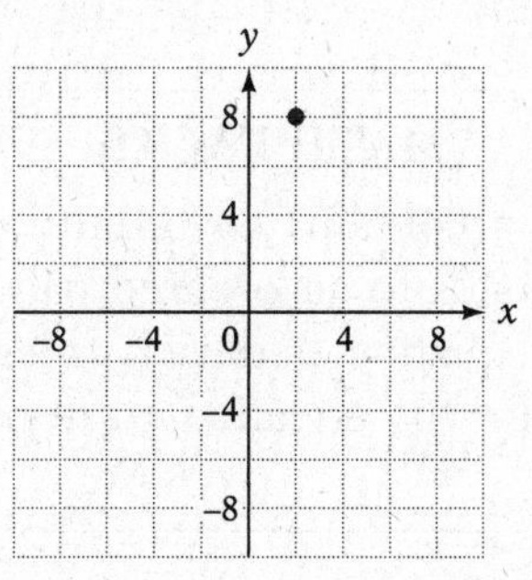

Do the same for the other two ordered pairs on the chart (3, 7) and (9, 1).

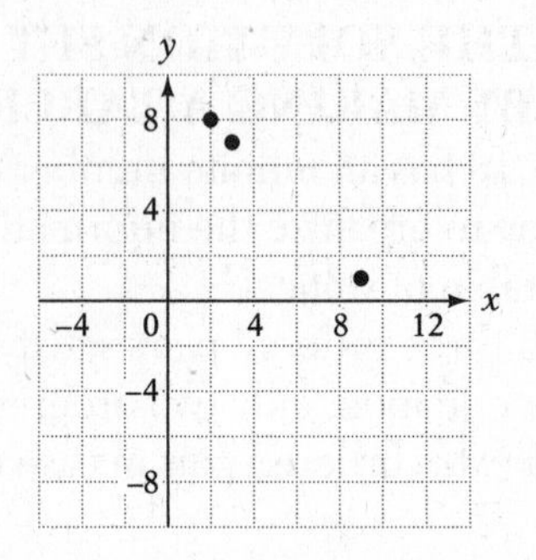

Draw a line through the three points. If the three points do not all lie on the same line, one of your ordered pairs is incorrect.

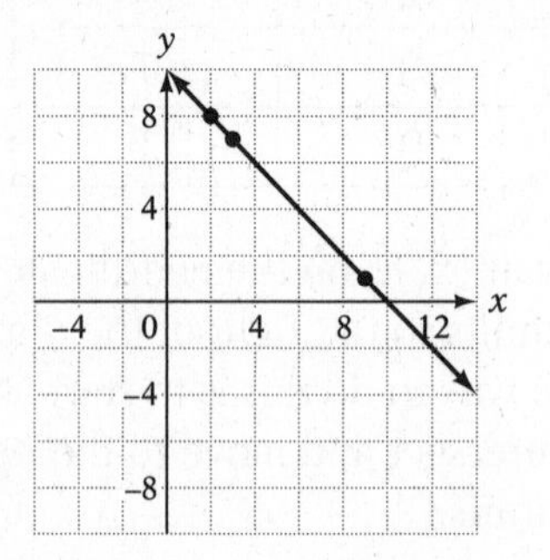

MATH FACTS

The line on a graph contains an infinite number of points. Each point corresponds to an ordered pair that is part of the solution set for the equation, and each ordered pair that is part of the solution set for the equation corresponds to a point on the line.

5.2 GRAPHING A LINEAR EQUATION USING THE INTERCEPT METHOD

When a linear equation is written in the form $ax + by = c$, the graph of the equation can be created quickly by finding the x-intercept and the y-intercept.

For the equation $2x - 3y = 12$:

- To find the y-intercept, substitute 0 for x and solve for y.

$$2(0) - 3y = 12$$
$$\frac{-3y}{-3} = \frac{12}{-3}$$
$$y = -4$$

The y-intercept is $(0, -4)$.

- To find the x-intercept, substitute 0 for y and solve for x.

$$2x - 3(0) = 12$$
$$2x = 12$$
$$x = 6$$

The x-intercept is $(6, 0)$.

- Plot (0, –4) and (6, 0) on a set of coordinate axes and draw the line that passes through both points.

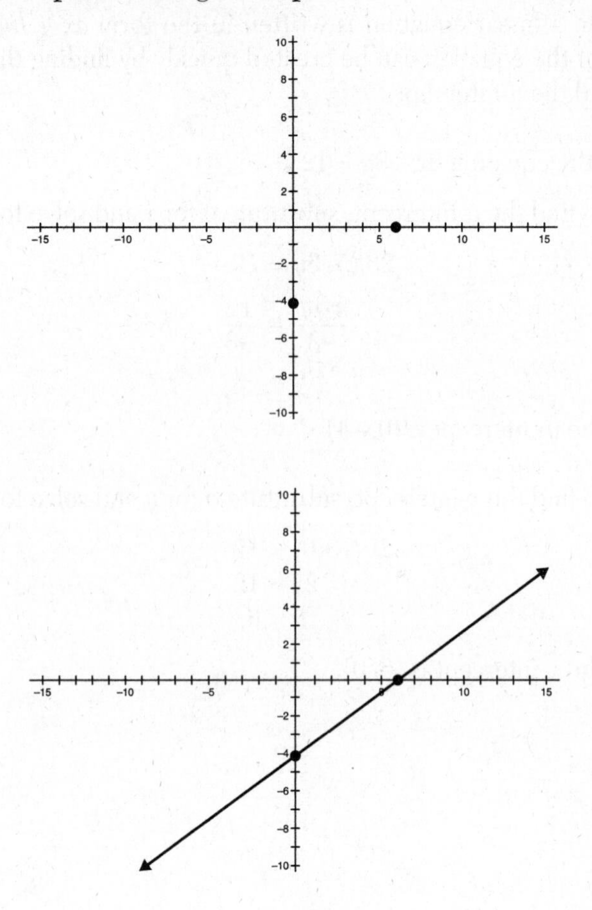

5.3 CALCULATING AND INTERPRETING SLOPE

The *slope* of a line is a number that measures how steep it is. A horizontal line has a slope of 0. A line with a positive slope goes up as it goes to the right. A line with a negative slope goes down as it goes to the right. The variable used for slope is the letter m.

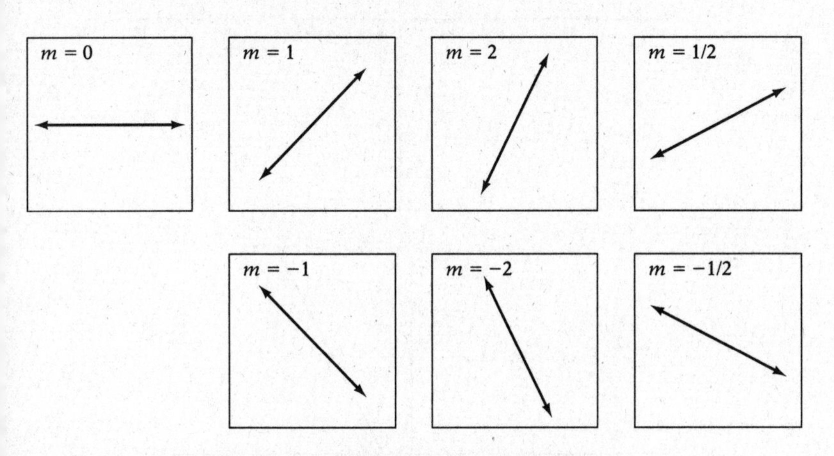

5.4 GRAPHING A LINEAR EQUATION IN SLOPE-INTERCEPT FORM

When the linear equation is in the form $y = mx + b$, it can be graphed very quickly. The y-intercept is $(0, b)$, and from that point to another point move one unit to the right and m units up.

For the equation $y = 2x - 3$, the y-intercept is $(0, -3)$. From that point, move 1 unit to the right and 2 units up. Draw the line through the two points $(0, -3)$ and $(1, -1)$.

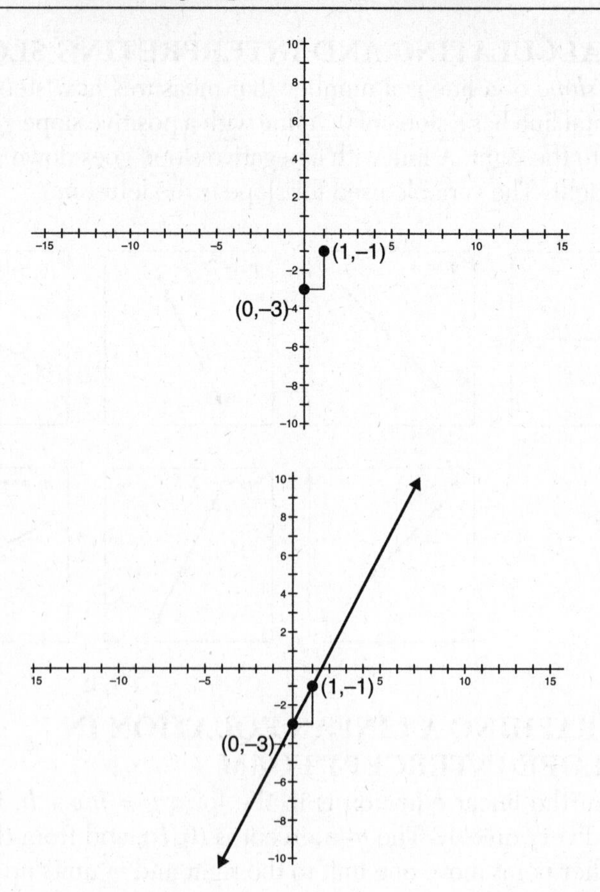

- If m is a fraction, n/d, it is more accurate to move d units to the right and n units up from the y-intercept.
- For the equation $y = \frac{2}{3}x + 2$, the y-intercept is (0, 2). From that point move 3 to the right and 2 up to get to the point (3, 4). Draw a line through these two points.

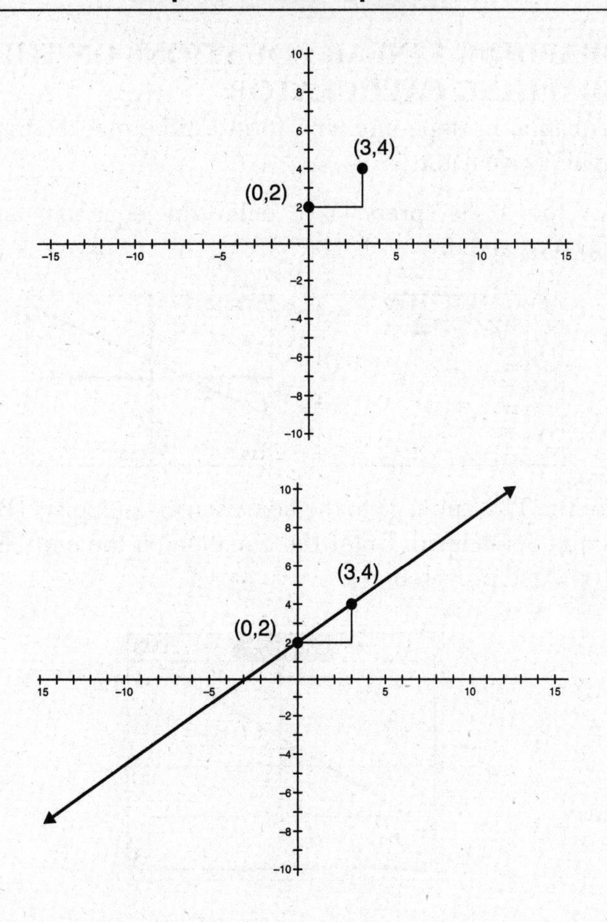

If the m value is negative, move down instead of up to get from the y-intercept to the next point.

5.5 GRAPHING LINEAR EQUATIONS ON THE GRAPHING CALCULATOR

An equation in slope-intercept form can be quickly graphed on the graphing calculator.

- For the TI-84, press [Y=], enter the equation, and press [ZOOM] and [6].

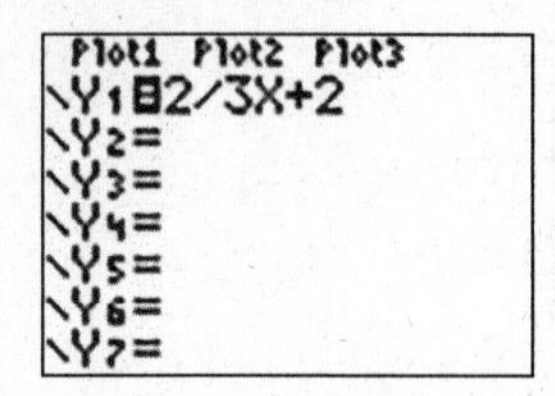

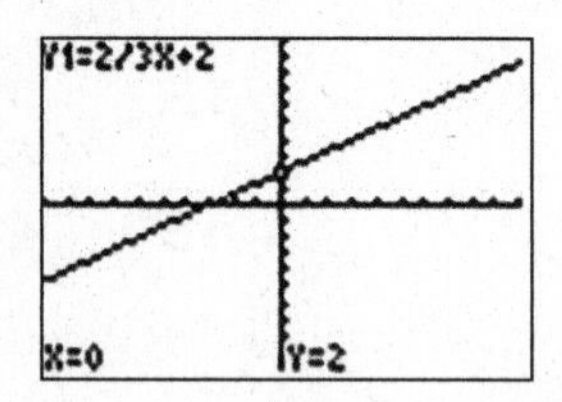

- For the TI-Nspire, go to the home screen and select [B] for the Graph Scratchpad. Enter the equation on the entry line after $f1(x)=$ and press [enter].

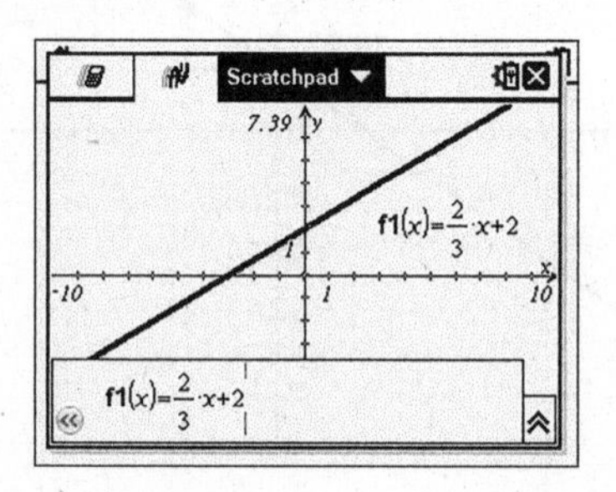

- If a linear equation with two variables is not originally in the slope-intercept form, $y = mx + b$, algebra can be used to rewrite it in slope-intercept form before graphing.

5.6 EQUATIONS OF VERTICAL AND HORIZONTAL LINES

The graph of the equation $y = k$ is a horizontal line through the point $(0, k)$. The graph of the equation $x = h$ is a vertical line through the point $(h, 0)$.

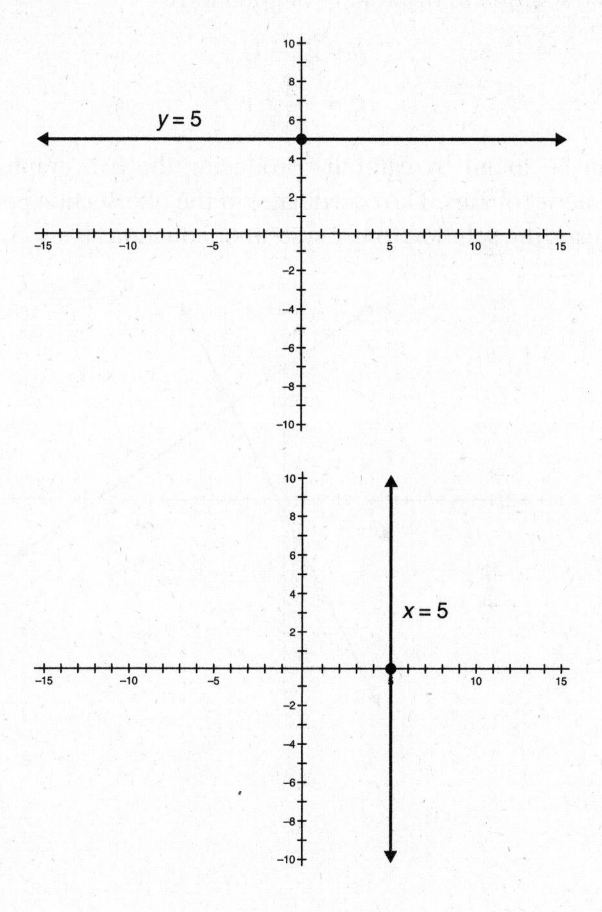

5.7 SOLVING SYSTEMS OF LINEAR EQUATIONS GRAPHICALLY

The solution to a system of equations is the set of coordinates of the point where the graphs of the two lines intersect.

- The solution to the system of equations

$$y = 2x - 1$$
$$y = -\frac{2}{3}x + 7$$

  can be found by carefully producing the two graphs on the same set of axes. The coordinates of the intersection point (3, 5) means the solution to the system of equations is $x = 3$, $y = 5$.

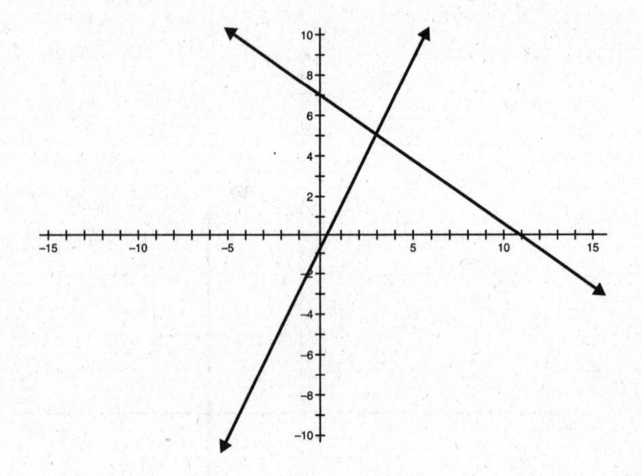

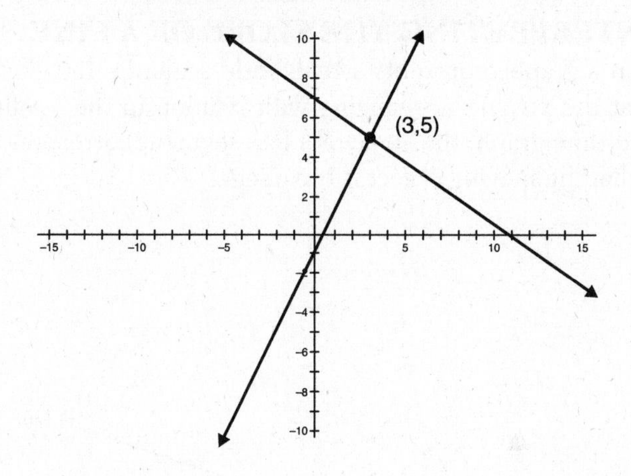

The graphing calculator can also determine the intersection point of two lines. On the TI-84, graph both lines and press [2ND], [TRACE], and [5] to find the intersection. On the TI-Nspire, graph both lines and press [menu], [6], and [4] to find the intersection.

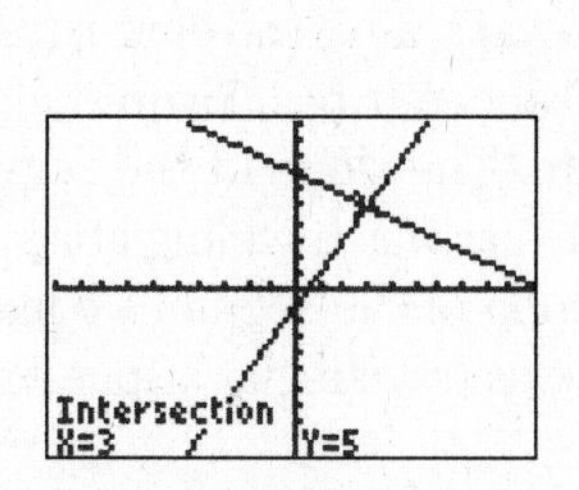

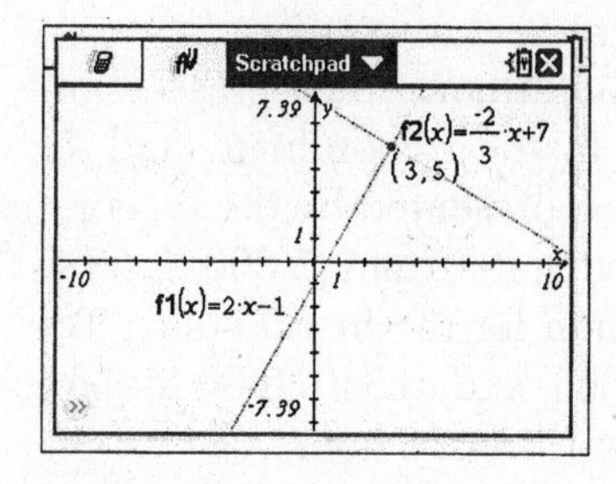

5.8 DETERMINING THE SLOPE OF A LINE

If two points on a line are (x_1, y_1) and (x_2, y_2), then the slope of the line can be determined by the formula

$$m = \frac{y_2 - y_1}{x_2 - x_1}$$

The slope of the line passing through (3,5) and (7,8) is

$$m = \frac{8-5}{7-3} = \frac{3}{4}$$

5.9 INTERPRETING THE SLOPE OF A LINE

When a graph represents a real-world scenario, the slope is the rate that the x value is changing with relation to the y value. In a distance-time graph, the slope of a line segment corresponds to the speed that the moving object is traveling.

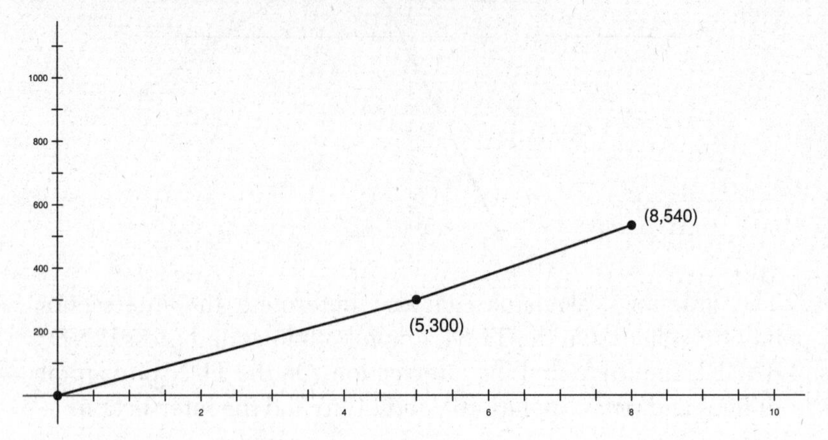

In the distance-time graph above, a car's time traveled is represented by the x-coordinate, and the distance it has covered at that time is represented by the y-coordinate. The slope of the segment connecting (0,0) and (5,300) is 60 so the car was traveling at a speed of 60 mph for the first 5 hours. The slope of the segment connecting (5,300) and (8,540) is 80 so the car was traveling at a speed of 80 mph for the last 3 hours.

5.10 FINDING THE EQUATION OF A LINE THROUGH TWO GIVEN POINTS

If two points are known, use the slope-intercept formula to find the slope of the line. Substitute the value you calculated for m and the x- and y-coordinates of either of the given two points into the equation $y = mx + b$ and solve for b. Substitute the values calculated for m and b into the equation $y = mx + b$.

- For the line through the two points (3, 5) and (9, 1), m is the slope of the line $= \dfrac{1-5}{9-3} = -\dfrac{4}{6} = -\dfrac{2}{3}$.
- Using the point (3, 5), substitute $x = 3$, $y = 5$, $m = -\dfrac{2}{3}$ into the equation $y = mx + b$ and solve for b.

$$5 = -\frac{2}{3}(3) + b$$
$$5 = -2 + b$$
$$+2 = +2$$
$$7 = b$$

$m = -\dfrac{2}{3}$ and $b = 7$ so the equation is $y = -\dfrac{2}{3}x + 7$.

Practice Exercises

1. What is the x-intercept of the graph of the solution set of the equation $2x + 5y = 20$?

(1) (10, 0)
(2) (4, 0)
(3) (0, 10)
(4) (0, 4)

2. This is a graph of the solution set of which equation?

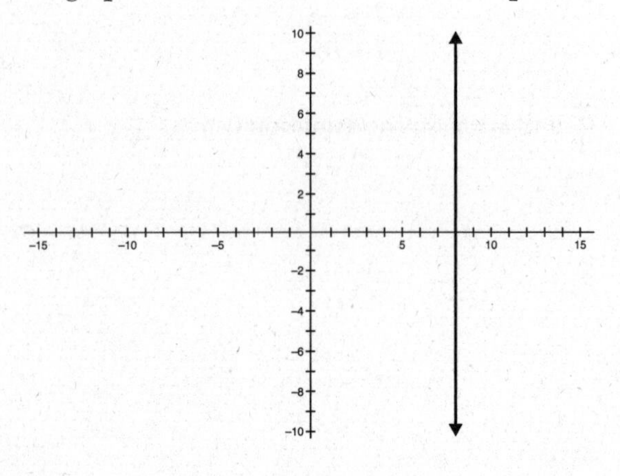

(1) $x = 8$
(2) $x = -8$
(3) $y = 8$
(4) $y = -8$

3. Below is the graph of the solution set of an equation. Based on this graph, which ordered pair does not seem to be part of the solution set of the equation $y = \frac{1}{3}x + 2$?

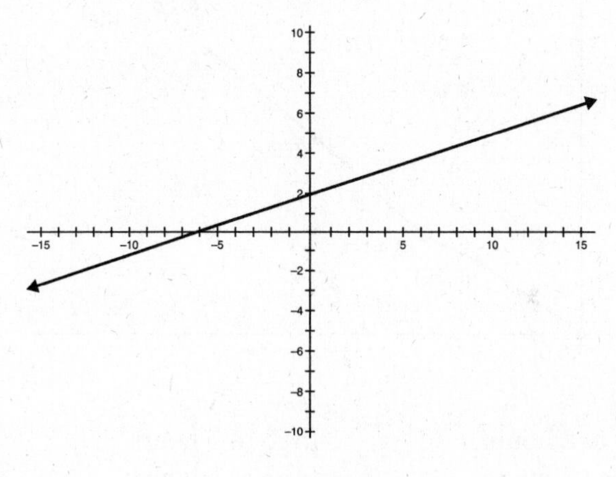

(1) (3, 3)
(2) (6, 4)
(3) (6, 8)
(4) (9, 5)

4. What is the slope of the line that passes through (–2, 1) and (8, 5)?

(1) $\frac{2}{5}$
(2) $-\frac{2}{5}$
(3) $\frac{5}{2}$
(4) $-\frac{5}{2}$

5. Below is a distance-time graph for a bicycle trip. During which time interval is the cyclist going the fastest?

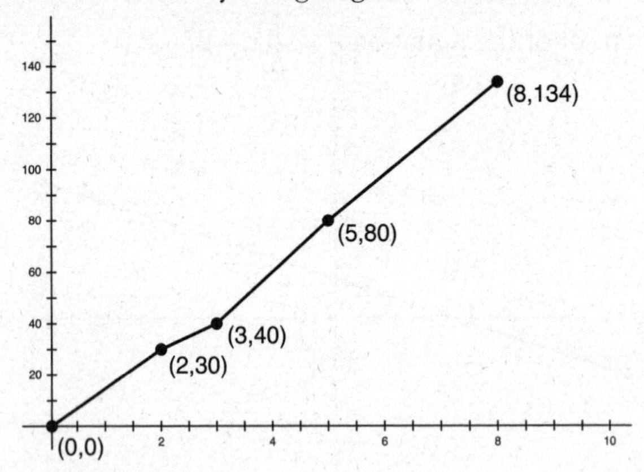

(1) 0 to 2 hours
(2) 2 to 3 hours
(3) 3 to 5 hours
(4) 5 to 8 hours

6. What is the slope of the line defined by the equation $y = -3x + 4$?

(1) 3
(2) –3
(3) 4
(4) –4

7. This is the graph of the solution set of which equation?

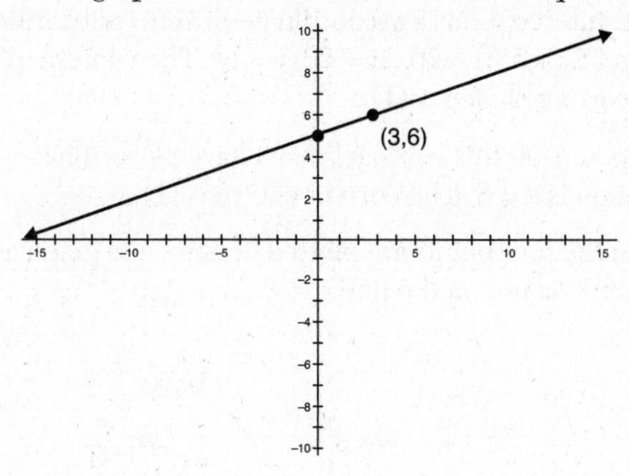

(1) $y = 5x + \frac{1}{3}$ (3) $y = 3x + 5$

(2) $y = 5x + 3$ (4) $y = \frac{1}{3}x + 5$

8. Find the equation of the line passing through the two points (0, –7) and (5, 8).

(1) $y = 3x + 7$ (3) $y = \frac{1}{3}x + 7$

(2) $y = 3x - 7$ (4) $y = \frac{1}{3}x - 7$

9. Find the equation of the line passing through the two points (4, –2) and (12, 4).

(1) $y = \frac{4}{3}x - 5$ (3) $y = \frac{3}{4}x - 5$

(2) $y = \frac{4}{3}x + 5$ (4) $y = \frac{3}{4}x + 5$

10. Find the equation of the line through the points (3, 5) and (3, 8).

(1) $x = -3$ (3) $y = -3$

(2) $x = 3$ (4) $y = 3$

Solutions

1. The x-intercept has a y-coordinate of zero. Substitute zero for y to get $2x + 5(0) = 20$, $2x = 20$, $x = 10$. The x-intercept is (10,0). The correct choice is **(1)**.

2. The points on this vertical line all have x-coordinates of 8. The equation is $x = 8$. The correct choice is **(1)**.

3. When the four points are plotted on the same graph as the line, only (6,8) is not on the line.

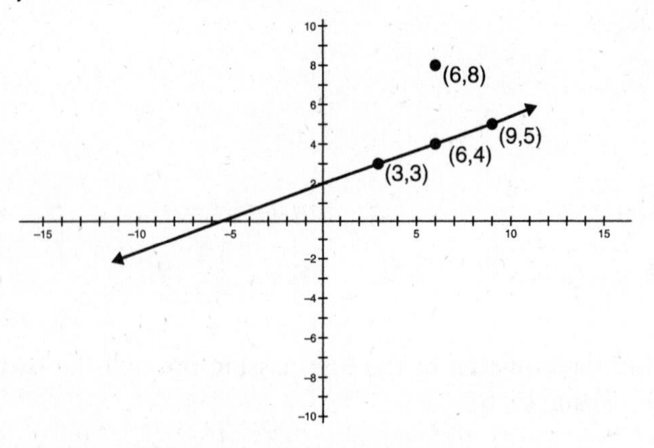

The correct choice is **(3)**.

4. Using the slope formula $m = \frac{y_2 - y_1}{x_2 - x_1}$, with the points (–2,1) and (8,5) becomes $m = \frac{5-1}{8-(-2)} = \frac{4}{10} = \frac{2}{5}$. The correct choice is **(1)**.

5. The interval between 0 and 2 has a slope of 30/2 = 15. The interval between 2 and 3 has a slope of 10/1 = 10. The interval between 3 and 5 has a slope of 40/2 = 20. The interval between 5 and 8 has a slope of 54/3 = 18. Since the slope represents the speed the bicycle is going, the interval between 3 and 5 hours is the fastest. It can also be seen from the graph that the interval between 3 and 5 hours looks the steepest. The correct choice is **(3)**.

6. When the equation for a line is in $y = mx + b$ form, the m is the slope. For this equation, the coefficient of the x is -3 so the slope of the line is -3. The correct choice is **(2)**.

7. According to the graph, the y-intercept is (0,5). The slope of the line through (0,5) and (3,6) is $m = \frac{6-5}{3-0} = \frac{1}{3}$. In $y = mx + b$ form, the m is the slope and the b is the y-coordinate of the y-intercept so the equation is $y = \frac{1}{3}x + 5$. The correct choice is **(4)**.

8. The slope of the line is $m = \frac{8-(-7)}{5-0} = \frac{15}{5} = 3$. Choose one of the points for the x and the y value and 3 for the m value. Using the point (5,8) the equation $y = mx + b$ becomes $8 = 3(5) + b$, $8 = 15 + b$, $b = -7$. The equation is $y = 3x - 7$. The correct choice is **(2)**.

9. The slope of the line is $m = \frac{4-(-2)}{12-4} = \frac{6}{8} = \frac{3}{4}$. Pick one of the points for x and y and substitute $\frac{3}{4}$ for m into $y = mx + b$. $4 = \frac{3}{4} \cdot 12 + b$, $4 = 9 + b$, $b = -5$. The equation is $y = \frac{3}{4}x - 5$. The correct choice is **(3)**.

10. Since the x-coordinates are the same, the line is a vertical line. Vertical lines have equations x = constant. Since the x-coordinates are both 3, the equation is $x = 3$. The correct choice is **(2)**.

6. GRAPHS OF QUADRATIC EQUATIONS

6.1 GRAPHING A QUADRATIC EQUATION WITH A CHART

A **quadratic equation** can be written in the form $y = ax^2 + bx + c$. The graph of a quadratic equation is always a **parabola**, which resembles the letter U. When the a value is positive, the parabola looks like a right-side-up U. When the a value is negative, the parabola looks like an upside-down U.

- The simplest way to create the graph of a quadratic equation is to choose at least five consecutive x values and create a chart.

For the equation $y = x^2 - 2x - 3$, the chart could look like this:

x	y
–2	$(-2)^2 - 2(-2) - 3 = 4 + 4 - 3 = 5$
–1	$(-1)^2 - 2(-1) - 3 = 1 + 2 - 3 = 0$
0	$(0)^2 - 2(0) - 3 = -3$
1	$(1)^2 - 2(1) - 3 = 1 - 2 - 3 = -4$
2	$(2)^2 - 2(2) - 3 = 4 - 4 - 3 = -3$

- The graph of these five ordered pairs looks like this:

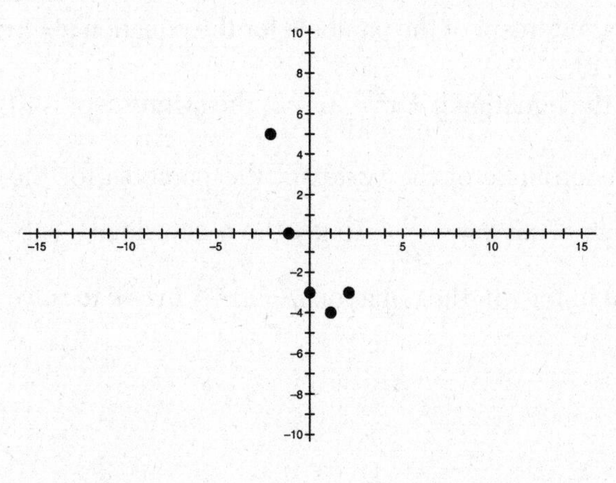

- *Connect* the points with a U-shaped parabola.

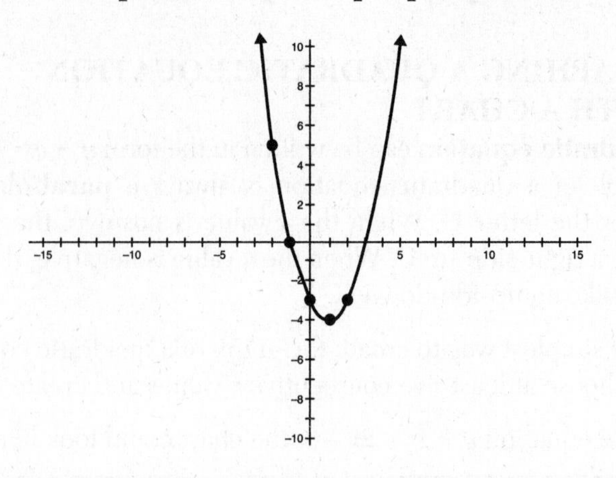

The low point of the parabola is called the **vertex**. In this example the vertex is (1, –4). The two x-intercepts of the parabola are (–1, 0) and (3, 0). The y-intercept of this parabola is (0, –3).

6.2 GRAPHING A PARABOLA BY FINDING THE VERTEX AND INTERCEPTS

The coordinates of the vertex, the y-intercept, and the x-intercepts can be calculated. These four points help produce an accurate graph of the parabola.

- The y-intercept of the parabola for the equation $y = ax^2 + bx + c$ is $(0, c)$.
- For the equation $y = x^2 - 2x - 3$, the y-intercept is (0, –3).

The x-coordinate of the vertex of the parabola for the equation $y = ax^2 + bx + c$ is $x = -\frac{b}{2a}$. To get the y-coordinate, substitute the $-\frac{b}{2a}$ value in for x in the equation $y = ax^2 + bx + c$ to solve for y.

- For the equation $y = x^2 - 2x - 3$, the x-coordinate of the vertex is $x = -\frac{-2}{2(1)} = \frac{2}{2} = 1$. The y-coordinate is $y = 1^2 - 2(1) - 3 = 1 - 2 - 3 = -4$. So the vertex is $(1, -4)$.

The x-intercepts of the parabola for the equation $y = ax^2 + bx + c$ can be found by solving the quadratic equation $0 = ax^2 + bx + c$.
For the equation $y = x^2 - 2x - 3$, this becomes

$$0 = x^2 - 2x - 3$$

This quadratic equation can be solved by factoring.

$$0 = (x + 1)(x - 3)$$
$$x + 1 = 0 \text{ or } x - 3 = 0$$
$$-1 = -1 \qquad +3 = +3$$
$$x = -1 \text{ or } \quad x = 3$$

The x-intercepts are $(-1, 0)$ and $(3, 0)$.

These four points help make an accurate sketch of the parabola.

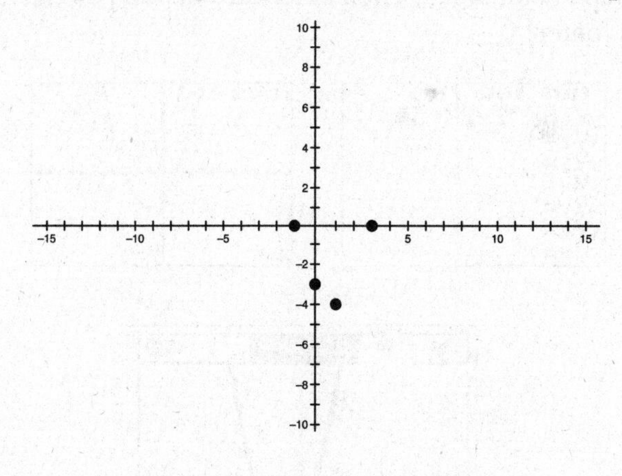

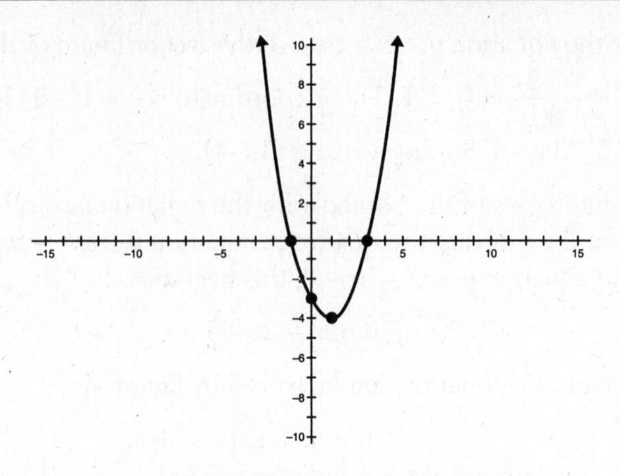

6.3 GRAPHING QUADRATIC EQUATIONS ON THE GRAPHING CALCULATOR

The graphing calculator can easily graph quadratic equations. On the TI-84 press [Y=] and enter the equation after "Y1=", and press [ZOOM] and [6]. On the TI-Nspire, from the home screen press [B] for the Graph Scratchpad. Then enter the equation on the entry line and press [enter].

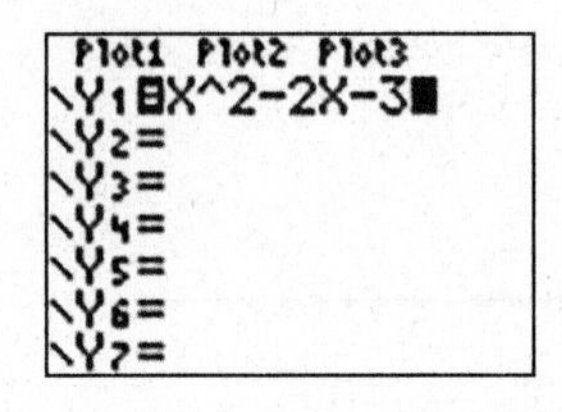

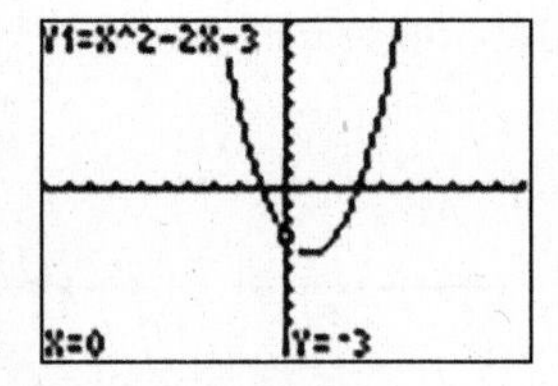

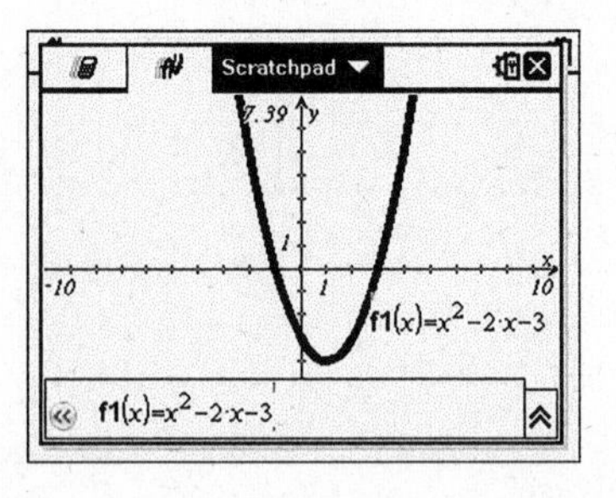

The graphing calculator can also determine the vertex and the intercepts using the min/max feature and the zeros feature.

- For the TI-84, press [2ND], [TRACE], and [3].

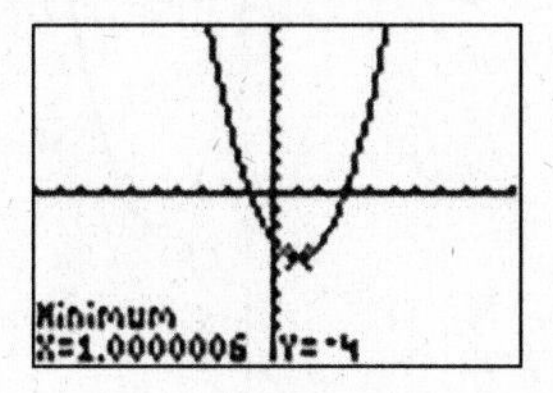

- For the TI-Nspire, press [menu], [6], and [2] to find the minimum point.

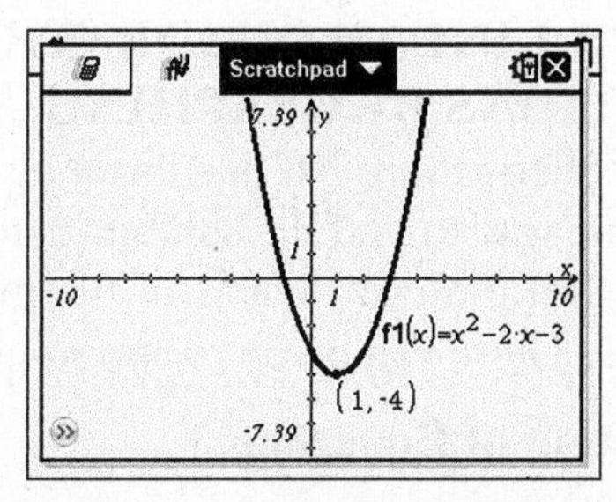

6.4 USING THE GRAPHING CALCULATOR TO SOLVE QUADRATIC EQUATIONS

The solutions to the equation $ax^2 + bx + c = 0$ are also the x-coordinates of the x-intercepts of the parabola defined by $y = ax^2 + bx + c$.

- To solve the equation $x^2 - 2x - 3 = 0$ with the graphing calculator, graph $y = x^2 - 2x - 3$ and then use the zeros feature to find the x-intercepts.
- For the TI-84, enter the equation $Y1 = x^2 - 2x - 3$ and [ZOOM] and [6] to graph. Then press [2ND], [TRACE], and [2] to find each zero.

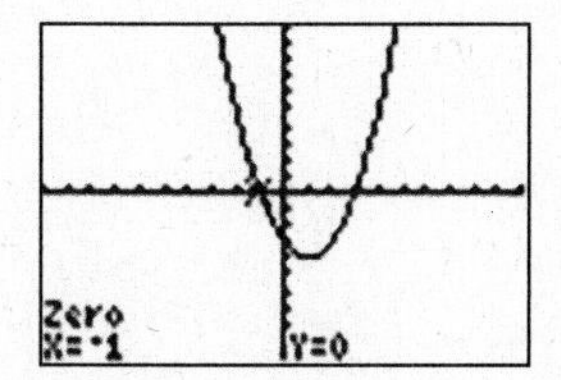

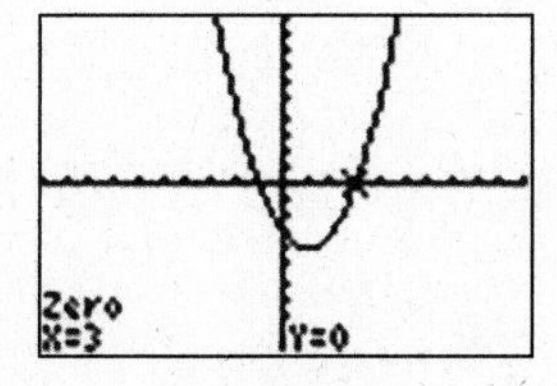

- For the TI-Nspire, enter the equation $f1(x) = x^2 - 6x + 8$ to graph it. Then press [menu], [6], and [1] to find each zero.

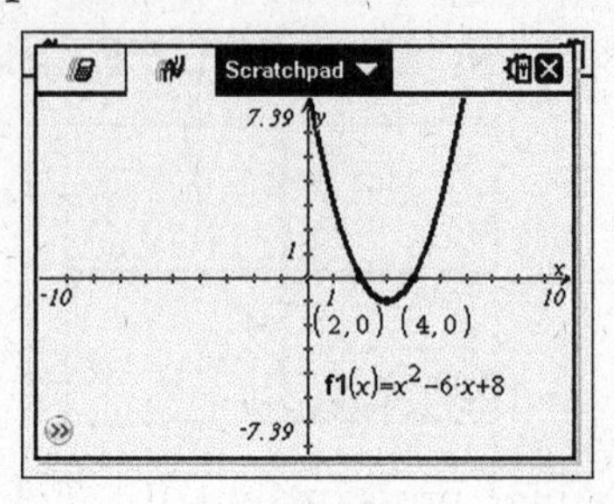

6.5 SOLVING A LINEAR-QUADRATIC SYSTEM OF EQUATIONS BY GRAPHING

When a system of equations has one linear equation and one quadratic equation, one way to find the solution is to graph the line and the parabola and find the coordinates of the intersection point, or intersection points. There can be up to two solutions.

- **For the System of Equations**

$$y = -2x + 1$$
$$y = x^2 - 2x - 3$$

- Create the graph for both equations either by hand or with the graphing calculator. The coordinates of the two intersection points are the two solutions to the system of equations. For this system the two solutions are $x = -2$, $y = 5$ and $x - 2$, $y = -3$.

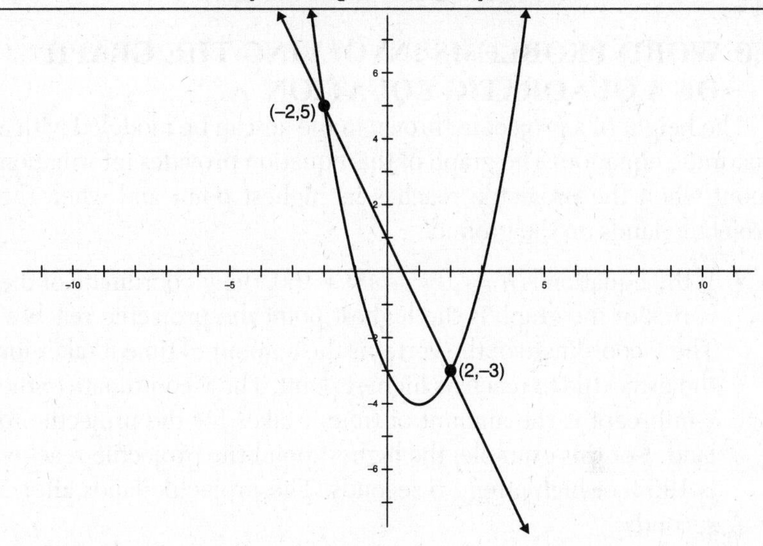

This can also be done on the graphing calculator.

- For the TI-84, graph the two equations and press [2ND], [TRACE], and [5] for each intersection point.

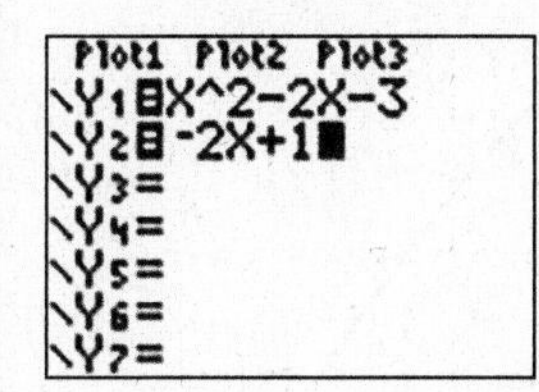

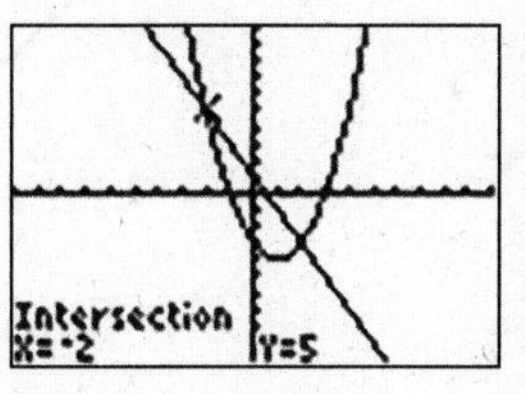

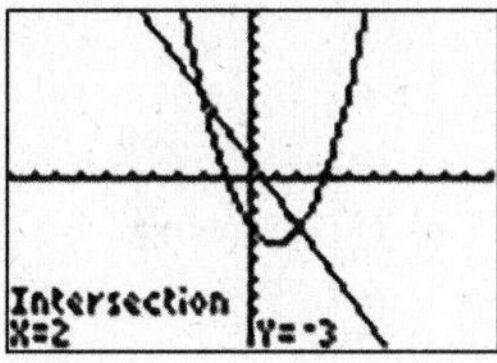

- For the TI-Nspire, graph the two equations and press [menu], [6], and [4] to find the intersection points.

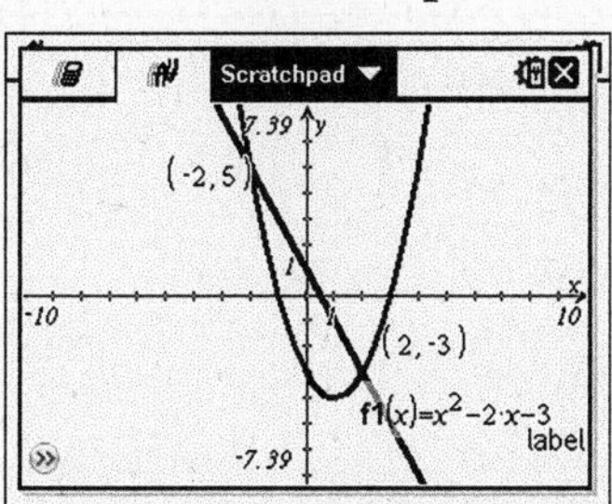

6.6 WORD PROBLEMS INVOLVING THE GRAPH OF A QUADRATIC EQUATION

The height of a projectile thrown in the air can be modeled with a quadratic equation. The graph of this equation provides information about when the projectile reaches its highest point and when the projectile lands on the ground.

- If the equation is $h = -16t^2 + 48t + 160$, the y-coordinate of the vertex of the graph is the highest point the projectile reaches. The x-coordinate of the vertex is the amount of time it takes for the projectile to reach its highest point. The x-coordinate of the x-intercept is the amount of time it takes for the projectile to land. For this example, the highest point the projectile reaches is 196 feet high after 1.5 seconds. The projectile lands after 5 seconds.

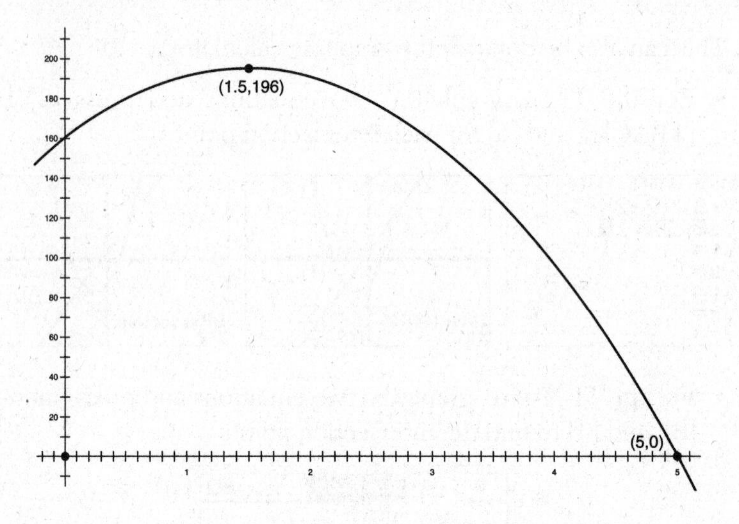

Practice Exercises

1. Which is a point on the graph of the solution set of $y = x^2 + 5x - 2$?

(1) (3, 19) (3) (3, 21)
(2) (3, 20) (4) (3, 22)

2. What are the coordinates of the vertex of the parabola defined by the equation $y = x^2 - 4x - 1$?

(1) (–2, 5) (3) (–2, –5)
(2) (2, 5) (4) (2, –5)

3. $x = -4$ is the x-coordinate of the vertex for the parabola defined by which equation?

(1) $y = x^2 + 8x + 3$
(2) $y = x^2 - 8x + 3$
(3) $y = x^2 + 4x + 3$
(4) $y = x^2 - 4x + 3$

4. What could be the equation that determines this parabola?

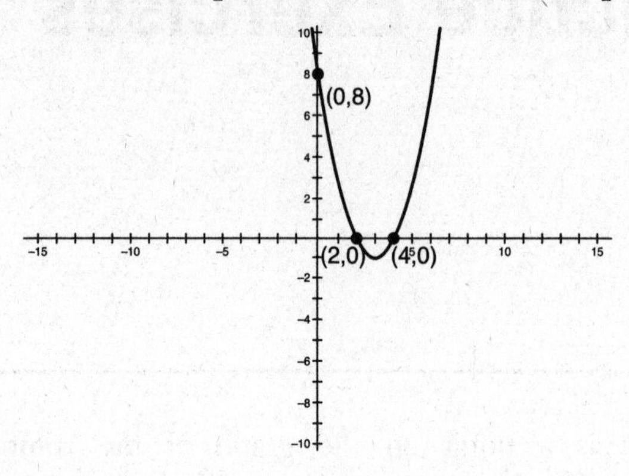

(1) $y = x^2 - 6x - 8$ (3) $y = x^2 + 6x - 8$
(2) $y = x^2 - 6x + 8$ (4) $y = x^2 + 6x + 8$

5. Which is the graph of $y = -x^2 + 2x + 3$?

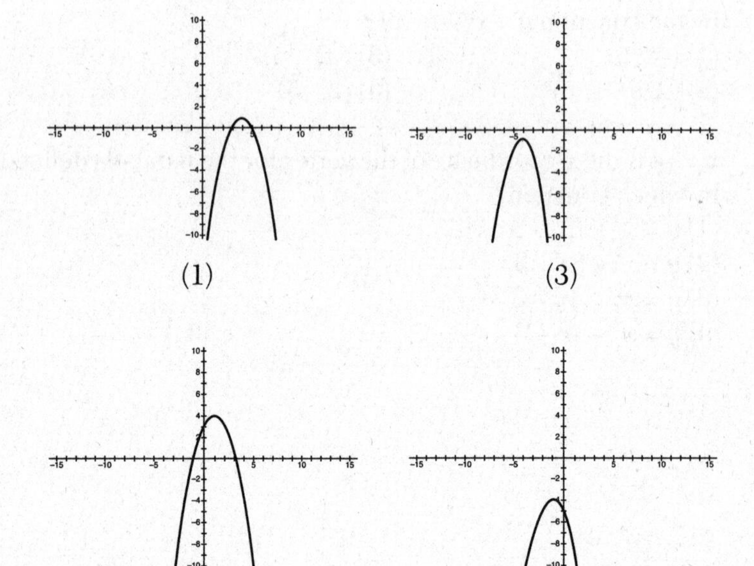

6. Which ordered pair is a solution to the system?

$$y = x^2$$
$$y = x + 2$$

(1) (1, 4) (3) (3, 4)
(2) (2, 4) (4) (4, 4)

7. Solve this system of equations using algebra.

$$y = x^2$$
$$y = 2x + 3$$

(1) (–1, 1) and (9, 3) (3) (–1, 1) and (–3, 9)
(2) (–1, 1) and (3, 9) (4) (1, –1) and (9, 3)

8. Which system of equations could be used to solve this graph?

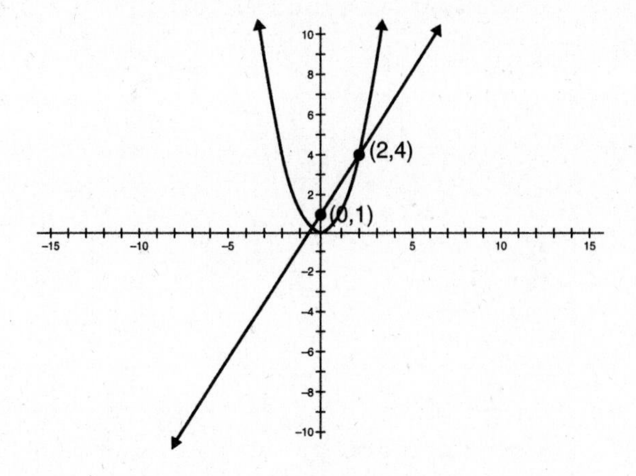

(1) $y = x^2$
$y = \frac{2}{3}x + 1$

(3) $y = x^2$
$y = \frac{3}{2}x + 1$

(2) $y = x^2$
$y = \frac{3}{2}x - 1$

(4) $y = x^2$
$y = \frac{2}{3}x - 1$

9. The x-intercepts of the parabola defined by which equation are the solutions to the equation $x^2 + 5x = 15$?

(1) $y = x^2 + 5x + 15$
(2) $y = x^2 + 5x$
(3) $y = x^2 + 5x - 15$
(4) $y = x^2 - 5x - 15$

10. Solve for all values of x, rounded to the nearest hundredth, $x^2 + 10x + 23 = 0$.

(1) –3.41, –6.58
(2) –3.59, –6.41
(3) –3.31, –6.64
(4) –3.62, –6.18

Solutions

1. For each choice, substitute the x-coordinate in for x and the y-coordinate in for y to see which makes the equation true. 22 does equal $3^2 + 5(3) - 2$ so the solution is (3,22). The correct choice is **(4)**.

2. The x-coordinate of the vertex of a parabola $y = ax^2 + bx + c$ is $x = \frac{-b}{2a}$. Since a is 1 and b is –4, the x-coordinate of the vertex is $x = -\frac{(-4)}{2(1)} = 2$. To get the y-coordinate of the vertex, substitute 2 for x into the equation and solve for y. $y = 2^2 - 4(2) - 1 = 4 - 8 - 1 = -5$. The vertex is (2, –5). The correct choice is **(4)**.

3. Graph each choice on the graphing calculator and use the minimum feature to find the vertex of each. The vertex of the parabola $y = x^2 + 8x + 3$ is (–4, –15), which has an x-coordinate of –4. The correct choice is **(1)**.

4. Since the x-intercepts are (2,0) and (4,0), the equation for the parabola is $y = a(x - 2)(x - 4) = a(x^2 - 6x + 8)$. Since the y-intercept is +8, $a = 1$ and the equation is $y = x^2 - 6x + 8$. The correct choice is **(2)**.

5. The x-coordinate of the vertex must be $x = -\frac{2}{2(1)} = 1$. The correct choice is **(2)**.

6. Substitute x^2 for y in the bottom equation to get $x^2 = x + 2$, $x^2 - x - 2 = 0$, $(x - 2)(x + 1) = 0$, $x = 2$ or $x = -1$. The x-coordinates of the intersection points are 2 and –1. This can also be done with the graphing calculator. The answer is (2,4). The correct choice is **(2)**.

7. Substitute x^2 for y in the bottom equation to get $x^2 = 2x + 3$, $x^2 - 2x - 3 = 0$, $(x - 3)(x + 1) = 0$. $x = 3$ and $x = -1$ are the x-coordinates of the two intersection points. The correct choice is **(2)**.

8. The parabola with vertex (0,0) passing through (2,4) is $y = x^2$. The line through (0,1) and (2,4) is $y = (3/2)x + 1$. The correct choice is **(3)**.

9. This equation can be rewritten as $x^2 + 5x - 15 = 0$. Equations in this form can be solved by finding the x-intercepts of the parabola $y = x^2 + 5x - 15$. The correct choice is **(3)**.

10. Using the graphing calculator, graph $y = x^2 + 10x + 23$ and use the "zero" feature to get $x = 3.59$ or $x = -6.41$. The correct choice is **(2)**.

7. LINEAR INEQUALITIES

7.1 ONE-VARIABLE LINEAR INEQUALITIES

A linear inequality is like a linear equation, but instead of an = sign there is either a >, <, ≥, or ≤ sign. Solving a one-variable linear inequality is almost the same as solving a linear equality. The only difference is that when multiplying or dividing both sides by a negative number to eliminate the coefficient, the direction of the inequality sign must be reversed.

$$-2x + 3 < 11$$

- Eliminate the constant 3 by subtracting it from both sides of the inequality.

$$\begin{aligned} -2x + 3 &< 11 \\ -3 &= -3 \\ -2x &< 8 \end{aligned}$$

- *Divide* both sides by -2 to isolate the x. Because you are dividing by a negative, the direction of the inequality sign must be reversed to keep the equation true. If the -2 were a $+2$, the direction of the inequality sign would not need to be reversed.

$$\frac{-2x}{-2} < \frac{8}{-2}$$

$$x > -4$$

The solution is $x > -4$.

7.2 GRAPHING TWO-VARIABLE INEQUALITIES

$y \leq 2x - 5$ is a two-variable inequality. To graph the two-variable inequality, first graph the line $y = 2x - 5$.

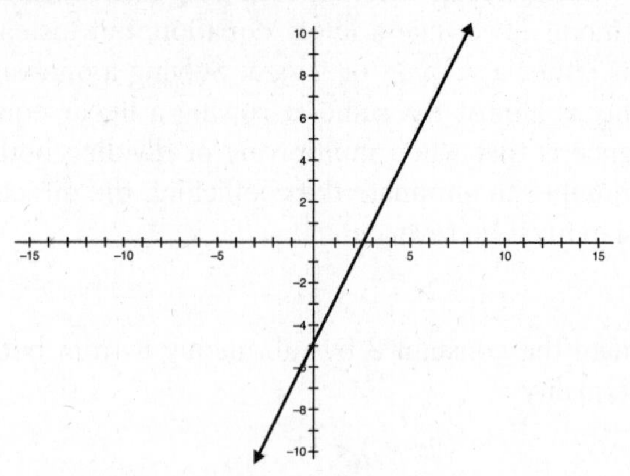

One side of the line needs to be shaded in. To determine which side of the line to shade, substitute the ordered pair (0, 0) into the inequality to test if it yields a true inequality.

$$0 < 2(0) - 5$$
$$0 < 0 - 5$$
$$0 < -5 \text{ is not true.}$$

Since (0, 0) does not satisfy the inequality, shade the side of the line that does not contain (0, 0).

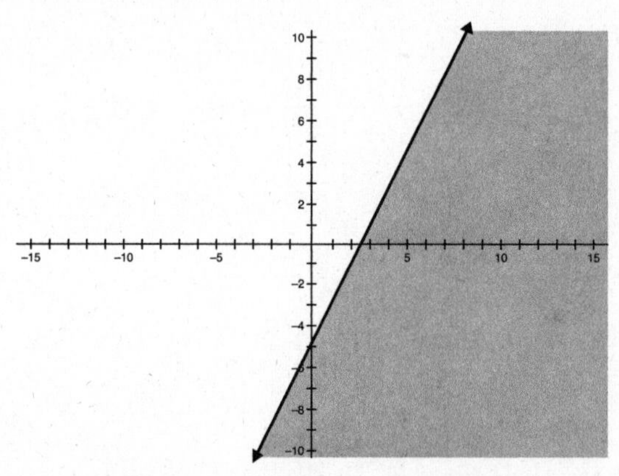

- If the inequality sign is a < or a >, the line must be a dotted line. If it is a ≤ or a ≥, it must be a solid line. If (0, 0) is on the line, an ordered pair that is not on the line must be used to test which side to shade.
- For the graph of $y > 3x$, graph the line $y = 3x$ and make it a dotted line.

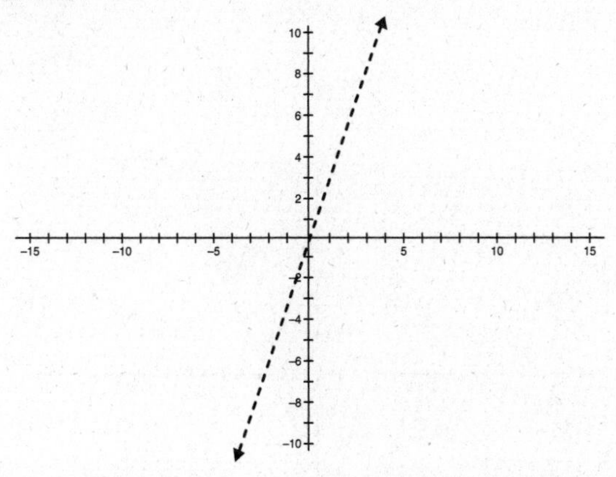

Since (0, 0) is on the line, test a point that is not on the line, like (2, 0).

$$0 > 3(2)$$

Since $0 > 6$ is not true, shade the side that does not contain (2, 0).

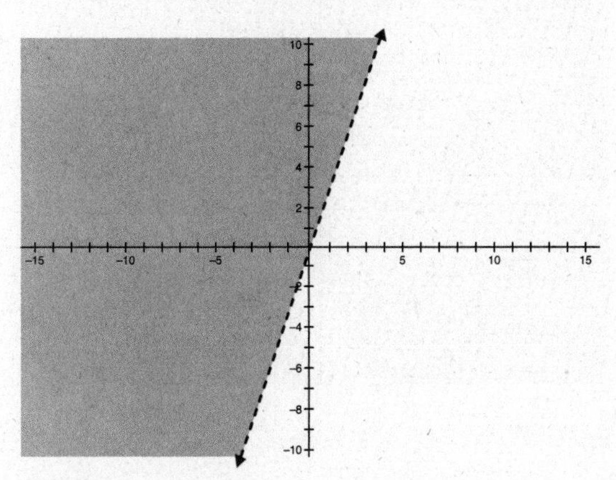

7.3 GRAPHING SYSTEMS OF LINEAR INEQUALITIES

To graph the solution set to a system of linear inequalities, graph both inequalities on the same graph and locate the portion of the graph that is shaded twice.

$$y > 2x - 5$$

$$y < -\frac{2}{3}x + 2$$

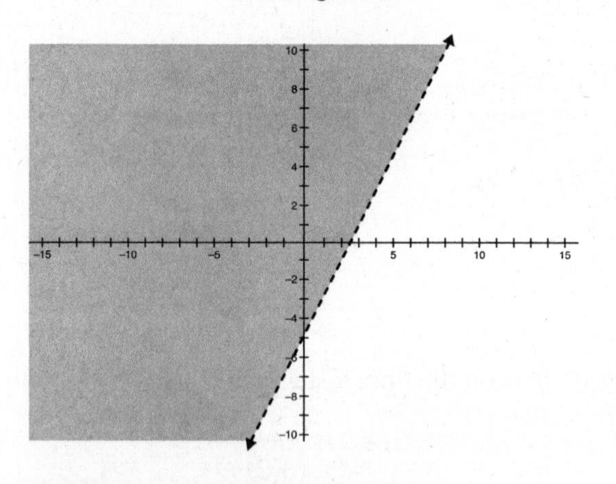

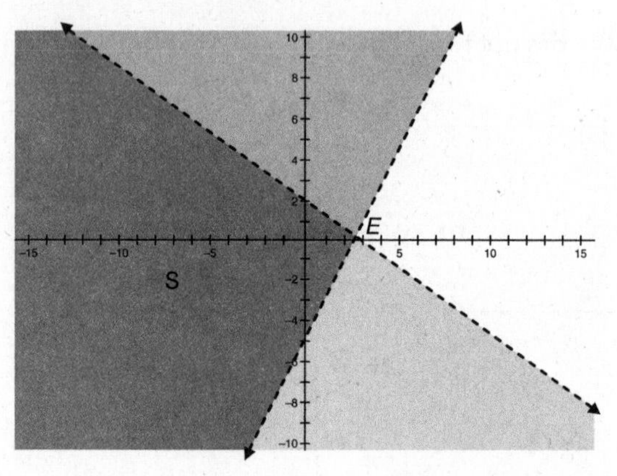

The portion of the graph that has an S is the solution set. Any point in that region will satisfy the system of inequalities. (–3, 1) is an example of a point that is in this double-shaded region and will satisfy both inequalities.

7.4 GRAPHING INEQUALITIES ON THE GRAPHING CALCULATOR

The TI-84 and the TI-Nspire can graph linear inequalities.

- On the TI-84, the shading can be set by moving the cursor to the icon to the left of the Y= and pressing [ENTER] until the icon shows the proper shading.

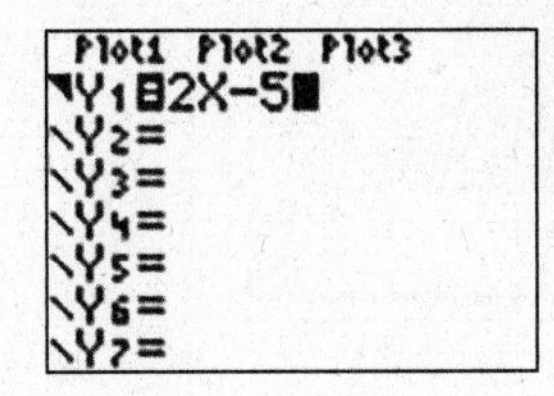

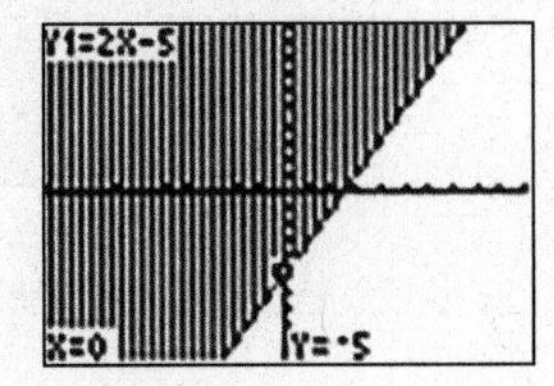

- On the TI-Nspire, delete the "=" from the equation in the entry line and choose the symbol to replace it with.

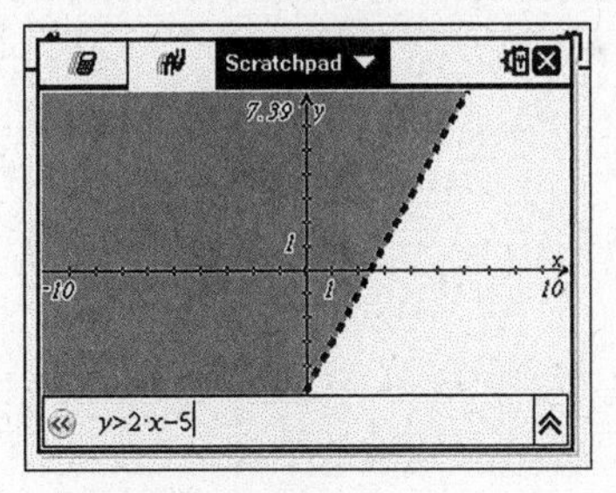

The graphing calculator can also graph systems of linear inequalities by graphing both inequalities on the same set of coordinate axes.

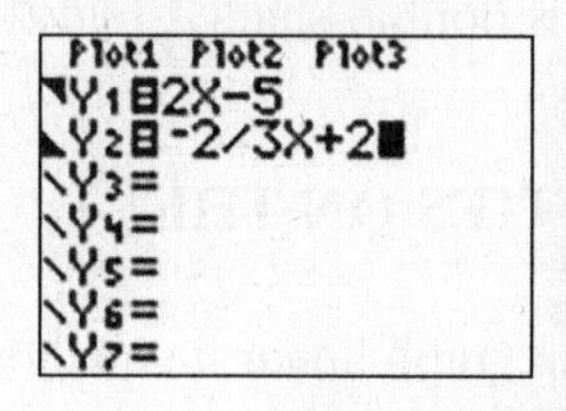

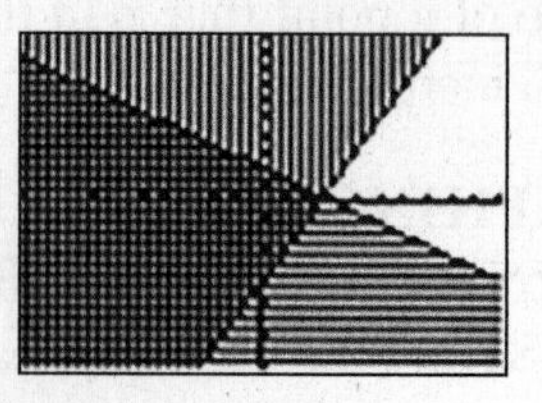

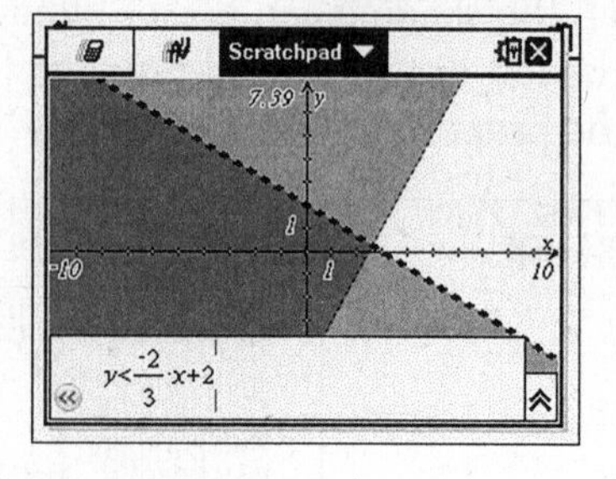

Practice Exercises

1. What is the solution set for $3x < -18$?
   (1) $x < -6$
   (2) $x > -6$
   (3) $x \geq -6$
   (4) $x \leq -6$

2. What is the solution set for $-4x \geq 20$?
   (1) $x \geq -5$
   (2) $x > -5$
   (3) $x \leq -5$
   (4) $x < -5$

3. What is the smallest integer that satisfies the equation $-6x < -18$?
   (1) 3
   (2) 4
   (3) 5
   (4) 6

4. Which is the graph of $y < x + 4$?

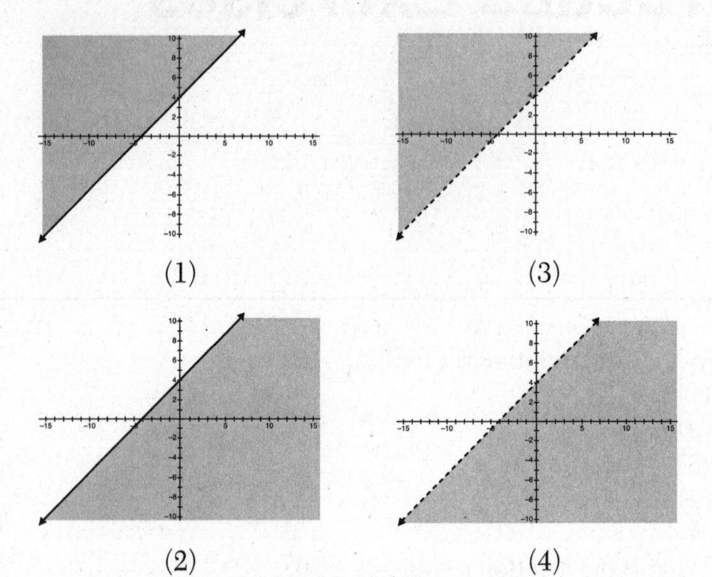

5. Below is the graph of $x + y \leq 8$. Which is a point in the solution set?

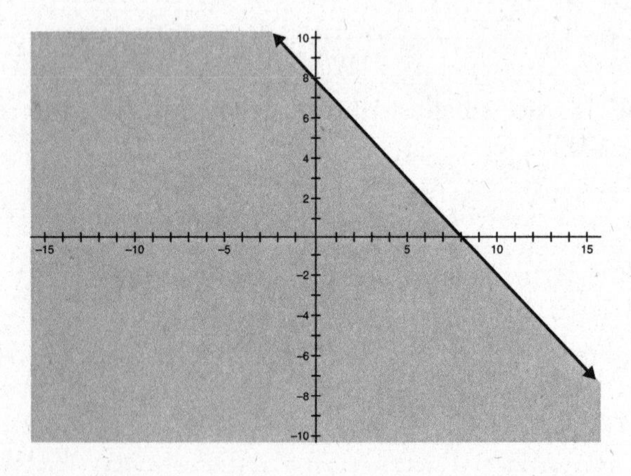

(1) (1, 8)
(2) (2, 7)
(3) (4, 5)
(4) (3, 5)

6. All of these ordered pairs are part of the shaded region for the graph of $2x + y \leq 12$ except

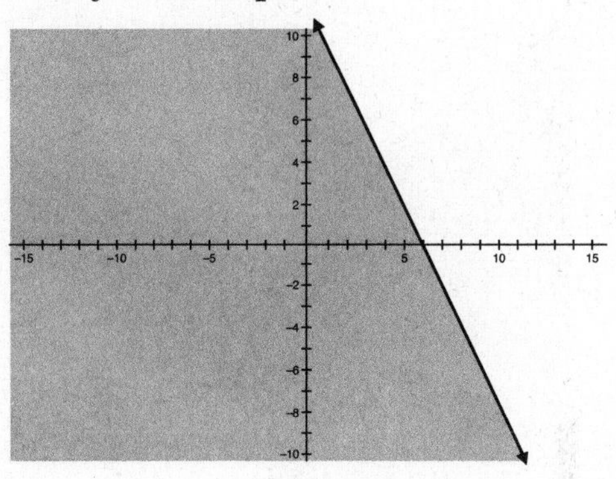

(1) (2, 8)
(2) (3, 6)
(3) (5, 3)
(4) (4, 4)

7. Which graph shows the solution to the system of inequalities?

$$y < 2x + 1$$

$$y > \frac{1}{3}x + 4$$

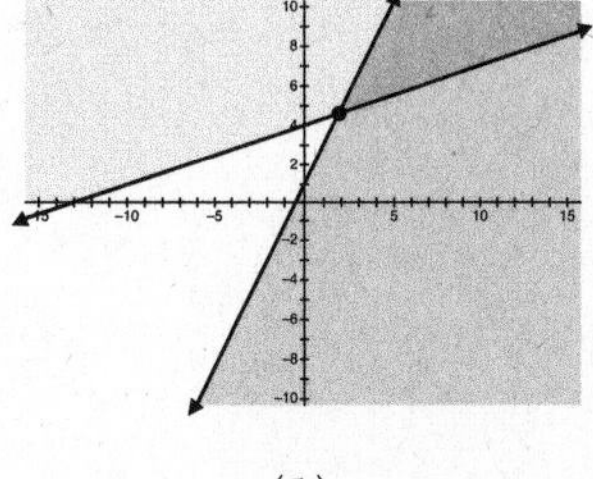

(1)

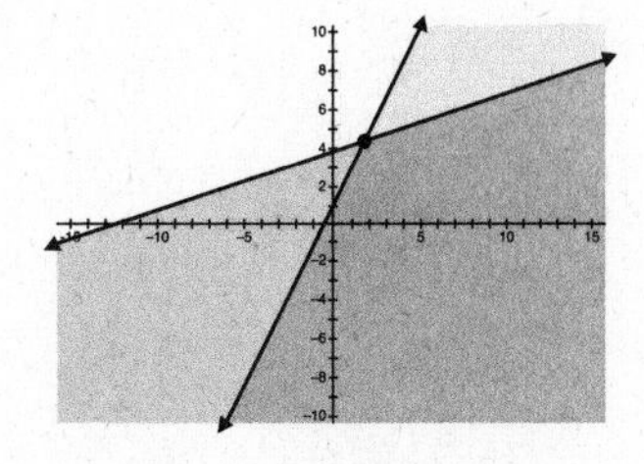

(3)

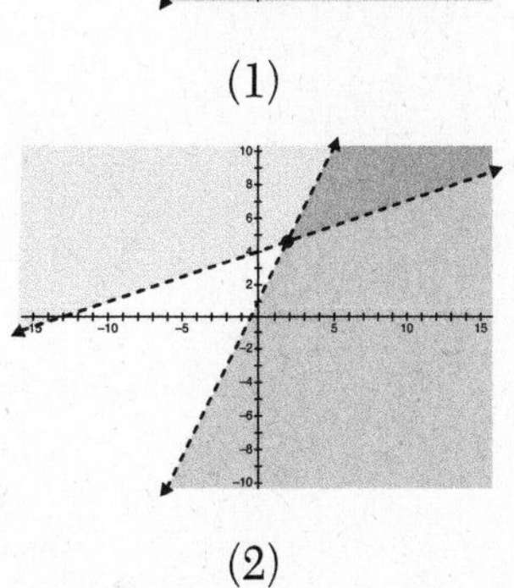

(2)

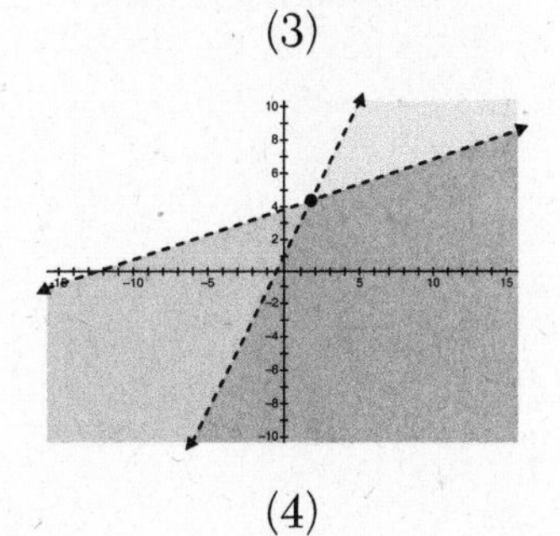

(4)

8. Which system of inequalities does the following graph show the solution for?

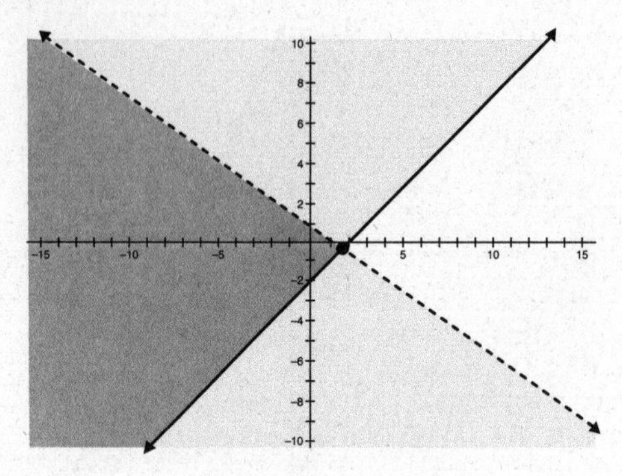

(1) $y \geq x - 2$

$y < \frac{2}{3}x + 1$

(2) $y \leq x - 2$

$y > -\frac{2}{3}x + 1$

(3) $y < x - 2$

$y \geq -\frac{2}{3}x + 1$

(4) $y > x - 2$

$y \leq -\frac{2}{3}x + 1$

9. Which graph has the solution set shaded in for the following system of inequalities?

$$y \leq -x + 6$$

$$y \geq \frac{1}{2}x - 1$$

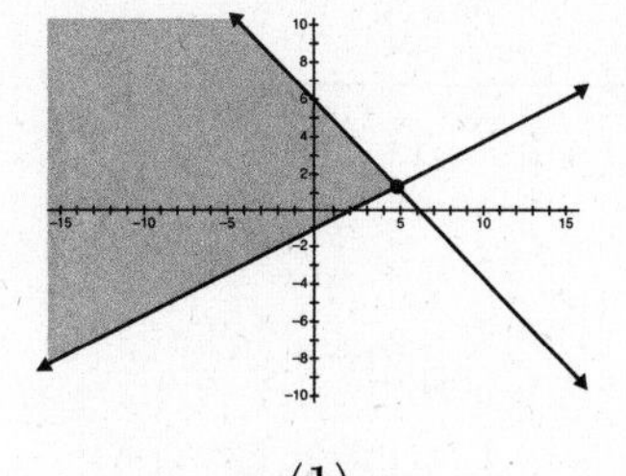

(1)

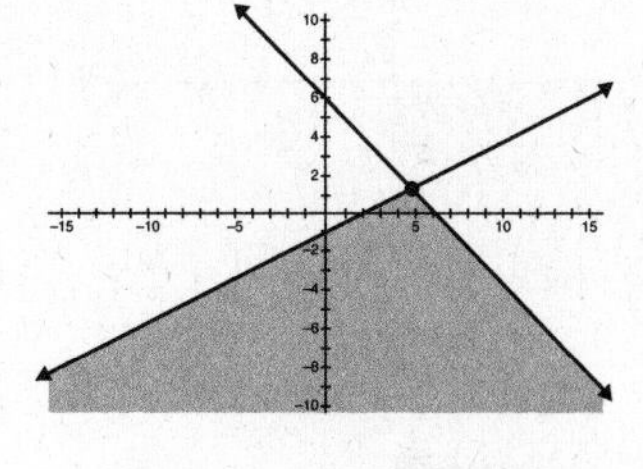

(3)

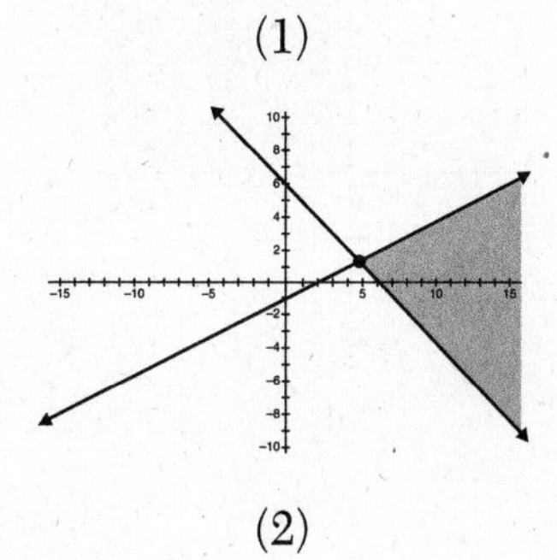

(2)

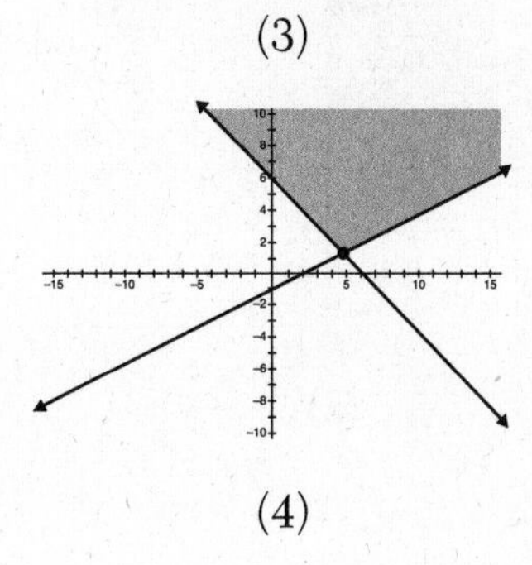

(4)

10. Which is the graph of the following system of inequalities?

$$y \geq 0$$
$$x \leq 0$$

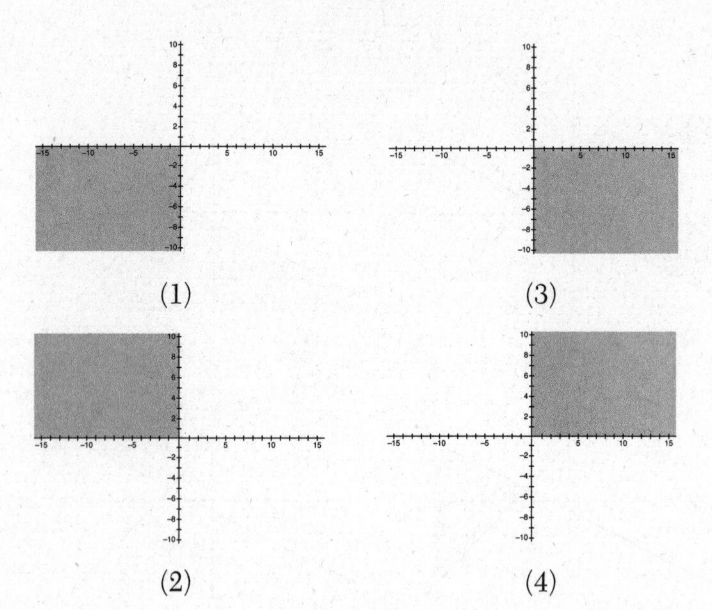

Solutions

1. Divide both sides of the inequality by 3 to get $x < -6$. The correct choice is **(1)**.

2. Divide both sides of the inequality by –4. Switch the direction of the inequality sign because you divided by a negative to get $x \leq -5$. The correct choice is **(3)**.

3. Divide both sides of the inequality by –6 and switch the direction of the inequality sign to get $x > 3$. The smallest integer that satisfies the equation is 4. The answer is not 3 since it is not true that $3 > 3$. The correct choice is **(2)**.

4. Because it is a < and not a ≤ sign, the line must be dotted. Test to see if (0,0) is in the solution set by checking if $0 < 0 + 4$ is true. Since it is true, the side of the line containing (0,0) must be shaded. The correct choice is **(4)**.

5. When all four choices are plotted on the graph, three of them are not in the shaded region. The point (3,5) is on the line, but since there is a ≤ sign, the line is part of the solution set. The correct choice is **(4)**.

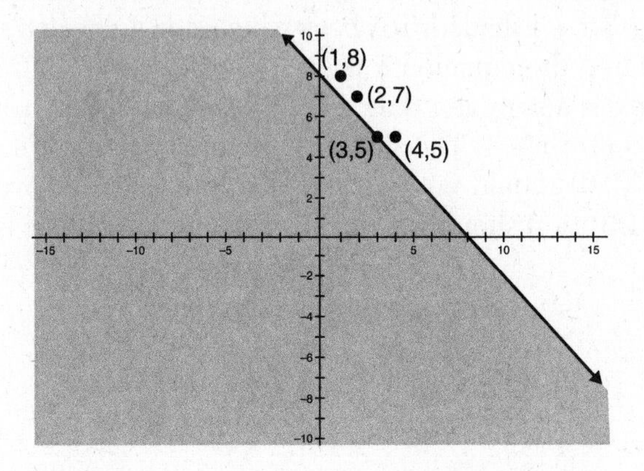

6. Three of the points are on the line, which is part of the solution set because it is a $\leq$ sign. The point (5,3) is not in the shaded area. The correct choice is **(3)**.

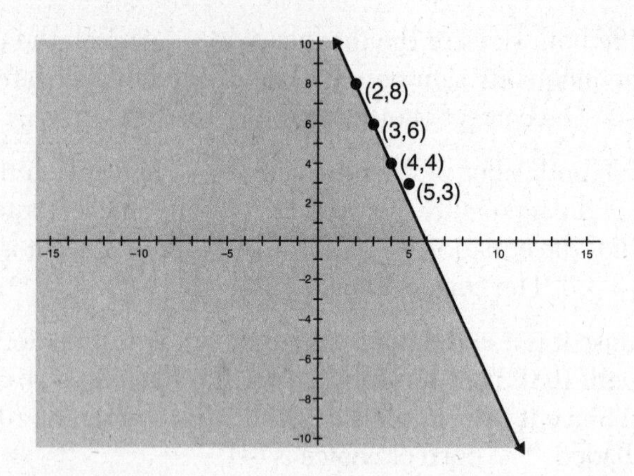

7. Both of the lines have to be dotted because of the inequality signs $<$ and $>$. This eliminates choices 1 and 3. Both choices 2 and 4 have the region below the line $y = 2x + 1$ shaded. The difference between choices 2 and 4 is that choice 2 has the region above $y = (1/3)x + 4$ shaded, and choice 4 has the region below $y = (1/3)x + 4$ shaded. To check which side is correct, substitute (0,0) into the equation $y > (1/3)x + 4$ to get $0 > (1/3)(0) + 4$. Since 0 is not greater than 4, (0,0) is not part of the solution set to $y > (1/3)x + 4$. The side of the line $y = (1/3)x + 4$ that contains (0,0) should not be shaded so the region above the line $y = (1/3)x + 4$ should be shaded. The correct choice is **(2)**.

8. Since the line $y = x - 2$ is solid, choices 3 and 4 can be eliminated. Substituting (0, 0) into both inequalities in choice 1 makes them both true while substituting (0, 0) into both inequalities in choice 2 makes them both false. The correct choice is **(1)**.

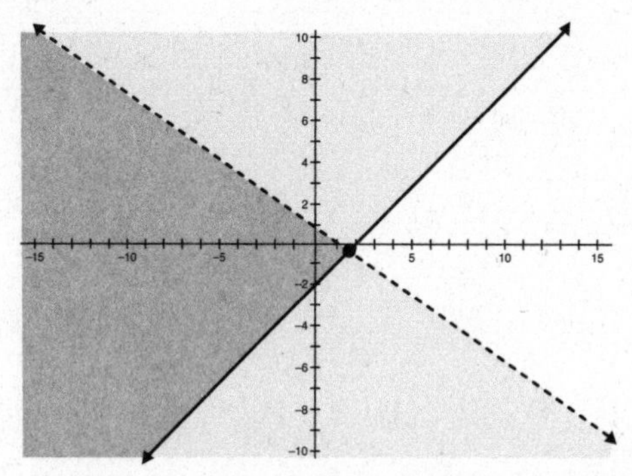

9. To check choice 1, pick a point in the shaded region for that choice and check to see if it satisfies both inequalities. Since (0,0) is a point in the shaded region, check to see if $0 \leq -0 + 6$ and $0 \geq (1/2) \cdot 0 - 1$. (0,0) does make both inequalities true. The correct choice is **(1)**.

10. The points on or above the x-axis make $y \geq 0$ true. The points on or to the left of the y-axis make $x \leq 0$ true. The points that satisfy both inequalities must be above the x-axis and to the left of the y-axis. The correct choice is **(2)**.

8. EXPONENTIAL EQUATIONS

8.1 EVALUATING EXPONENTIAL EXPRESSIONS

An **exponential equation** is one where one of the variables is an exponent. The equation $y = 3 \cdot 2^x$ is a two-variable exponential equation. Exponential equations can be written in the form $y = a \cdot b^x$.

- When evaluating an *exponential expression*, raise the base to the exponent before multiplying. For example, to evaluate $y = 3 \cdot 2^x$ when $x = 4$ it becomes

$$y = 3 \cdot 2^4$$
$$y = 3 \cdot 16 \text{ (NOT } y = 6^4\text{)}$$
$$y = 48$$

8.2 EXPONENTIAL GROWTH VS. EXPONENTIAL DECAY

In the exponential equation $y = a \cdot b^x$, the b is called the **base**. When b is greater than 1, the equation is an example of *exponential growth* since $a \cdot b^x$ grows as x grows. When b is between 0 and 1 the equation is an example of *exponential decay* since $a \cdot b^x$ gets smaller as x grows.

$y = 3 \cdot 2^x$ is an example of exponential growth since $b = 2$, which is greater than 1.

$y = 3 \cdot (0.7)^x$ is an example of exponential decay since $b = 0.7$, which is between 0 and 1.

8.3 GRAPHS OF EXPONENTIAL EQUATIONS

A table of values for a two-variable exponential equation, like $y = 3 \cdot 2^x$, looks like this:

x	y
–3	$3 \cdot 2^{-3} = 3 \cdot \frac{1}{2^3} = 3 \cdot \frac{1}{8} = \frac{3}{8}$
–2	$3 \cdot 2^{-2} = 3 \cdot \frac{1}{2^2} = 3 \cdot \frac{1}{4} = \frac{3}{4}$
–1	$3 \cdot 2^{-1} = 3 \cdot \frac{1}{2^1} = 3 \cdot \frac{1}{2} = \frac{3}{2}$
0	$3 \cdot 2^0 = 3 \cdot 1 = 3$
1	$3 \cdot 2^1 = 3 \cdot 2 = 6$
2	$3 \cdot 2^2 = 3 \cdot 4 = 12$
3	$3 \cdot 2^3 = 3 \cdot 8 = 24$

Remember that raising a number to a negative power, like a^{-x}, is equivalent to $\frac{1}{a^x}$. For example, $2^{-3} = \frac{1}{2^3} = \frac{1}{8}$.

The graph for this equation has a shape like this:

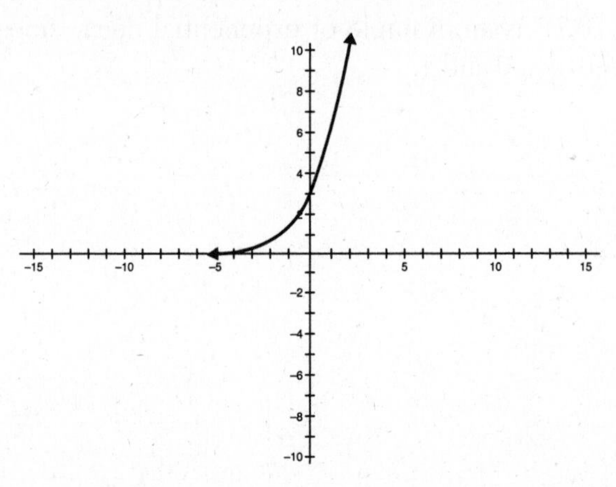

- For exponential decay, like $y = 3 \cdot \left(\frac{1}{2}\right)^x$, the graph has a shape like this:

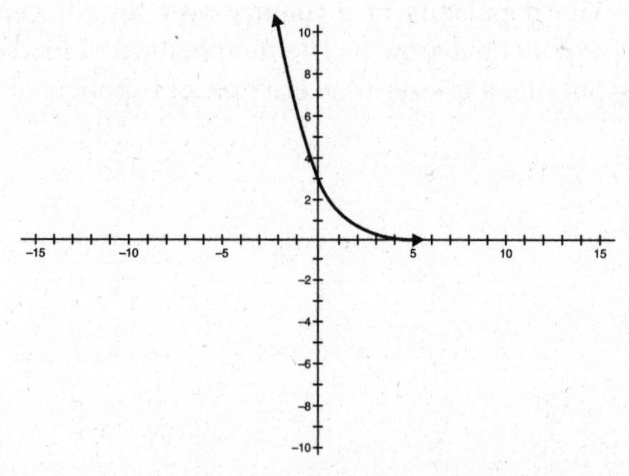

Exponential equations can also be graphed on the graphing calculator.

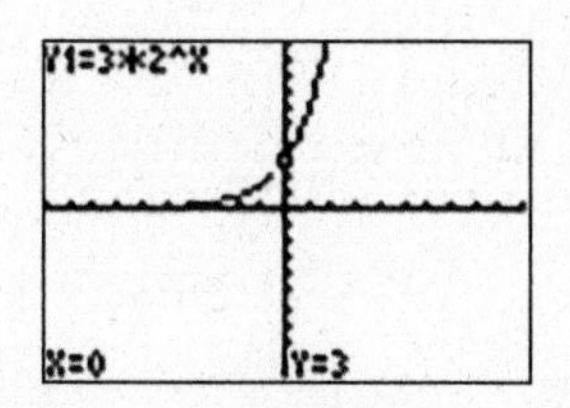

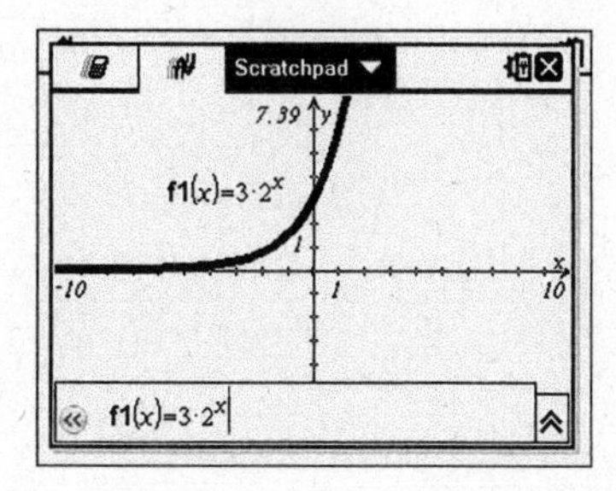

8.4 REAL-WORLD SCENARIOS INVOLVING EXPONENTIAL EQUATIONS

Many real-world scenarios can be modeled with exponential equations. The population of a country over time is generally an example of exponential growth. The temperature of food over time after being put into a freezer is an example of exponential decay.

Practice Exercises

1. If $x = 2$ and $y = 3^x$, solve for y.
(1) 8
(2) 9
(3) 10
(4) 11

2. If $x = 3$ and $y = 2 \cdot 3^x$, solve for y.
(1) 216
(2) 27
(3) 54
(4) 18

3. Which ordered pair is in the solution set of $y = 5 \cdot 2^x$?
(1) (0, 0)
(2) (2, 25)
(3) (3, 40)
(4) (3, 100)

4. Which is the graph of $y = 2^x$?

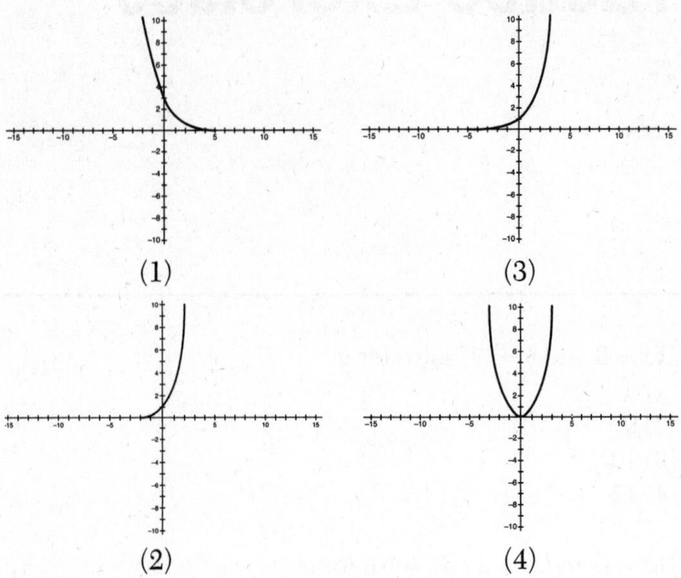

5. Below is the graph of which equation?

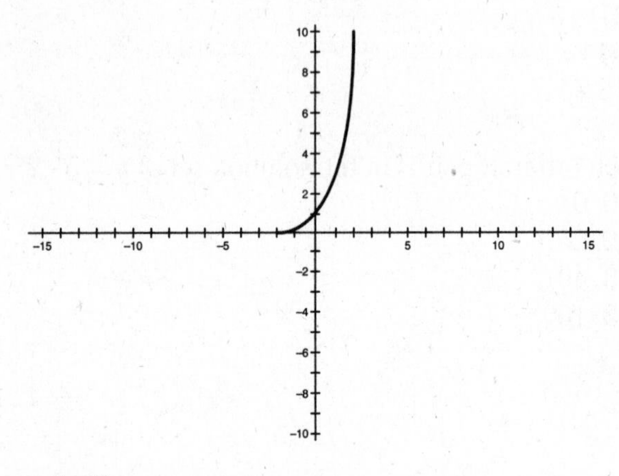

(1) $y = \left(\frac{1}{3}\right)^x$

(2) $y = 10^x$

(3) $y = 4^x$

(4) $y = 3^x$

6. In what interval is the graph of $y = 1.5^x$ increasing?

(1) Always | (3) When $x \geq 0$
(2) Never | (4) When $x \leq 0$

7. Below is the graph of $y = b^x$. What is true about the value of b?

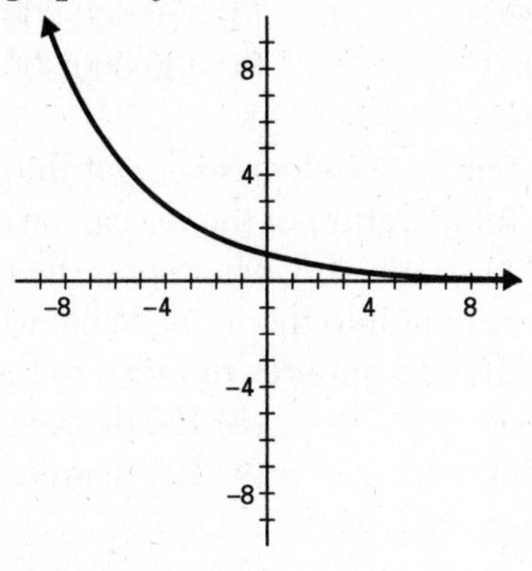

(1) b must be greater than 1
(2) b must be less than 1
(3) b must be less than 0
(4) b must be less than –1

8. What type of equation has a graph like the one below?

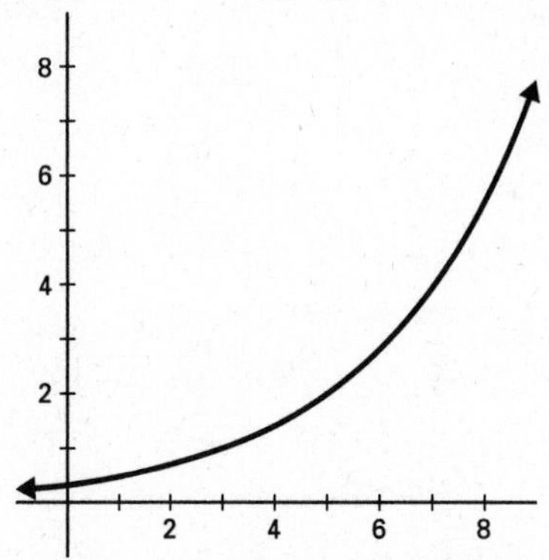

(1) Linear | (3) Quadratic
(2) Exponential | (4) None of the above

9. The population of a country can be modeled with the equation $P = 250 \cdot 1.07^t$, where P is the population in millions and t is the number of years since 2010. According to this model, rounded to the nearest ten million, what will the population of this country be in 2019?

(1) 450,000,000
(2) 460,000,000
(3) 470,000,000
(4) 480,000,000

10. A cup of tea that is 200 degrees is put into a room that is 80 degrees. The temperature of the tea can be calculated with the formula $t = 200 \cdot 0.9^m + 80$, where m is the number of minutes since the tea was put into the room. What will the temperature of the tea be after 15 minutes rounded to the nearest degree?

(1) 116 degrees
(2) 118 degrees
(3) 120 degrees
(4) 121 degrees

Solutions

1. Substitute 2 for x and the equation becomes $y = 3^2 = 9$. The correct choice is **(2)**.

2. Substitute 3 for x and the equation becomes $y = 2 \cdot 3^3 = 2 \cdot 27 = 54$. The correct choice is **(3)**.

3. Substitute each ordered pair into the equation to see which makes it true. For the ordered pair (3, 40), $5 \cdot 2^3 = 5 \cdot 8 = 40$. The correct choice is **(3)**.

4. The graph for $y=2^x$ must pass through the points (0,1), (1,2), and (2,4). It also passes through (–1,1/2) and (–2,1/4). The correct choice is **(3)**.

5. Since this graph passes through (0,1), (1,3), and (2,9), it is the equation $y = 3^x$. The correct choice is **(4)**.

6. An exponential graph with a base greater than 1 is always increasing. The correct choice is **(1)**.

7. When an exponential graph is decreasing, the base is between 0 and 1. The correct choice is **(2)**.

8. This graph has the standard shape of an exponential curve based on an equation $y = b^x$ with $b > 1$. An exponential curve like this starts off relatively flat for x values close to zero and then increases rapidly. The correct choice is **(2)**.

9. Since 2019 is 9 years past 2010, substitute 9 for t to get $P = 250 \cdot 1.07^9 = 459.6$. In millions, this is approximately 460,000,000. The correct choice is **(2)**.

10. Substitute 15 for m to get $t = 200 \cdot 0.9^{15} + 80 = 121.1782$ or 121.2, which rounds to 121. The correct choice is **(4)**.

9. CREATING AND INTERPRETING EQUATIONS

9.1 CREATING AND INTERPRETING LINEAR EQUATIONS

Many real-world situations can be modeled with a linear equation in the form $y = mx + b$. The m and the b have to be replaced with the appropriate values from the situation. In general, the b value is the starting value and the m value is the amount the total changes each time the x variable increases.

- If a carnival costs $10 admission and $3 for each ride, the 10 and the 3 can be used in a linear equation. Since the 10 is the starting amount, it would take the place of the b in the equation. Since 3 is the amount the total increases by for each new ride, it would take the place of m in the equation.

  The equation could be written $y = 3x + 10$, where y is the total and x is the number of rides. Instead of x and y, the total could be represented by the variable T, whereas the number of rides could be represented by the variable R to form the equation $T = 3R + 10$.

  When given an equation modeling a real-world situation, it is also possible to interpret what the values for m and b represent. The b represents the starting value, and the m represents the amount that the total changes for each increase in x.
- If the equation for the cost of a pizza with N toppings is $C = 2N + 12$, you could be asked to interpret what the 2 and the 12 represent. Since the 2 is in the place of the m in $y = mx + b$, it represents the cost of each topping. Since the 12 is in the place of the b in $y = mx + b$, it represents the cost of the pizza before any toppings are added.

9.2 CREATING AND INTERPRETING EXPONENTIAL EQUATIONS

An **exponential equation** is a good model for many real-world situations including population growth, compound interest, and liquid cooling. Exponential equations have the form $y = a \cdot (1 + r)^x$, where the a represents the starting value and the r represents the growth rate.

- If the population of a town is 10,000 people and the annual growth rate is 7%, then the equation that relates total population, P, to the number of years that have passed, T, is $P = 10{,}000 \cdot 1.07^T$. Since 10,000 is the starting value, it takes the place of the a and since the growth rate is .07, it takes the place of the r in the equation $y = a \cdot (1 + r)^x$.

- If an exponential equation that models a real-world situation is given, it is possible to interpret what the numbers in the equation represent.

- If in another town, the equation relating their population to the number of years that have passed is $P = 20{,}000 \cdot 1.09^T$, you could be asked to interpret what the numbers 20,000 and .09 represent. In this case, the 20,000 represents the starting population and the .09 represents the growth rate.

- When a ball is dropped, each bounce is 80% as high as the previous bounce. If the ball is dropped from a window 50 feet above the ground, the equation that relates the height of the bounce (H) to the number of bounces (B) will be $H = 50 \cdot 0.80^B$. Since the starting height is 50, it takes the place of the a, and since 0.80 can be expressed as $(1 - 0.20)$, the r value is -0.20, which is the growth rate. A -0.20 growth rate can also be called a *decay* rate of 0.20.

Practice Exercises

1. It costs \$10 to go to the movies and \$3 for each bag of popcorn. Which equation relates the total cost (C) to the number of bags of popcorn purchased (P)?
(1) $C = 3P + 10$
(2) $C = 10P + 3$
(3) $P = 3C + 10$
(4) $P = 10C + 3$

2. A tablet computer costs \$400 and \$2 for each app. Which equation relates the total cost (C) to the number of apps purchased (A)?
(1) $A = 2 + 400C$
(2) $A = 400 + 2C$
(3) $C = 2 + 400A$
(4) $C = 400 + 2A$

3. A cable TV plan costs \$80 a month plus \$10 extra for each premium channel. Which equation relates the monthly bill (B) to the number of premium channels ordered (C)?
(1) $B = 80C + 10$
(2) $B = 10C + 80$
(3) $C = 80B + 10$
(4) $C = 10B + 80$

4. Lydia wants to buy a DVD player and some DVDs. The equation that relates the total cost for the DVD player and N DVDs is $P = 20N + 200$. What does the number 200 in the equation represent?
(1) The cost of the DVD player
(2) The cost of each DVD
(3) The cost of all N DVDs
(4) The total cost of the DVD player and all N DVDs

5. Amelia buys an empty sticker album and some sticker sheets. The equation that relates the total cost for the empty sticker album and N sticker sheets is $P = 0.75N + 3.00$. What does the number 0.75 in the equation represent?
(1) The cost of the empty sticker album
(2) The cost of each sticker sheet
(3) The cost of all N sticker sheets
(4) The total cost of the empty sticker album and all N sticker sheets

6. There were 900 birds in a forest. Each year the bird population increases by 12%. Which equation relates the bird population (P) to the number of years that have passed?
(1) $P = 900(1.12)^t$
(2) $P = 900(0.12)^t$
(3) $P = 900(0.88)^t$
(4) $t = 900(1.12)^P$

7. A bouncing ball is dropped from 20 feet high. After each bounce, the height of the next bounce is 65% as high as the last bounce. Which equation relates the height of the bounce (H) to the number of bounces that have happened (N)?
(1) $H = 20(0.35)^N$
(2) $H = 20(1.65)^N$
(3) $H = 20(0.65)^N$
(4) $N = 20(0.65)^H$

8. The population (P) of a town after t years can be modeled with the equation $P = 20{,}000(1.07)^t$. What does the 20,000 represent?
(1) The growth rate
(2) The percent increase each year
(3) The population after t years
(4) The starting population of the town

9. After Allie takes some medicine, each hour the number of milligrams of medicine (M) remaining in her body after t minutes can be modeled with the equation $M = 200(1 - 0.27)^t$. Which number represents the decay rate?
(1) 0.27
(2) 200
(3) 0.73
(4) 146

10. Mason puts money into a bank that offers interest compounded annually. The formula relating the amount of money in the bank (A) to the number of years it has been in the bank (t) is $A = 800(1.2)^t$. What is the interest rate the bank offers?
(1) 1.2%
(2) 2%
(3) 20%
(4) 120%

Solutions

1. P bags of popcorn cost \$3P. The entrance price is \$10 so the total price is $3P + 10$. The correct choice is **(1)**.

2. A apps cost \$2A. The computer costs \$400 so the total price is $2A + 400$ or $400 + 2A$. The correct choice is **(4)**.

3. C premium channels cost \$10C. The monthly cost is \$80 so the total price is $10C + 80$. The correct choice is **(2)**.

4. The 20 is the cost for each DVD and the 200 is the price of the DVD player. In general, the constant represents the fixed cost. The correct choice is **(1)**.

5. The 3 is the part that does not depend on the value of N. The $0.75N$ increases by 0.75 each time N increases by 1 so the 0.75 represents the cost of each sticker sheet. The correct choice is **(2)**.

6. In an exponential equation of the form $P = a \cdot (1 + r)^t$, the r value is the percent increase (or decrease if r is negative), and the a is the initial value. When $r = 0.12$ and $a = 900$, the equation is $P = 900(1.12)^t$. The correct choice is **(1)**.

7. The height of each bounce is 0.65 multiplied by the height of the previous bounce. The initial height is 20 feet. After one bounce, the second is $20 \cdot 0.65$. After two bounces it has a height of $20 \cdot 0.65 \cdot 0.65$. In general, after N bounces, the height will be $H = 20 \cdot 0.65^N$. The correct choice is **(3)**.

8. In an equation of the form $P = a \cdot (1 + r)^t$, the r is the percent increase each year and the a is the initial value at $t = 0$. The 20,000 then represents the starting population of the town. The correct choice is **(4)**.

9. In an equation of the form $M = a \cdot (1 - r)^t$, the r is the decay rate. Since the $(1 - 0.27)$ is being raised to the t power in this equation, the decay rate is 0.27. The correct choice is **(1)**.

10. The interest rate is the percent increase each year. The percent increase in an equation of the form $A = P \cdot (1 + r)^t$ is the r variable. Since $1.2 = 1 + 0.2$, the r value is 0.2, which is 20%. The correct choice is **(3)**.

10. FUNCTIONS

10.1 DIFFERENT REPRESENTATIONS OF FUNCTIONS

A **function** is like a machine that takes a number as an input and outputs a number. Functions are often named with lowercase letters like f and g.

- If a function f takes the number 2 as an input and outputs the number 7, we say $f(2) = 7$. The number in the parentheses is the number that is input into the function. The number after the equals sign is the number that is output from the function.

A function can be represented in several different ways. Here are the most common ways:

1. As a list of ordered pairs

- If the function f is defined as $f = \{(1,4), (2,7), (3,10), (4,13), (5,16)\}$, the numbers in the parentheses represent an input value and an output value for the function. In this example, the point (1,4) in the definition means that if 1 is put into the function, 4 is output from the function, or $f(1) = 4$. Likewise, $f(2) = 7$, $f(3) = 10$, $f(4) = 13$, and $f(5) = 16$.

2. As an equation

- A function can be defined by an equation that allows you to calculate the output value for a given input value. An example is the definition $f(x) = 3x + 1$. With this definition, it is possible to calculate the output value for any input value. For example, to calculate $f(10)$, substitute the number 10 into the equation $f(x) = 3x + 1$ to become $f(10) = 3(10) + 1 = 30 + 1 = 31$ so $f(10) = 31$.

3. As a graph

- If the function is defined as a graph, determine the value of $f(2)$ by finding a point on the graph that has an x-coordinate of 2. The y-coordinate of that point is the value of $f(2)$. Since the point with x-coordinate of 2 is (2,7) the value of $f(2) = 7$.

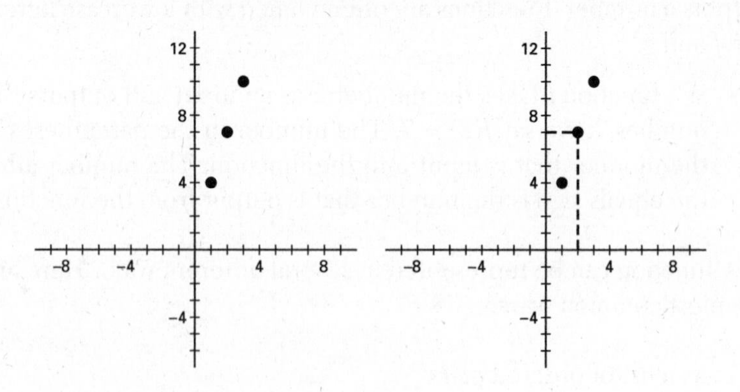

10.2 DOMAIN AND RANGE OF FUNCTIONS

The possible input values of a function are called the *domain* of the function. The possible output values are called the *range* of the function.

For the function $f = \{(1, 4), (2, 7), (3, 10), (4, 13), (5, 16)\}$, the domain is the set of input values $\{1, 2, 3, 4, 5\}$. The range is the set of output values $\{4, 7, 10, 13, 16\}$.

- When a function is described as a graph, the domain is the set of x-coordinates of all the points on the graph, and the range is the set of y-coordinates of all the points on the graph.

For this graph, the domain is {1, 2, 3, 4}, and the range is {3, 5, 8}.

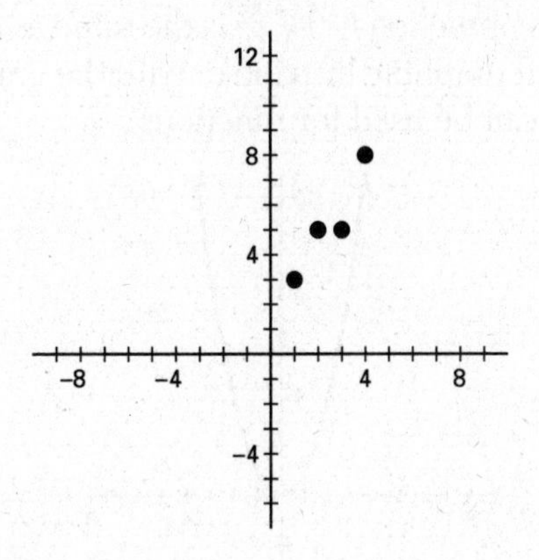

For this graph, the domain is $2 \leq x \leq 5$, and the range is $3 \leq y \leq 7$.

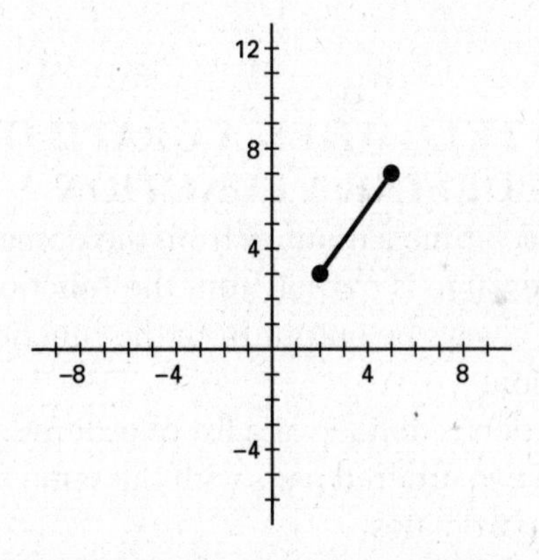

In a real-world situation, the domain is often a special subset of the real numbers. If the function has as its input the number of cars a salesman sells in a month, the domain would be the set of non-negative integers {0, 1, 2, ...} since you cannot sell fractions of a car or negative cars.

10.3 GRAPHING FUNCTIONS

The graph of the function $f(x) = x^2$ is the same as the graph $y = x^2$. All the methods for graphing by hand or with the graphing calculator described earlier can be used for functions.

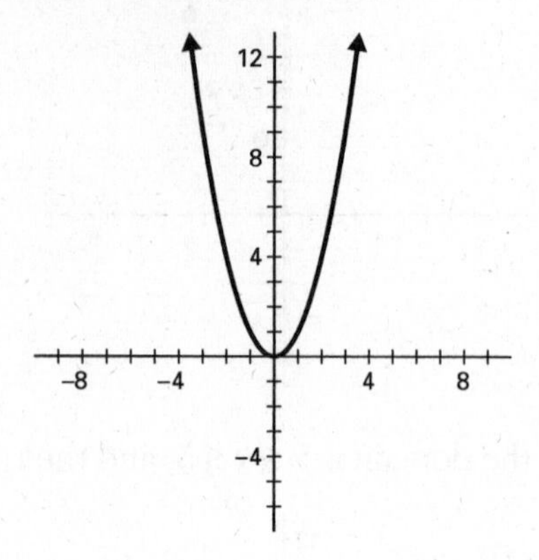

10.4 HOW TO TELL WHEN A GRAPH OR A LIST CANNOT DEFINE A FUNCTION

In a function, each time a number from the domain is put into the function, the same value is output from the function. So if $f(2) = 7$, the number 7 will always be output from the function whenever 2 is put into the function.

- When a function is defined as a list of ordered pairs, then there will never be two ordered pairs with the same x-coordinate, but different y-coordinates.

$f = \{(1, 4), (2, 7), (2, 10), (3, 13)\}$ is not the definition of a function since $f(2)$ can be 7 or 10.

$g = \{(1, 4), (2, 7), (3, 7), (4, 13)\}$ is the definition of a function. The fact that $g(2) = 7$ and $g(3) = 7$ does not contradict the definition of a function. As long as there are no repeats of x-coordinates, there can be repeats of y-coordinates.

A graph will not be the graph of a function if there are two points that have the same x-coordinate, but different y-coordinates. Graphs that cannot be functions fail the **vertical line test**, which means that at least one vertical line will pass through two or more points. All the points it passes through will have the same x-coordinates but different y-coordinates.

These graphs cannot be graphs of functions since they fail the vertical line test at at least one location.

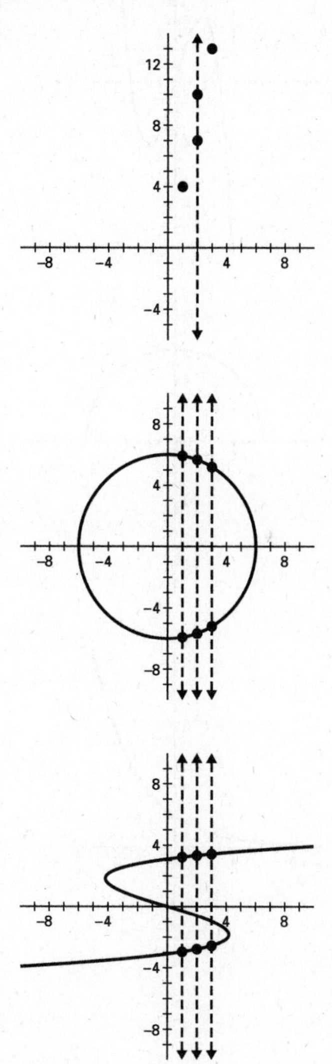

These graphs can be graphs of functions since they pass the vertical line test at all locations.

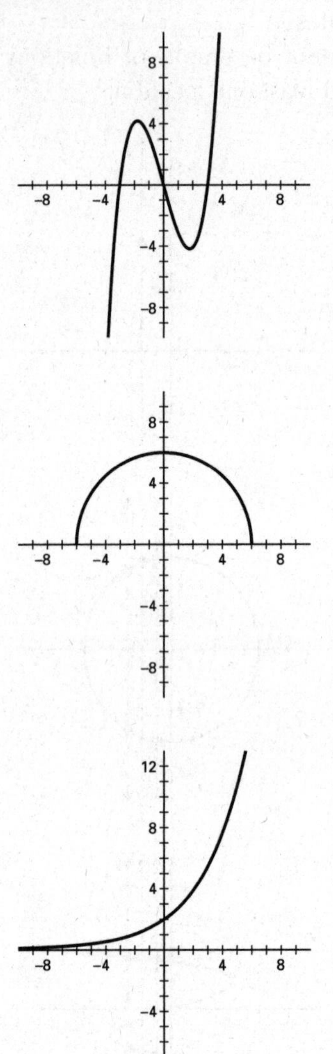

10.5 GRAPHING TRANSFORMED FUNCTIONS

- If the graph of $y = f(x)$ is already known, the graph of $y = f(x + a)$, $y = f(x - a)$, $y = f(x) + a$, and $y = f(x) - a$ can be easily graphed by knowing the four basic transformations.
- If the graph of $y = f(x)$ looks like this:

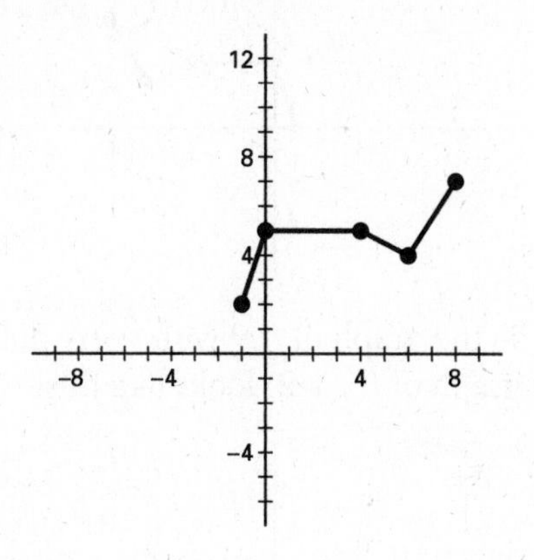

then the graph of

1. $f(x) + a$ will be the graph of $f(x)$ with every point shifted *up* by a units. The graph of $f(x) + 2$ looks like this:

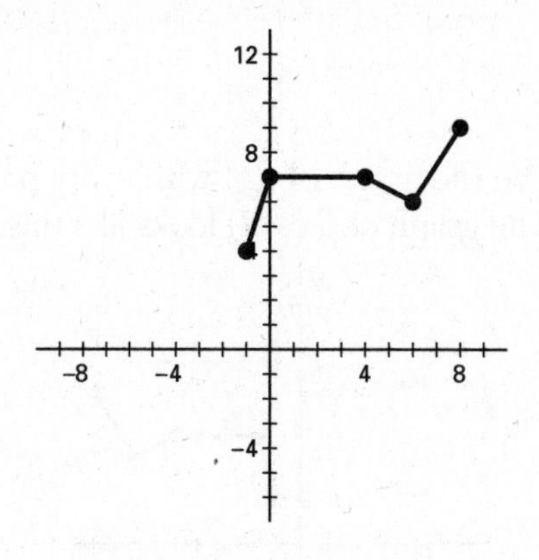

2. $f(x) - a$ will be the graph of $f(x)$ with every point shifted *down* by a units. The graph of $f(x) - 2$ looks like this:

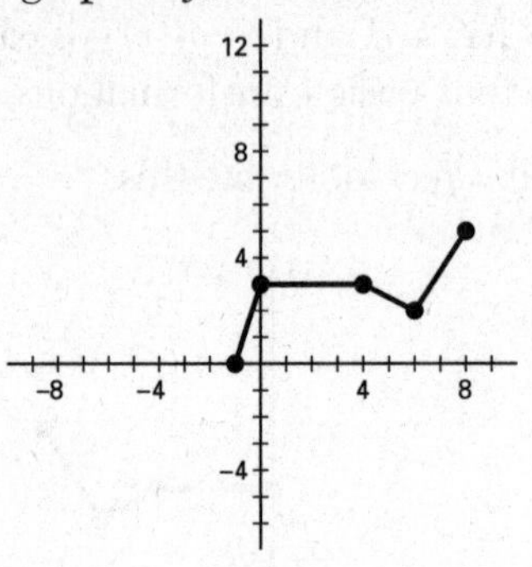

3. $f(x + a)$ will be the graph of $f(x)$ with every point shifted *left* by a units. The graph of $f(x + 2)$ looks like this:

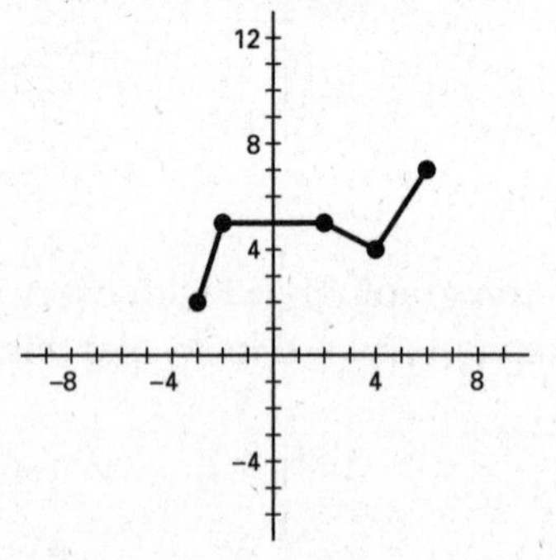

4. $f(x - a)$ will be the graph of $f(x)$ with every point shifted *right* by a units. The graph of $f(x - 2)$ looks like this:

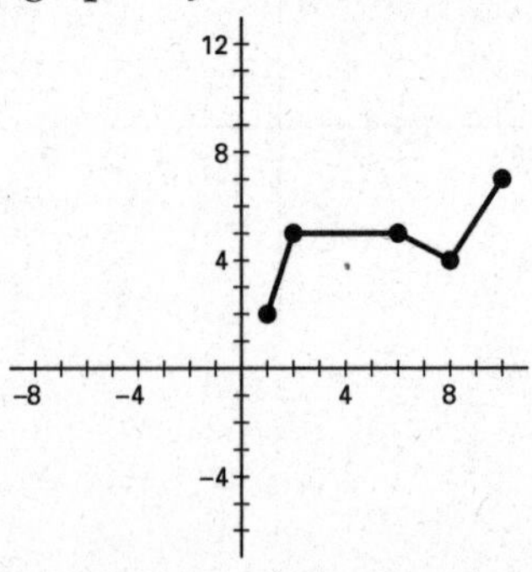

Practice Exercises

1. If a function f is defined as $f = \{(1, 2), (2, 3), (3, 1), (4, 4)\}$, what is $f(2)$?
(1) 1
(2) 2
(3) 3
(4) 4

2. Which of the following *cannot* be the definition of a function?
(1) $f = \{(1, 5), (2, 7), (2, 8), (4, 9)\}$
(2) $f = \{(1, 2), (2, 2), (3, 2), (4, 2)\}$
(3) $f = \{(0, 0), (1, 1), (-1, 1), (2, 4), (-2, 4)\}$
(4) $f = \{(6, 1)\}$

3. What is the domain of the function defined as $f = \{(1, 4), (3, 7), (4, 8), (5, 8)\}$?
(1) $\{4, 7, 8\}$
(2) $\{1, 3, 4, 5, 7, 8\}$
(3) $\{1, 3, 4, 5\}$
(4) $\{4\}$

4. Below is the graph of $y = f(x)$. What is the value of $f(3)$?

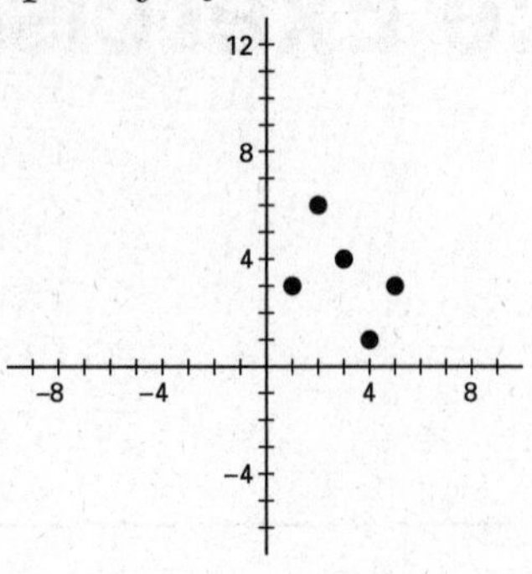

(1) 1
(2) 2
(3) 3
(4) 4

5. Which is the graph of a function?

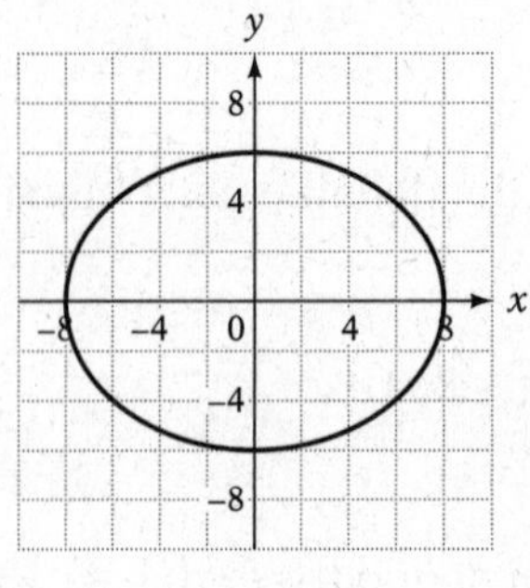

(1)

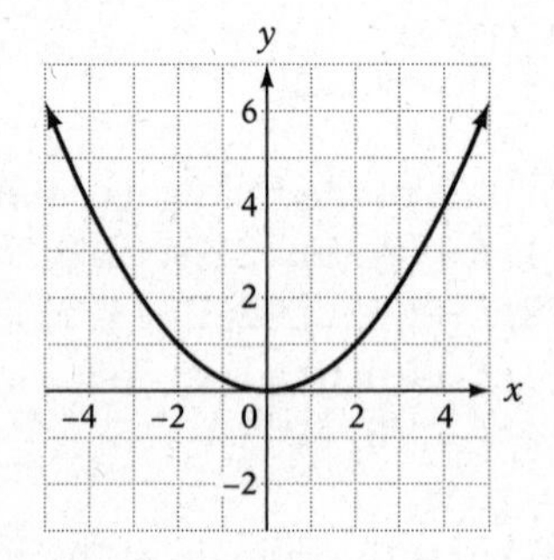

(3)

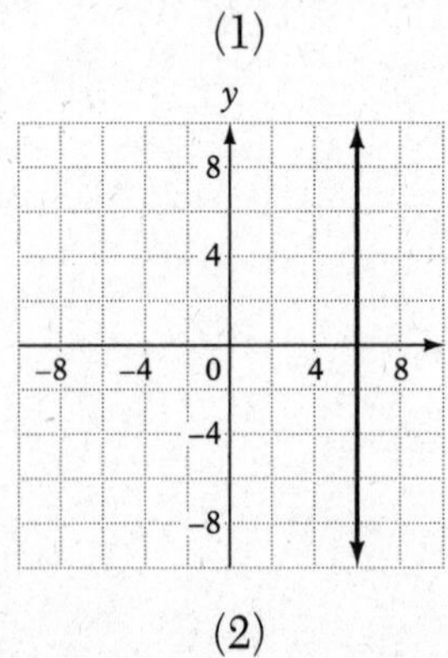

(2)

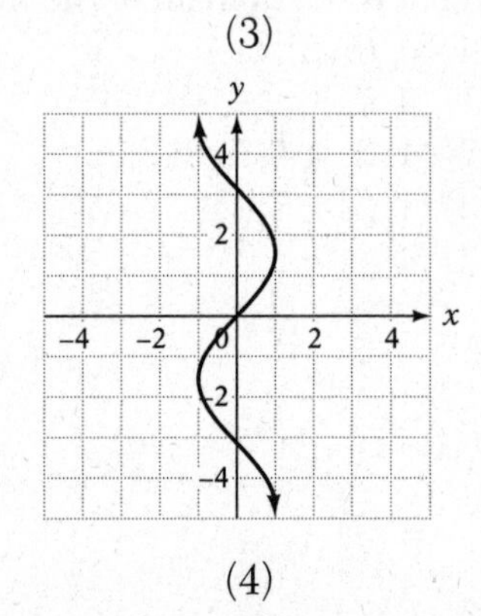

(4)

6. If $g(x) = -x^2 + 7x + 1$ what is $g(2)$?
(1) 11 (3) 27
(2) 19 (4) 35

7. If below is the graph of $y = f(x)$, which is the graph of $y = f(x) - 5$?

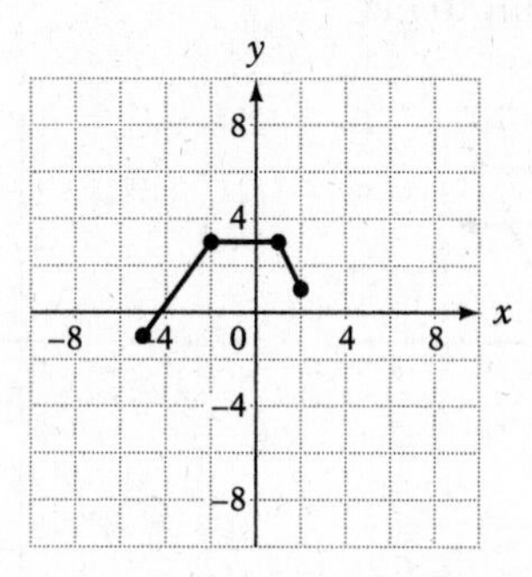

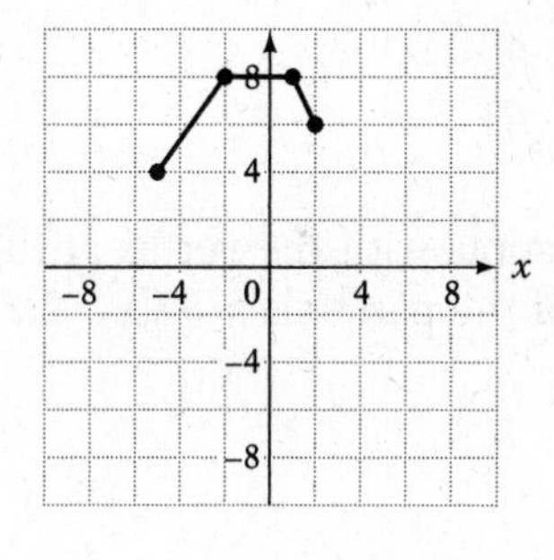

(1)

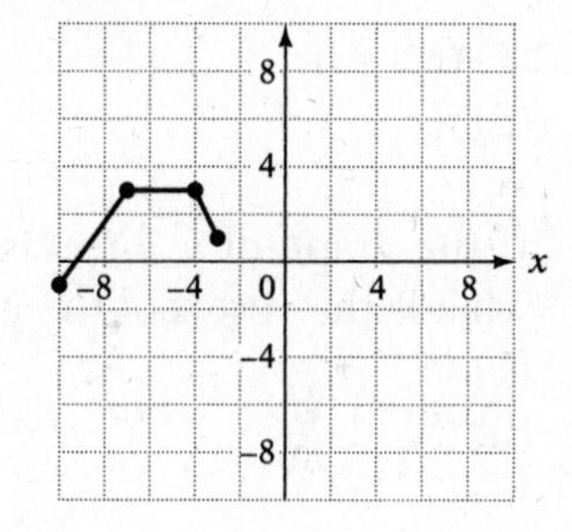

(3)

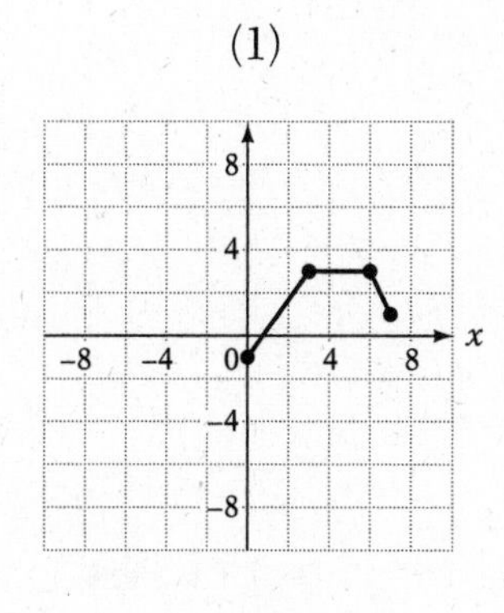

(2)

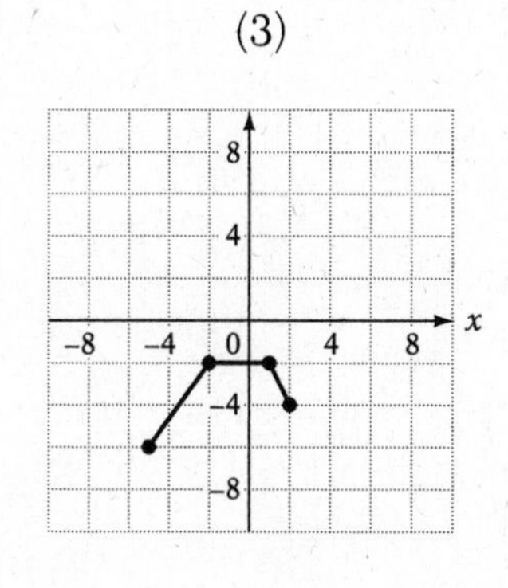

(4)

8. If $f(x) = 5x - 2$, what is $f(x + 1)$?

(1) $5x - 2$
(2) $5x - 3$
(3) $5x + 1$
(4) $5x + 3$

9. Below is the graph of $f(x)$ on the left and $g(x)$ on the right. Which is equivalent to $g(x)$?

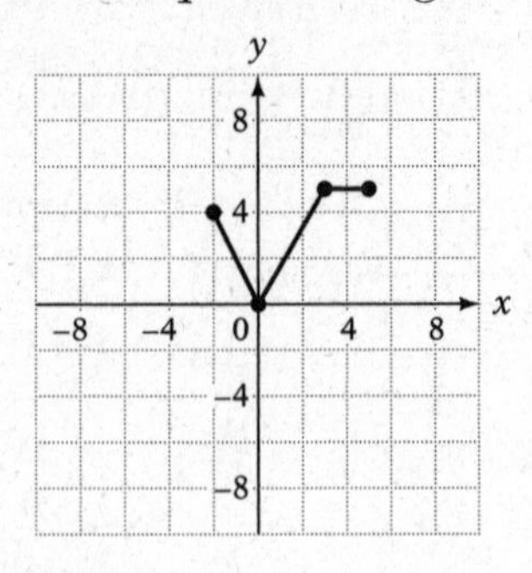

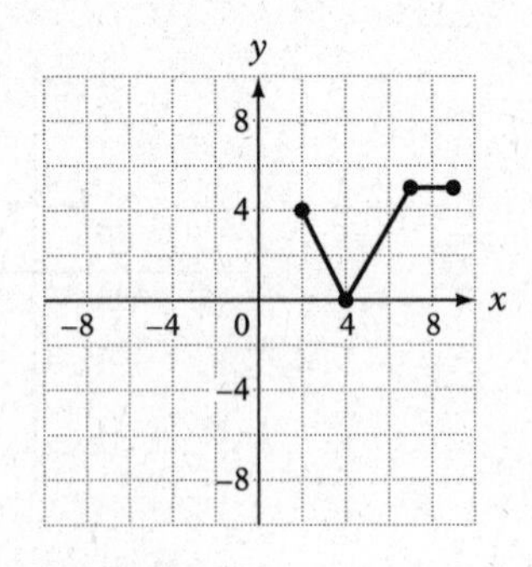

(1) $f(x) + 4$
(2) $f(x) - 4$
(3) $f(x + 4)$
(4) $f(x - 4)$

10. If the graph of $y = f(x)$ is a parabola with the vertex at (5, 1), what is the vertex of the graph of the parabola $y = f(x - 2)$?

(1) (5, 3)
(2) (5, –1)
(3) (7, 1)
(4) (3, 1)

Solutions

1. Since the ordered pair (2, 3) is in the set, $f(2) = 3$. The correct choice is **(3)**.

2. A set of ordered pairs cannot be the definition of a function if there are two ordered pairs with the same x-coordinate, but different y-coordinates. In choice 1, there are two ordered pairs with x-coordinates of 2, (2,7) and (2,8), so that cannot be the definition of a function. The correct choice is **(1)**.

3. The domain is the set of x-coordinates. In this example, the x-coordinates are 1, 3, 4, and 5. The correct choice is **(3)**.

4. There is a point at (3,4) so $f(3) = 4$. The correct choice is **(4)**.

5. Choices 1, 2, and 4 all fail the vertical line test since at least one vertical line can pass through more than one point on them. Choice 3 can be the graph of a function since there is no possible vertical line that would pass through more than one point on it. The correct choice is **(3)**.

6. Substitute 2 for x to get $g(2) = -2^2 + 7 \cdot 2 + 1 = -4 + 14 + 1 = 11$. The correct choice is **(1)**.

7. Substitute $x + 1$ for x to get $f(x + 1) = 5(x + 1) - 2 = 5x + 5 - 2 = 5x + 3$. The correct choice is **(4)**.

8. The graph of $y = f(x) = -5$ is the same as the graph of $y = f(x)$ with each point shifted 5 units down. The correct choice is **(4)**.

9. The graph of $g(x)$ is the same as the graph of $y = f(x)$ with each point shifted 4 units to the right. $g(x)$ must be equivalent to $f(x - 4)$. The correct choice is **(4)**.

10. The graph of $f(x - 2)$ is the same as the graph of $f(x)$ with each point shifted two units to the right. If the vertex of the parabola for $f(x)$ is at (5,1), the vertex of the parabola for $f(x - 2)$ will be two units to the right of (5,1), which is at (7,1). The correct choice is **(3)**.

11. SEQUENCES

11.1 TYPES OF SEQUENCES

A **sequence** is a series of numbers that can be predicted by some kind of pattern. Two of the most common types of sequences are **arithmetic sequences** and **geometric sequences**.

- The sequence 2, 5, 8, 11, 14, … is an example of an arithmetic sequence since each term after the first can be obtained by adding the same number, 3, to the previous term.
- The sequence 2, 6, 18, 54, 162, … is an example of a geometric sequence since each term after the first can be obtained by multiplying the same number, 2, by the previous term.
- There are two notations that are used to describe the terms of a sequence. If the sequence is 2, 5, 8, 11, 14, … the first term can be described as either $a_1 = 2$ or $a(1) = 2$. The second notation is similar to function notation.

11.2 DESCRIBING A SEQUENCE WITH A DIRECT FORMULA

The terms of the sequence 2, 5, 8, 11, … can be described by the formula $a_n = 2 + 3(n - 1)$ or $a(n) = 2 + 3(n - 1)$. If you substitute $n = 1$ into the formula, it becomes $a_1 = 2 + 3(1 - 1) = 2 + 3(0) = 2 + 0 = 2$.

- For any arithmetic sequence, the direct formula for the nth term is $a_n = a_1 + d(n - 1)$ where a_1 is the first term of the sequence and d is the common difference between two consecutive terms. For the sequence 2, 5, 8, 11, …, $a_1 = 2$ and $d = 3$.
- For any geometric sequence, the direct formula for the nth term is $a_n = a_1 \cdot r^{n-1}$, where a_1 is the first term of the sequence and r is the common ratio between two consecutive terms. For the sequence 2, 6, 18, 54, …, $a_1 = 2$ and $r = 3$ since if you divide any term by the previous term you get 3.

11.3 DESCRIBING A SEQUENCE WITH A RECURSIVE FORMULA

Another way to describe the nth term of a sequence is to relate it to the previous term. The term before the a_n term is called the a_{n-1} term. A *recursive formula* for a sequence first defines the exact value of one or more of the terms and then describes how to obtain new terms from the previous ones.

- For the sequence 2, 5, 8, 11, 14, … the recursive formula is
  $a_1 = 2$
  $a_n = 3 + a_{n-1}$ for $n > 1$

From the set of two equations, the entire sequence can be calculated. For example, when $n = 2$, $a_2 = a_1 + 3 = 2 + 3 = 5$. From this, the value of a_3 can be calculated, and this pattern can continue to obtain other n values. The recursive formulas are not convenient for large values of n, however.

- For the sequence 2, 6, 18, 54, 162, …, the recursive formula is
  $a_1 = 2$
  $a_n = 3 \cdot a_{n-1}$ for $n > 1$

Practice Exercises

1. What type of sequence is 3, 7, 11, 15, …?
(1) Increasing arithmetic
(2) Decreasing arithmetic
(3) Increasing geometric
(4) Decreasing geometric

2. What type of sequence is 4, 8, 16, 32, …?
(1) Increasing arithmetic
(2) Decreasing arithmetic
(3) Increasing geometric
(4) Decreasing geometric

3. What is the next number in the sequence 20, 10, 5, $\frac{5}{2}$?

(1) $\frac{5}{4}$ (3) $\frac{5}{16}$

(2) $\frac{5}{8}$ (4) $\frac{5}{32}$

4. Find the value of a_2 in the sequence defined by

$$a_1 = 4$$
$$a_n = 3 + a_{n-1} \text{ for } n > 1$$

(1) 3
(2) 5
(3) 7
(4) 12

5. The sequence 5, 11, 17, 23, ... can be generated by which definition?

(1) $a_1 = 5$
$a_n = 6a_{n-1}$ for $n > 1$

(2) $a_1 = 5$
$a_n = \frac{11}{5} a_{n-1}$ for $n > 1$

(3) $a_1 = 5$
$a_n = -6 + a_{n-1}$ for $n > 1$

(4) $a_1 = 5$
$a_n = 6 + a_{n-1}$ for $n > 1$

6. The sequence 3, 15, 75, 375, ... can be generated by which definition?

(1) $a_1 = 3$
$a_n = 5\,a_{n-1}$ for $n > 1$

(2) $a_1 = 3$
$a_n = 12 + a_{n-1}$ for $n > 1$

(3) $a_1 = 3$
$a_n = \frac{1}{5} a_{n-1}$ for $n > 1$

(4) $a_1 = 3$
$a_n = -12 + a_{n-1}$ for $n > 1$

7. A ball is dropped from a window 50 feet above the ground. Each bounce is $\frac{4}{5}$ the height of the previous bounce. Which definition would generate the height of the bounces?

(1) $a_1 = 50$
$a_n = -10 + a_{n-1}$ for $n > 1$

(2) $a_1 = 50$
$a_n = 10 + a_{n-1}$ for $n > 1$

(3) $a_1 = 50$
$a_n = \frac{4}{5}a_{n-1}$ for $n > 1$

(4) $a_1 = 50$
$a_n = \frac{5}{4}a_{n-1}$ for $n > 1$

8. What is the fifth term of the sequence generated by the definition $a_n = 10 - 4(n - 1)$?
(1) 18
(2) 12
(3) 2
(4) –6

9. What definition would produce the sequence 4, 13, 22, 31, …?
(1) $a_n = 4 + 9n$
(2) $a_n = 4 + 9(n - 1)$
(3) $a_n = 9 + 4n$
(4) $a_n = 9 + 4(n - 1)$

10. Which expression could be used to find the 20th term of the sequence 5, 9, 13, 17, 21, …?
(1) 5 + 4(20)
(2) 5 + 4(19)
(3) 4 + 5(20)
(4) 4 + 5(19)

Solutions

1. Since 7 = 3 + 4 and 11 = 7 + 4 and 15 = 11 + 4, this is an increasing arithmetic sequence. The correct choice is **(1)**.

2. Since $8 = 2 \cdot 4$ and $16 = 2 \cdot 8$ and $32 = 2 \cdot 16$, this is an increasing geometric sequence. The correct choice is **(3)**.

3. Each term is equal to $\frac{1}{2}$ multiplied by the previous term. So the next term is $\frac{1}{2} \cdot \frac{5}{2} = \frac{5}{4}$. The correct choice is **(1)**.

4. $a_2 = 3 + a_{2-1} = 3 + a_1 = 3 + 4 = 7$. The correct choice is **(3)**.

5. The first term is a_1, so $a_1 = 5$. Each term is equal to 6 plus the previous term, so $a_n = 6 + a_{n-1}$. The correct choice is **(4)**.

6. The first term is a_1, so $a_1 = 3$. Each term is equal to 5 multiplied by the previous term, so $a_n = 5 \cdot a_{n-1}$. The correct choice is **(1)**.

7. The initial height is 50, so $a_1 = 50$. Each bounce is 4/5 the height of the previous bounce so $a_n = 4/5 \cdot a_{n-1}$. The correct choice is **(3)**.

8. Substitute 5 for n to get $a_5 = 10 - 4(5 - 1) = 10 - 4(4) = 10 - 16 = -6$. The correct choice is **(4)**.

9. Substitute 1 for n in each equation. Only the second equation gets $a_1 = 4$. Also, an arithmetic sequence has the form $a_n = a_1 + d\,(n - 1)$, where d is the common difference and a_1 is the first term. Since $a_1 = 4$ and $d = 9$, the equation is $a_n = 4 + 9(n - 1)$. The correct choice is **(2)**.

10. An arithmetic sequence with first term a_1 and common difference of d has the equation $a_n = a_1 + (n - 1)d$. a_1 is 5 in this example, and $d = 4$. Therefore, the equation is $a_n = 5 + 4(n - 1)$. Substitute 20 for n to get a $20 = 5 + 4(19)$. The correct choice is **(2)**.

12. REGRESSION CURVES

12.1 THE LINE OF BEST FIT

A **line of best fit** is a line that comes as close as possible to a set of points on a graph. In the scatter plot below, there is no line that could pass through all ten points. Of all the possible lines, though, there is one that is a better fit than the others and this is the line of best fit.

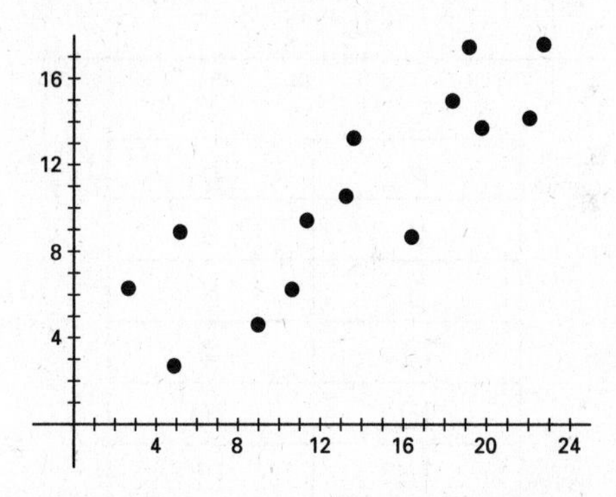

The equation for the line of best fit can be determined quickly on a graphing calculator.

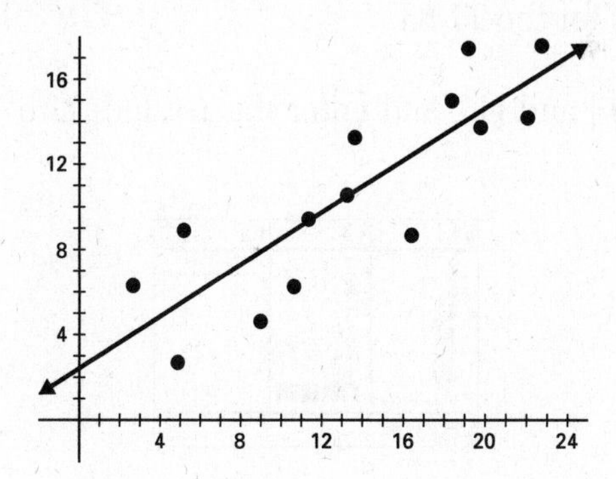

Below is a scatter plot and the chart on which it was based.

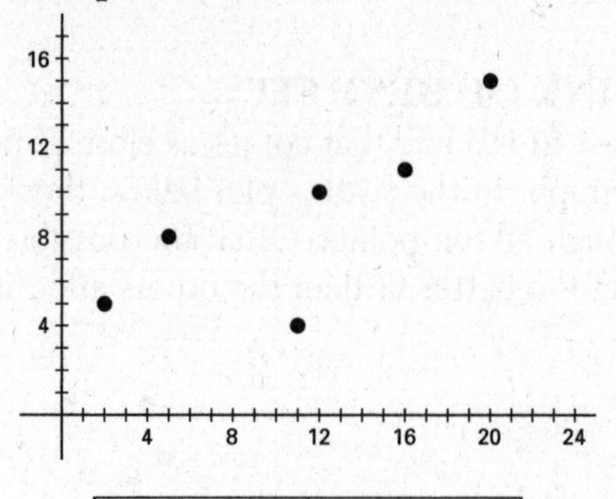

x	y
2	5
5	8
11	4
12	10
16	11
20	15

Instructions for the TI-84:

Press [STAT] and [1], and enter the x values into L1 and the y values into L2.

L1	L2	L3 2
2	5	------
5	8	
11	4	
12	10	
16	11	
20	15	

L2(7) =

Press [STAT], [right], [4], and [ENTER] for the equation of the line of best fit.

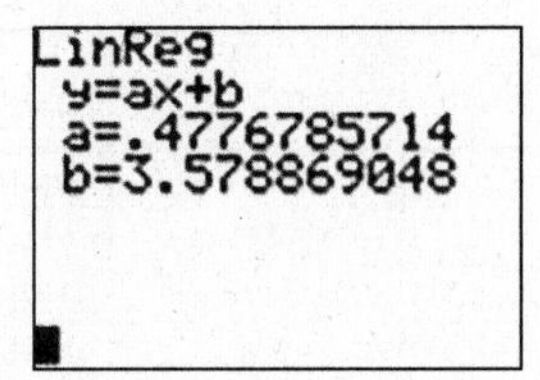

Instructions for the TI-Nspire:

From the home screen select the Add Lists and Spreadsheet icon. In the header row, label column A with an x and column B with a y. Enter the x values into cells A1 to A6. Enter the y values into cells B1 to B6.

	A x	B y	C	D
1	2	5		
2	5	8		
3	11	4		
4	12	10		
5	16	11		

Press [menu], [4], [1], and [3] and select x in the X List field and y in the Y List field.

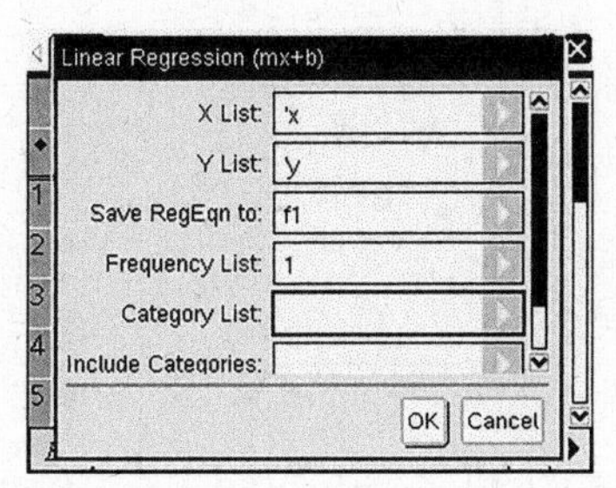

Press the OK button.

1.1 *Unsaved

	A x	B y	C	D
◆				=LinRegM
1	2	5	Title	Linear Re..
2	5	8	RegEqn	m*x+b
3	11	4	m	0.477679
4	12	10	b	3.57887
5	16	11	r^2	0.617042

D1 ="Linear Regression (mx+b)"

The equation for the line of best fit is $y = 0.477679x + 3.57887$.

12.2 THE CORRELATION COEFFICIENT

The **correlation coefficient**, denoted by the variable r, is a number between –1 and 1. When a line of best fit has a positive slope and passes exactly through each of the points, the correlation coefficient is 1. When a line of best fit has a negative slope and passes exactly through each of the points, the correlation coefficient is –1. All correlation coefficients are between –1 and 1.

- The line of best fit for the scatter plot below is close to 1 because it has a positive slope and comes close to the points on the scatter plot. In this case $r = 0.9$

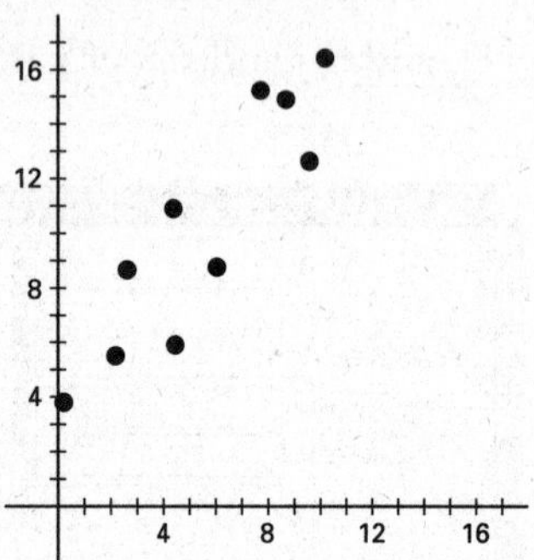

- The line of best fit for the following scatter plot is close to –1 because it has a negative slope and comes close to the points on the scatter plot, but not as close as the line in the previous example did to the points on that graph. In this case the $r = -0.8$

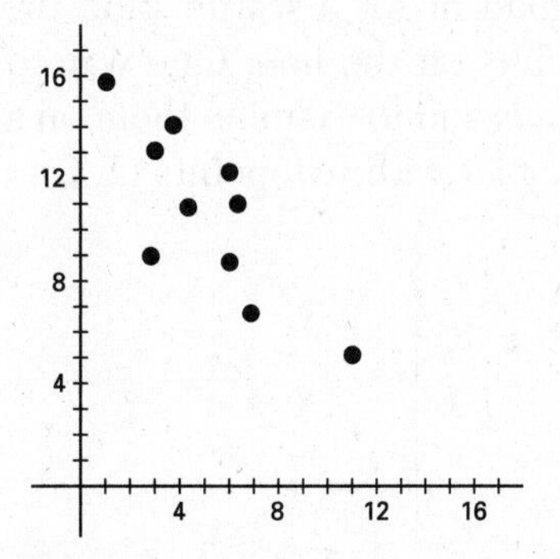

The graphing calculator can display the correlation coefficient. On the TI-84 it will be displayed along with the line of best fit only if the "diagnostics" are turned on. Press [2ND] and [0] to access the catalog and scroll to the DiagnosticOn command to do this. On the TI-Nspire it will display the correlation coefficient along with the equation of the line of best fit.

```
CATALOG
 DependAuto
 det(
 DiagnosticOff
▸DiagnosticOn
 dim(
 Disp
 DispGraph
```

```
LinReg
y=ax+b
a=.4776785714
b=3.578869048
r²=.6170415349
r=.7855199138
```

The correlation coefficient for the example from Section 12.1 was $r = 0.78552$.

1.1 *Unsaved

	A x	B y	C	D
◆				=LinRegM
2	5	8	RegEqn	m*x+b
3	11	4	m	0.477679
4	12	10	b	3.57887
5	16	11	r²	0.617042
6	20	15	r	0.78552

D6 =0.78551991376656

12.3 RESIDUAL PLOTS

When a line is a good fit for a scatter plot, the points in the plot are close to the points on the line. One way to measure this is to calculate the *residuals* and to examine them on a *residual plot*.

Here is a scatter plot with five points (1, 2), (2, 4), (3, 4), (4, 6), (5, 7).

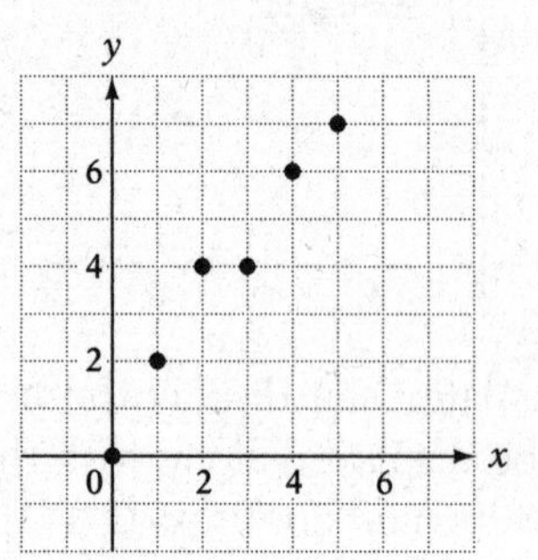

The line of best fit for this scatter plot is $y = 1.2x + 1$.

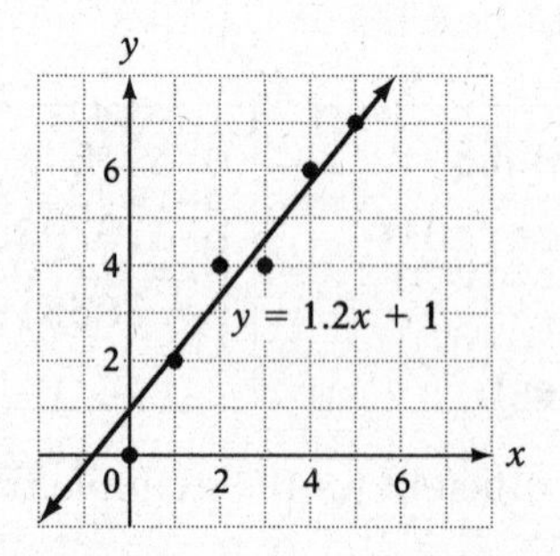

Only one of the points is on the line. For the others, two are above the line and two are below. Residuals measure how far above or below the line the points are. The information can be collected on a chart.

x	y	$1.2x + 1$	Residuals (y column) – ($1.2x + 1$ column)
1	2	2.2	–0.2
2	4	3.4	0.6
3	4	4.6	–0.6
4	6	5.8	0.2
5	7	7	0

The two points above the line have positive residuals. The two points below the line have negative residuals.

A residual plot is a scatter plot with the x-coordinates from the x column in the chart and the y-coordinates from the residuals column in the chart.

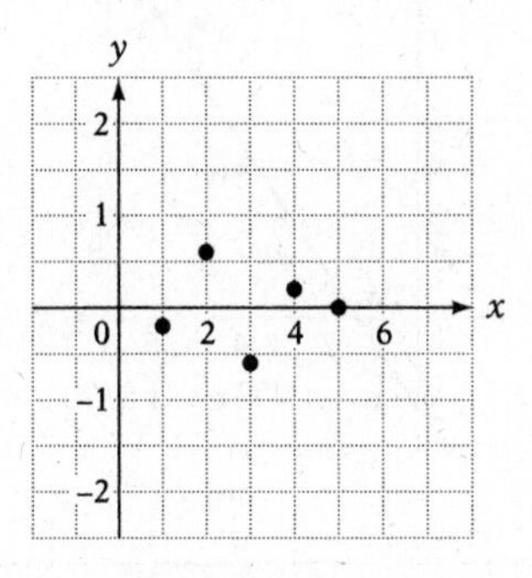

MATH FACTS

When the original scatter plot resembles a line, the residual plot will be a random scattering of points with no apparent pattern. When the original scatter plot resembles some other kind of curve, the residual plot will not be a bunch of random points but will look like a line or some kind of curve like a line or a U shape.

Example 1

Construct a residual plot for the scatter plot based on the following chart. What does this residual plot suggest about the original scatter plot?

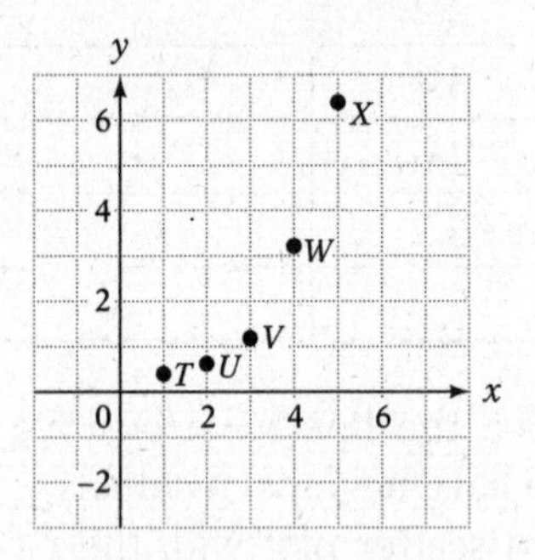

Solution: The line of best fit is $y = 1.453x - 1.987$.

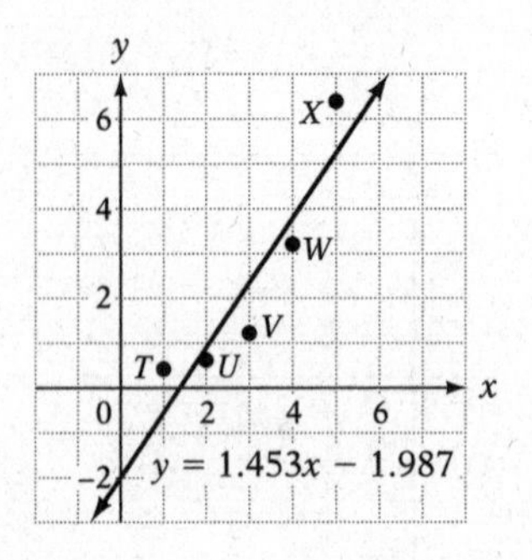

x	y	$1.453x - 1.987$	Residuals (y column) – ($1.2x + 1$ column)
1	0.42	–0.534	0.954
2	0.62	0.919	–0.299
3	1.22	2.372	–1.152
4	3.21	3.825	–0.615
5	6.39	5.278	1.112

The residual plot looks like this:

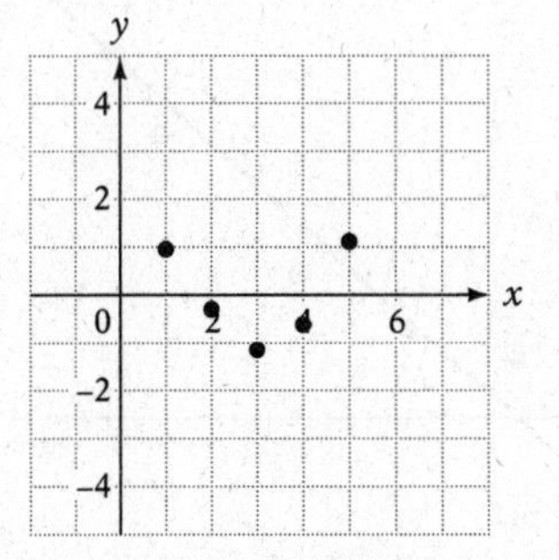

Because this has a U shape and is not just a random scattering of points, it suggests that the original scatter plot did not resemble a line.

12.4 PARABOLAS AND EXPONENTIALS OF BEST FIT

Even though every scatter plot has a line of best fit, sometimes even the best possible line isn't a very good fit.

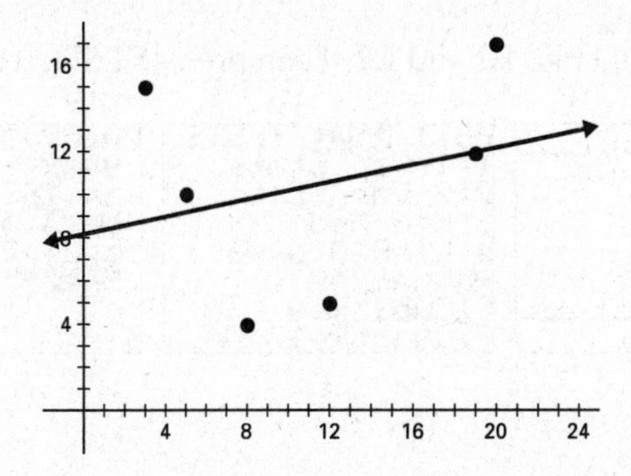

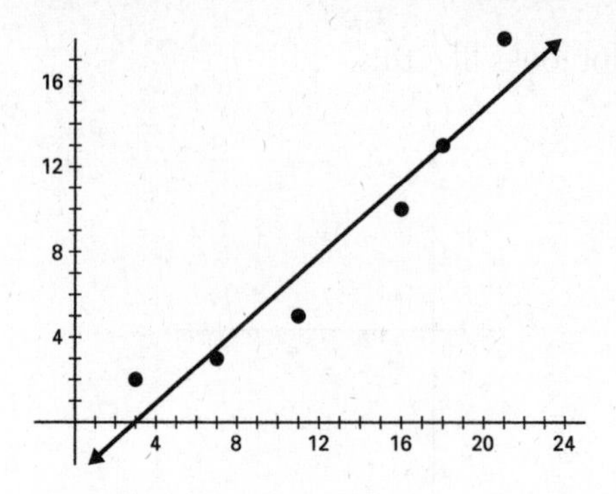

If the points in a scatter plot look more like a parabola or an exponential curve, the most appropriate curve to model it may not be a line. For parabolas, there is a parabola of best fit, and for exponential curves, there is an exponential curve of best fit. Both can be calculated on the graphing calculator. Select quadratic regression for the parabola of best fit and exponential regression for the exponential curve of best fit.

For the TI-84:

Enter the data into L1 and L2. Then press [STAT], [right], and [5].

```
L1     L2     L3    2
3      15     ------
5      10
8      4
12     5
19     12
20     12
------
L2(7) =
```

```
EDIT CALC TESTS
1:1-Var Stats
2:2-Var Stats
3:Med-Med
4:LinReg(ax+b)
5:QuadReg
6:CubicReg
7↓QuartReg
```

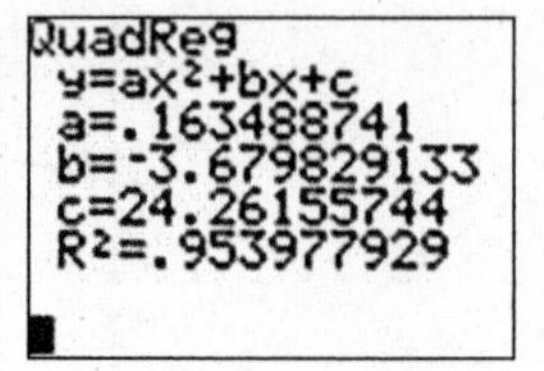

For the TI-Nspire:

To find the parabola of best fit, after entering the data into column A and column B, press [menu], [4], [1], and [6].

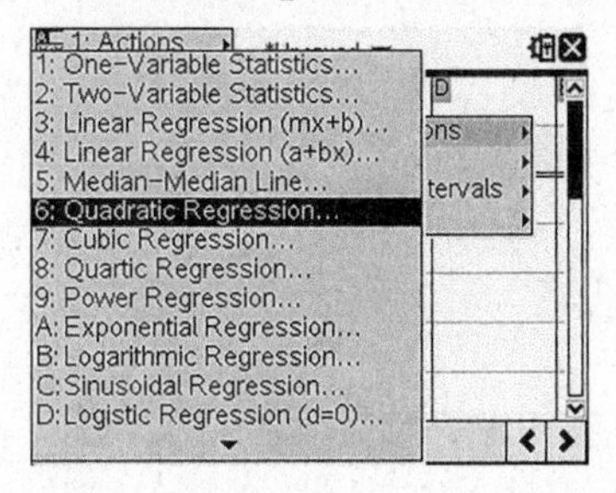

1.1 *Unsaved

	A x	B y	C	D
◆				=QuadReg
1	3	15	Title	Quadrati...
2	5	10	RegEqn	a*x^2+b*...
3	8	4	a	0.163489
4	12	5	b	-3.67983
5	19	12	c	24.2616

D1 ="Quadratic Regression"

To find the exponential curve of best fit on the TI-84, enter the data into L1 and L2 and press [STAT], [right], and [0].

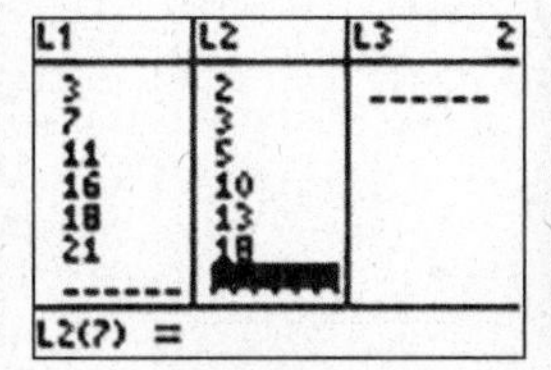

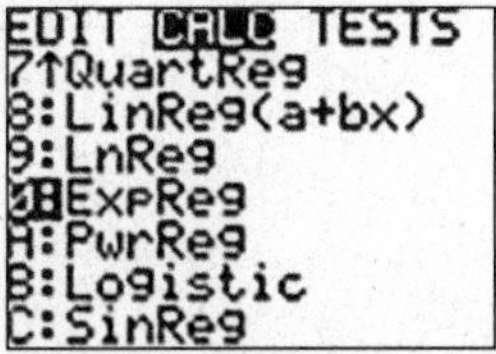

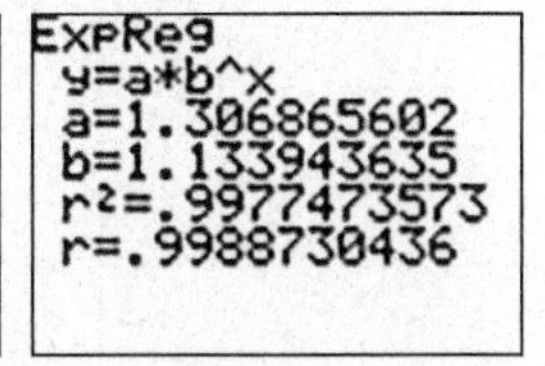

To find the exponential curve of best fit on the TI-Nspire, enter the data into column A and column B, press [menu], [4], [1], and [A].

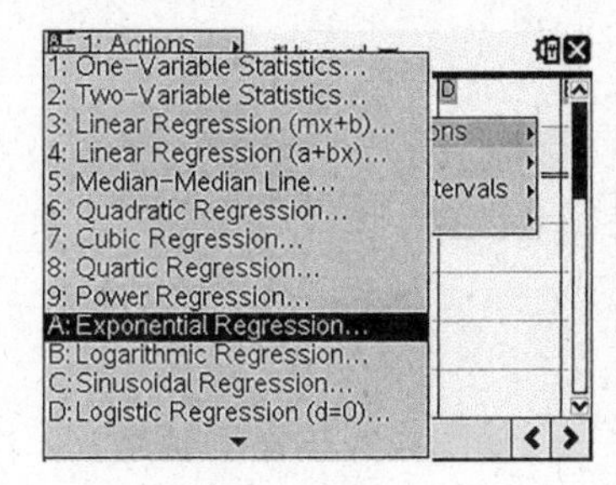

1.1 *Unsaved

	A x	B y	C	D
◆				=ExpReg(
1	3	2	Title	Exponen...
2	7	3	RegEqn	a*b^x
3	11	5	a	1.30687
4	16	10	b	1.13394
5	18	13	r²	0.997747

D1 ="Exponential Regression"

The parabola of best fit for the first scatter plot is $y = 0.1634989x^2 - 3.67983x + 24.2616$.

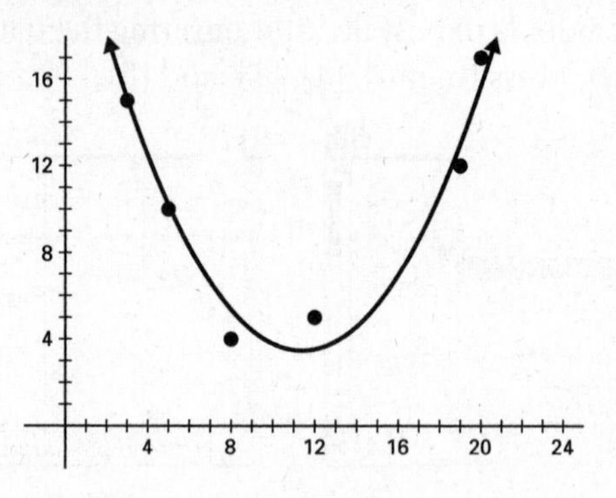

The exponential curve of best fit for the second scatter plot is $y = 1.30687 \cdot 1.13394^x$.

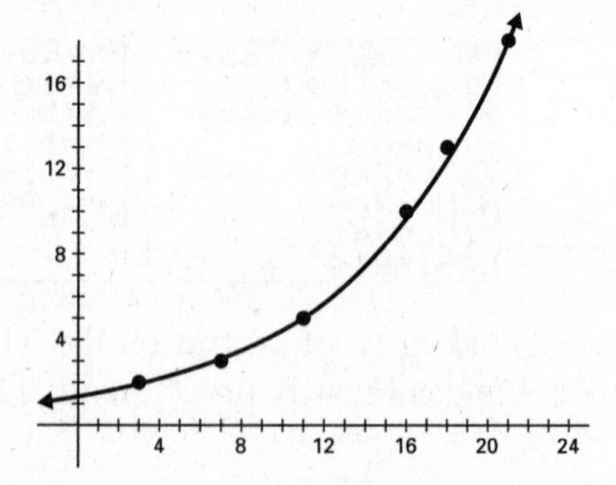

Practice Exercises

1. Calculate the equation for the line of best fit for the following set of data in $y = mx + b$ form. Round m and b to the nearest tenth.

x	y
1	3
2	5
3	4
4	6
5	8

(1) $y = 1.1x + 1.9$
(2) $y = 1.4x + 1.7$
(3) $y = 1.7x + 1.4$
(4) $y = 1.9x + 1.1$

2. Calculate the equation for the line of best fit for the following set of data in $y = mx + b$ form. Round m and b to the nearest tenth.

x	y
10	33
20	20
30	10
40	14
50	6

(1) $y = 34.6x - 0.6$
(2) $y = -0.6x + 34.6$
(3) $y = 31.8x - 0.7$
(4) $y = -0.7x + 31.8$

3. What is the equation for the line of best fit for the points on this scatter plot?

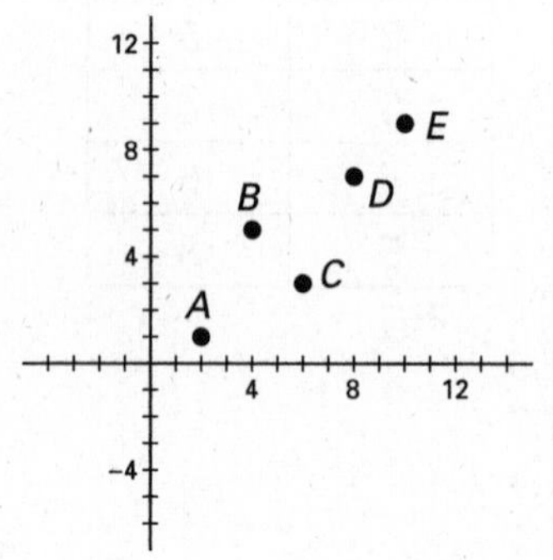

(1) $y = -0.5x + 7$
(2) $y = 0.7x - 0.5$
(3) $y = -0.4x + 0.9$
(4) $y = 0.9x - 0.4$

4. Of these four choices, which line appears to be the best fit for this scatter plot?

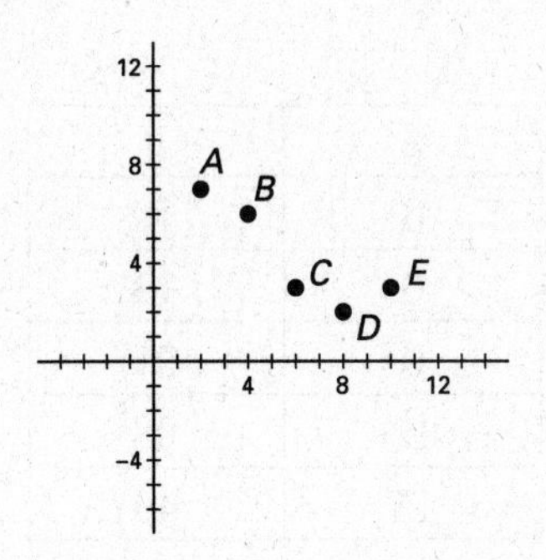

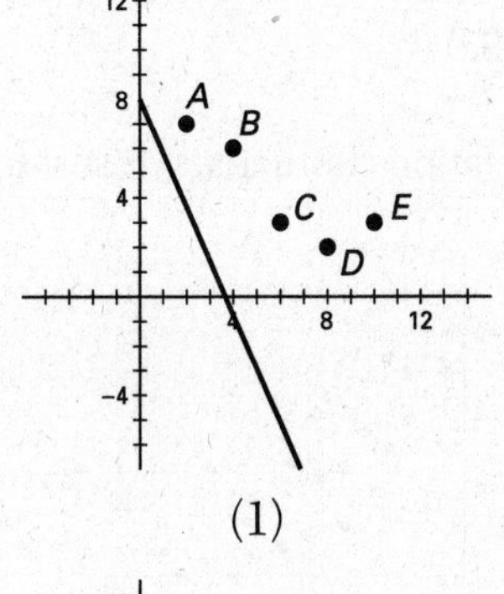

(1)

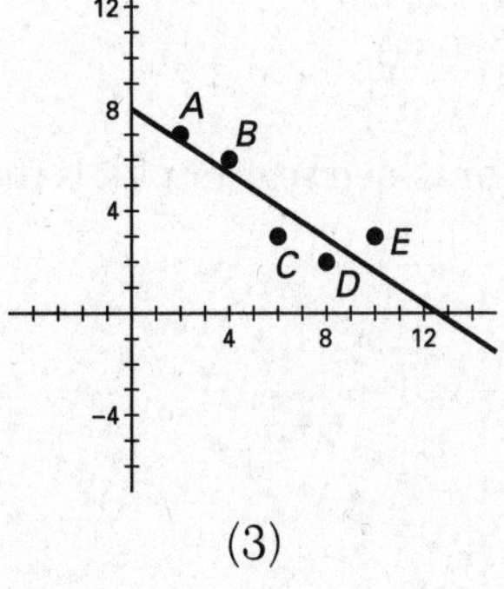

(3)

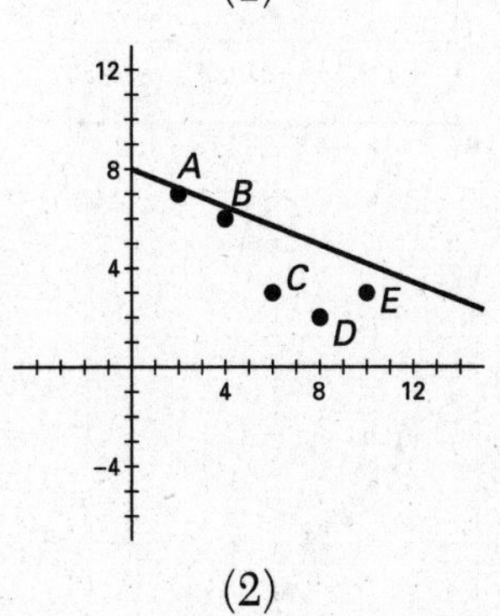

(2)

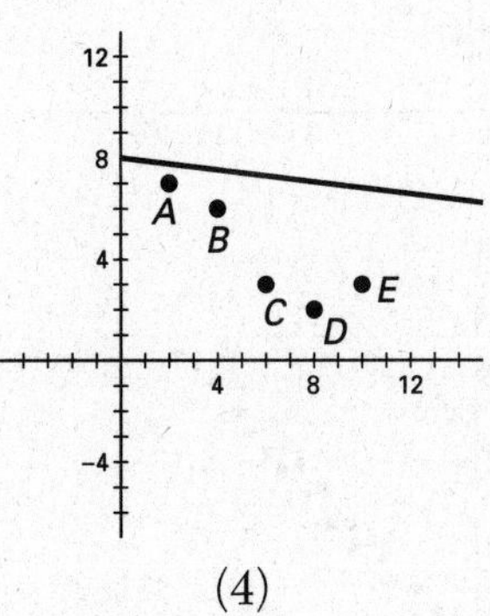

(4)

5. What is the correlation coefficient (r), rounded to the nearest hundredth, for the line of best fit for the data on the table below?

x	y
3	10
6	13
9	27
12	38
15	40

(1) 0.97
(2) 0.96
(3) 0.95
(4) 0.94

6. For which scatter plot is the correlation coefficient closest to 1?

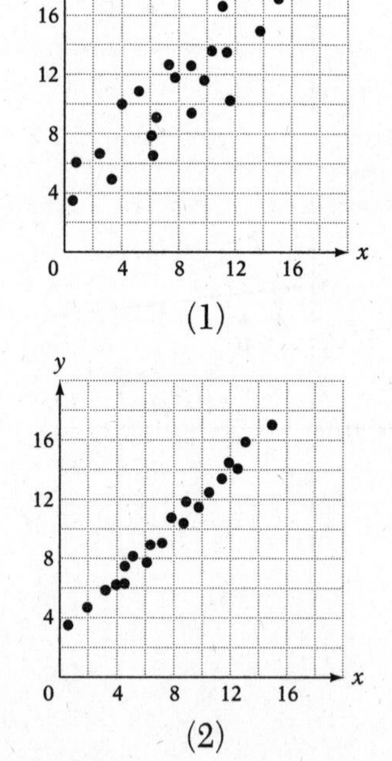

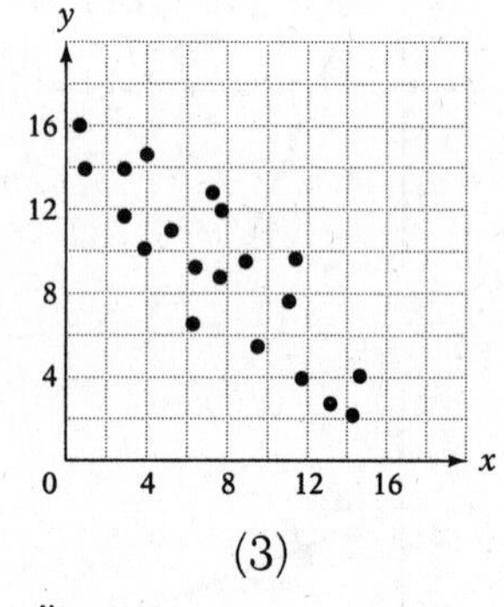

(3)

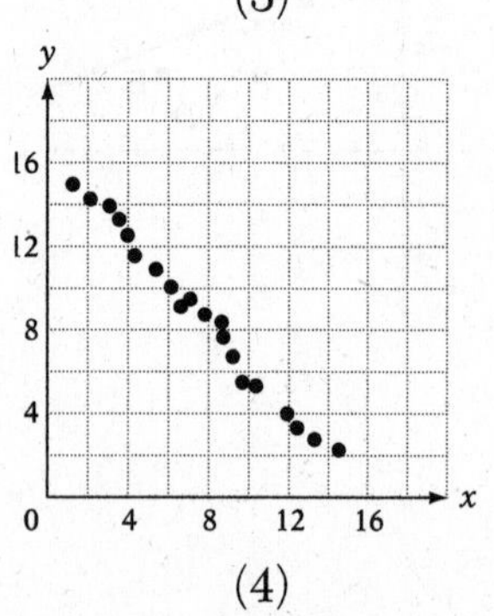

(4)

7. Of the four choices, which is closest to the correlation coefficient for this scatter plot?

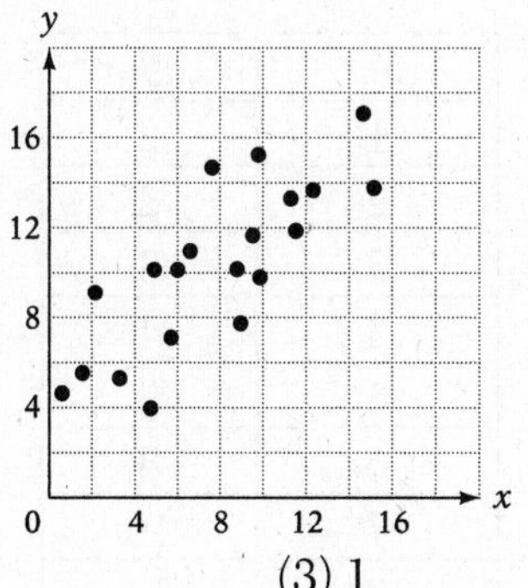

(1) 0.8
(2) –0.8
(3) 1
(4) –1

8. Find the equation of the parabola of best fit for the data on the table below.

x	y
1	2
2	2
3	3
4	4
5	6
6	8
7	10
8	14

(1) $y = 0.31x^2 - 67x + 3.53$
(2) $y = 0.24x^2 - 0.52x + 2.23$
(3) $y = 0.58x^2 - 0.41x + 4.12$
(4) $y = 0.73x^2 - 0.72x + 6.87$

9. Find the equation of the exponential curve of best fit for the data on the table below.

x	y
1	2
2	2
3	3
4	4
5	6
6	8
7	10
8	14

(1) $y = 1.28 \cdot 1.35^x$

(2) $y = 1.35 \cdot 1.28^x$

(3) $y = 1.47 \cdot 1.39^x$

(4) $y = 1.39 \cdot 1.47^x$

10. For which scatter plot would an exponential curve of best fit be most appropriate?

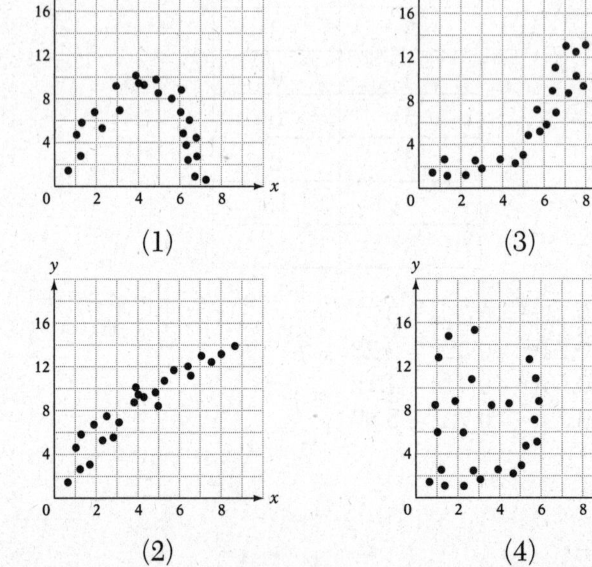

Solutions

1. Enter the data into the graphing calculator and do linear regression to get the equation $y = 1.1x + 1.9$. The correct choice is **(1)**.

2. Enter the data into the graphing calculator and do linear regression to get the equation $y = -0.6x + 34.6$. The correct choice is **(2)**.

3. Enter 2, 4, 6, 8, 10 for the x values and 1, 5, 3, 7, 9 for the y values into the graphing calculator and do linear regression. The equation is $y = 0.9x - 0.4$. The correct choice is **(4)**.

4. Of the four choices, choice 3 seems to have the line closer to the points than the other three choices. The correct choice is **(3)**.

5. Enter the data into the graphing calculator and do linear regression to get an r value of 0.97. The correct choice is **(1)**.

6. To have an r value close to +1, the points must lie close to a line with a positive slope. Choices 1 and 2 both resemble lines with positive slopes. In choice 2 the points seem to fall closer to a straight line than choice 1. The correct choice is **(2)**.

7. Since this scatter plot resembles a line with a positive slope, choices 2 and 4 can be eliminated because r must be positive. To have an r value of +1 the points would have to lie perfectly on a line, so choice 3 can be eliminated too. The correct choice is **(1)**.

8. Enter the data into the graphing calculator and do quadratic regression to get $y = 0.24x^2 - 0.52x + 2.23$. The correct choice is **(2)**.

9. Enter the data into the graphing calculator and do exponential regression to get $y = 1.28 \cdot 1.35^x$. The correct choice is **(1)**.

10. Of the four choices, choice 3 looks the most like an exponential curve. Choice 1 looks like a parabola. Choice 2 looks like a line. Choice 4 looks like a bunch of random points. The correct choice is **(3)**.

13. STATISTICS

13.1 MEAN, MEDIAN, AND MODE

In a set of numbers, the **mean**, also known as the **average**, is the sum of the numbers divided by how many numbers there are. For the set 70, 75, 78, 80, 82, the mean can be calculated as $\frac{70+75+78+80+82}{5} = \frac{385}{5} = 77$.

The **median** is the middle number (or average of the two middle numbers if there are an even amount of numbers) when the numbers are arranged from least to greatest. In the set 70, 75, 78, 80, 82, the median is the number 78. If the set were 70, 75, 78, 80, 82, 85, the median would be the average of 78 and 80, which is $\frac{78+80}{2} = 79$.

The **mode** is the number that appears most frequently. In the set 70, 75, 78, 80, 80, the mode is 80 since there are two 80s and only one of each of the other numbers.

13.2 FIRST QUARTILE AND THIRD QUARTILE

The median is greater than 50% of the numbers in the list. The number that is greater than just 25% of the numbers on the list is called the *first quartile*. The number that is greater than 75% of the numbers on the list is called the *third quartile*. To find the first quartile, find the median of all the numbers less than the median of the list. To find the third quartile, find the median of all the numbers greater than the median of the list. The interquartile range is the difference between the third quartile and the first quartile.

- For the list 40, 43, 45, 47, 48, 52, 57, 60, 61, 64, 68:
- The numbers less than the median 52 are 40, 43, 45, 47, 48. The median of these five numbers is 45, which is the first quartile.
- The numbers greater than the median 52 are 57, 60, 61, 64, 68. The median of these five numbers is 61, which is the third quartile. The interquartile range is $61 - 45 = 16$.

13.3 BOX PLOTS

A box plot is a picture that shows five different metrics, minimum, first quartile, median, third quartile, and maximum. To make a box plot, draw a segment connecting the maximum and the minimum. Then draw a rectangle with sides at the first quartile and third quartile. Finally, draw a line through the rectangle at the median.

- For the following numbers, the box plot looks like this:

10, 10, 10, 12, 12, 12, 12, 17, 25, 25, 25, 25, 30, 30, 30

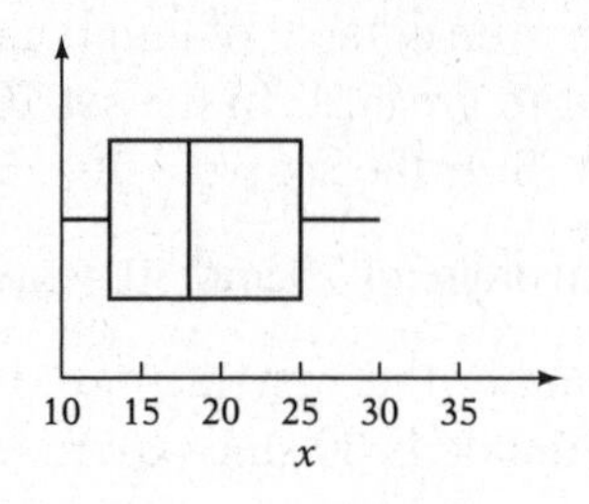

13.4 USING THE GRAPHING CALCULATOR TO DETERMINE MAXIMUM, MINIMUM, MEDIAN, FIRST QUARTILE, AND THIRD QUARTILE

For the TI-84:

The graphing calculator can calculate the five measures of central tendency. First enter all the numbers into L1 by pressing [STAT] and [1] for Edit.

To find the minimum, first quartile, median, third quartile, and maximum of the seven numbers 10, 4, 8, 12, 6, 16, 14, enter them into L1. Then press [STAT] and [1] for 1-Var Stats and press [ENTER].

L1	L2	L3 2
10		------
4		
8		
12		
6		
16		
14		

L2(1)=

EDIT CALC TESTS
1:1-Var Stats
2:2-Var Stats
3:Med-Med
4:LinReg(ax+b)
5:QuadReg
6:CubicReg
7↓QuartReg

The screen will display

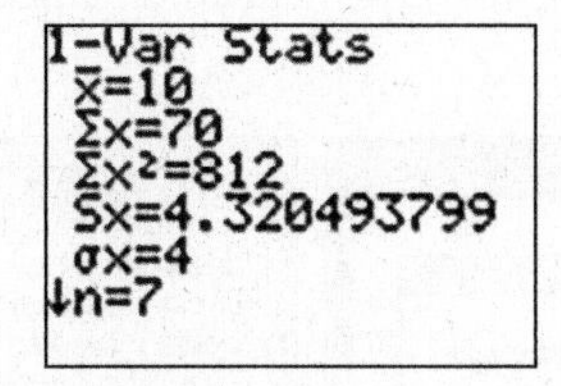

The $\overline{x} = 10$ is for the mean. The $n = 7$ means that there were seven elements in the list. For the minimum, first quartile, median, third quartile, and maximum, press the down arrow five times.

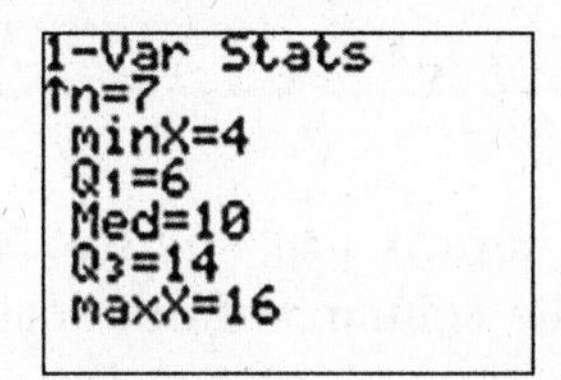

minX is for the minimum, Q1 is for the first quartile, Med is for the median, Q3 is for the third quartile, and maxX is for the maximum element.

For the TI-Nspire:

From the home screen, select the Add Lists & Spreadsheet icon. Name column A x and fill in cells A1 through A7 with the numbers 10, 4, 8, 12, 6, 16, 14.

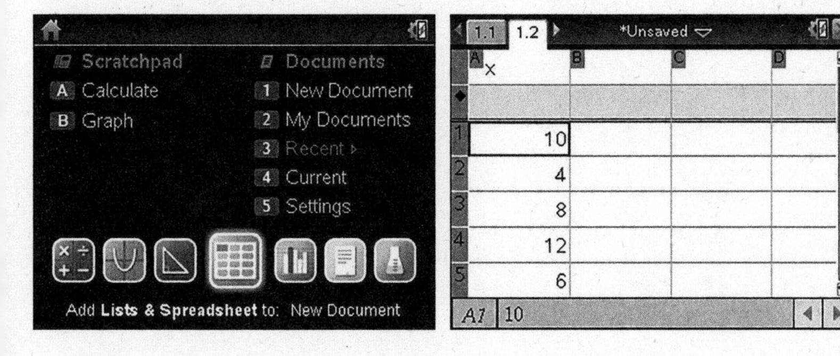

Press [4], [1], and [1] for One-Variable Statistics.

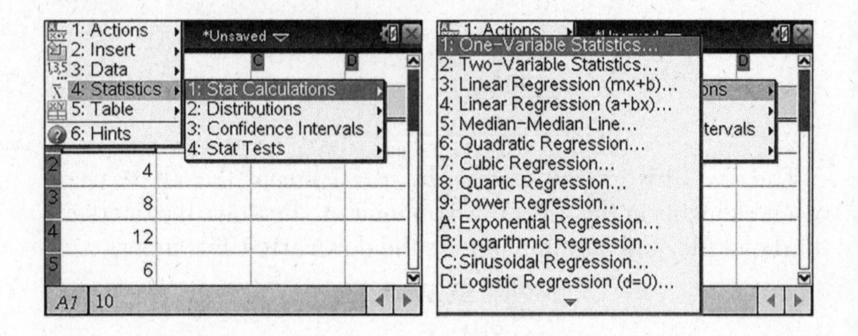

Press [OK] since there is just one list. Set the X1 List to x since that was what the column with the data was named in the spreadsheet.

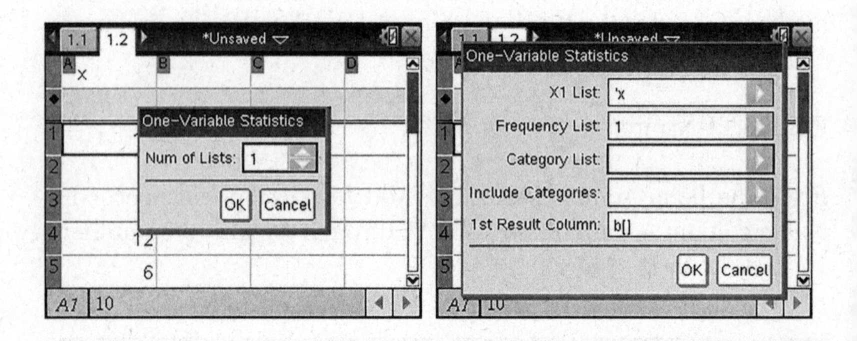

In cells C2 through C13, the one-variable statistics will be displayed. The median is the $\bar{x}$-bar. n is for the number of numbers. MinX is the smallest number. Q1X is the first quartile. MedianX is the median. Q3X is the third quartile. MaxX is the largest number.

1.1 1.2 *Unsaved

	A x	B	C	D
◆			=OneVar('	
1	10	Title	One–Var...	
2	4	x̄	10.	
3	8	Σx	70.	
4	12	Σx²	812.	
5	6	sx := sn-1...	4.32049	

C1 ="One–Variable Statistics"

1.1 1.2 *Unsaved

	A x	B	C	D
◆			=OneVar('	
6	16	σx := σnx...	4.	
7	14	n	7.	
8		MinX	4.	
9		Q₁X	6.	
10		MedianX...	10.	

C10 =10.

1.1 1.2 *Unsaved

	A x	B	C	D
◆			=OneVar('	
11		Q₃X	14.	
12		MaxX	16.	
13		SSX := Σ...	112.	
14				
15				

C15

Practice Exercises

1. Find the mean of this set of numbers {4, 5, 8, 8, 8, 10, 10, 13, 15, 17, 23}.
   (1) 8 (3) 10
   (2) 9 (4) 11

2. Find the median of this set of numbers {4, 5, 8, 8, 8, 10, 10, 13, 15, 17, 23}.
   (1) 8 (3) 11
   (2) 10 (4) 15

3. Find the interquartile range of this set of numbers {4, 5, 8, 8, 8, 10, 10, 13, 15, 17, 23}.
   (1) 7 (3) 9
   (2) 8 (4) 10

4. For the first four days of a five-day vacation, the mean temperature was 80 degrees. What must the temperature be on the fifth day in order for the mean temperature to be 82 degrees?
   (1) 88 (3) 90
   (2) 89 (4) 91

5. For which data set is the median greater than the mean?
   (1) {4, 7, 10, 13, 16} (3) {8, 9, 10, 11, 12}
   (2) {8, 9, 10, 18, 19} (4) {1, 2, 10, 11, 12}

6. What is the mode of the data in this histogram?

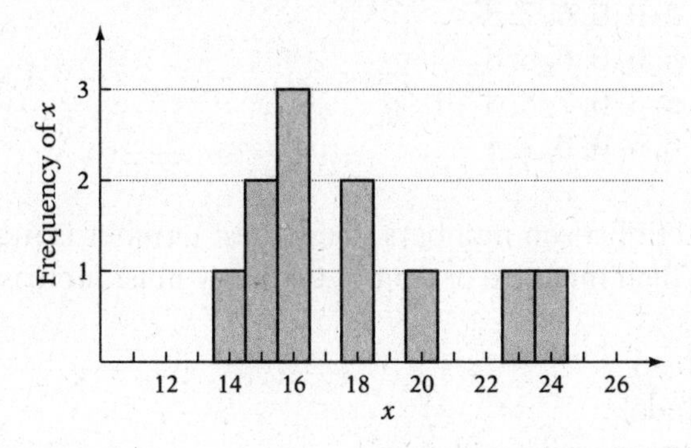

(1) 23
(2) 24
(3) 15
(4) 16

7. What is the median of the data in this box plot?

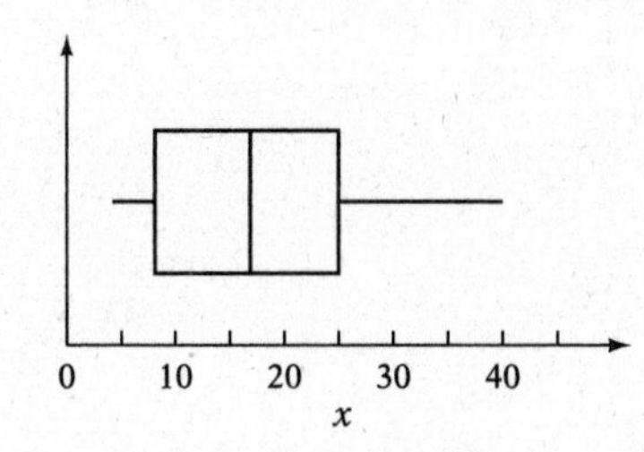

(1) 17
(2) 15.2
(3) 8
(4) 25

8. What is true about this data set: 1, 2, 10, 11, 11?
(1) The median is greater than the mean.
(2) The median is equal to the mean.
(3) The median is less than the mean.
(4) The median is greater than the mode.

9. For which data set is the interquartile range equal to 0?
(1) 2, 3, 6, 6, 6, 7, 8
(2) 2, 6, 6, 6, 6, 6, 8
(3) 1, 2, 3, 6, 7, 8, 9
(4) 2, 6, 6, 6, 6, 7, 8

10. In a set of seven numbers, the largest number is increased by 10. Which measure of central tendency must increase because of this?
(1) Mean
(2) Mode
(3) First quartile
(4) Interquartile range

Solutions

1. To get the mean, add the 11 values and divide by 11. (4 + 5 + 8 + 8 + 8 + 10 + 10 + 13 + 15 + 17 + 23)/11 = 121/11 = 11. The correct choice is **(4)**.

2. The median is the middle number after the numbers have been arranged from least to greatest. These numbers are already arranged from least to greatest so the middle number is the sixth number, which is 10. The correct choice is **(2)**.

3. The interquartile range is the difference between the first quartile and the third quartile. The median is the sixth number so the first quartile is the median of the first five numbers. The first five numbers are 4, 5, 8, 8, 8 with a median of 8. The third quartile is the median of the last five numbers. The last five numbers are 10, 13, 15, 17, 23, with a median of 15. The first quartile is 8. The third quartile is 15. The interquartile range is 15 – 8 = 7. The correct choice is **(1)**.

4. The sum of a set of numbers is the product of the average and the number of numbers. If the first four days had an average of 80, the sum of all four temperatures was 320. If the first five days need to have an average of 82, the sum of all five temperatures needs to be $82 \cdot 5 = 410$. The fifth temperature needs to be 410 – 320 = 90. The correct choice is **(3)**.

5. The median for all four data sets is 10. The mean of the numbers in choice 4 is (1 + 2 + 10 + 11 + 12)/5 = 7.2, which is less than the median. All the other choices have the mean greater than or equal to the median. The correct choice is **(4)**.

6. The mode is the most frequent number. In a histogram the most frequent number is the tallest bar, which is 16 in this example. The correct choice is **(4)**.

7. The vertical line in the middle of the rectangle represents the median in a box plot. Since this is at 17, the median is 17. The correct choice is **(1)**.

8. The median of the data set is 10. The mode is 11. The mean is 7. The median is greater than the mean. The correct choice is **(1)**.

9. The interquartile range is zero if the first quartile is equal to the third quartile. This is the case for choice 2 where both the first quartile and the third quartile are 6. The correct choice is **(2)**.

10. When the largest number is increased by 10, the sum of all the numbers is increased by 10 so the mean will increase. The other measures will not be affected by increasing the largest number. The correct choice is **(1)**.

Glossary of Terms

Addition property of equality A property of algebra that states that when equal values are added to both sides of a true equation, the equation continues to be true. To solve the equation $x - 2 = 5$, add 2 to both sides of the equation by using the addition property of equality.

Arithmetic sequence A number sequence in which the difference between two consecutive terms is a constant. The sequence 2, 5, 8, 11, 14, ... is an arithmetic sequence because the difference between consecutive terms is always 3.

Axis of symmetry An imaginary vertical line that passes through the vertex of a parabola. The equation for the axis of symmetry of a parabola is defined by $y = ax^2 + bx + c$ is $x = \frac{-b}{2a}$.

Base The number being raised to a power in an exponential expression. In the expression $2 \cdot 3^x$, the 3 is the base.

Binomial A polynomial with only two terms. $3x + 5$ is a binomial.

Box plot A graphical way to summarize data. The five numbers represented by the minimum, first quartile, median, third quartile, and maximum are graphed on a number line. A line segment connects the minimum to the first quartile. A rectangle is drawn around the first quartile and third quartile with a vertical line at the median. A line segment connects the third quartile to the maximum.

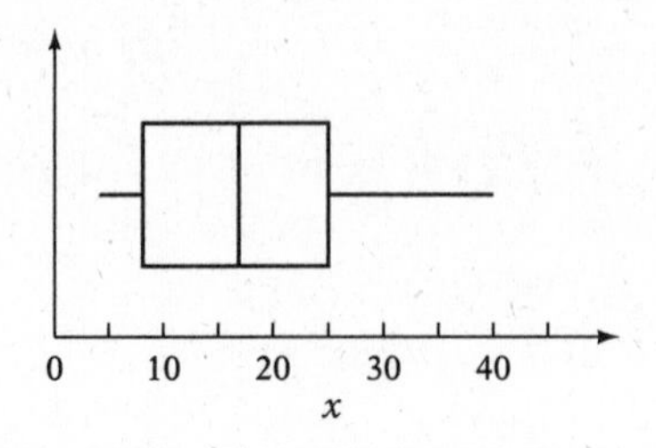

Closed form defined sequence A formula that defines the *n*th term of a sequence. The formula $a_n = 3 + 2(n - 1)$ is a closed form definition of a sequence. To get the 50th term of the sequence, substitute 50 for n in the definition.

Coefficient A number multiplied by a variable expression. In the expression $5x + 2$, 5 is the coefficient of x.

Common difference In an arithmetic sequence, the difference between consecutive terms. The common difference in the sequence 2, 5, 8, 11, 14, … is 3.

Common ratio In a geometric sequence, the ratio between consecutive terms. The common ratio in the sequence 2, 6, 18, 54, 162, … is 3 since $162/54 = 54/18 = 18/6 = 6/2 = 3$.

Commutative property of addition or multiplication The law from arithmetic that states that the order in which two numbers are added or multiplied does not matter. Because of the commutative property of addition, $5 + 2 = 2 + 5$.

Completing the square A method of solving a quadratic equation that involves turning one side of the equation into a perfect square trinomial.

Constant A number that does not have a variable part. In the expression $5x + 2$, 2 is a constant.

Correlation coefficient A number represented by r that measures how well a curve of best fit matches the points in a scatter plot.

When the correlation coefficient is very close to 1 or to –1, the curve is a very good fit.

Degree of a polynomial The exponent on the highest power of a polynomial. In the polynomial $x^3 - 2x^2 + 5x - 2$, the degree is 3 since the highest power is a 3.

Difference of perfect squares Factoring a quadratic binomial in the form $x^2 - a^2$ into $(x - a)(x + a)$. For example, $x^2 - 9 = (x - 3)(x + 3)$.

Distributive property of multiplication over addition The rule that allows expressions of the form $a(b + c)$ to become $a \cdot b + a \cdot c$. For example, $2(3x + 5) = 6x + 10$.

Division property of equality A property of algebra that states that when both sides of a true equation are divided by the same non-zero number, the equation continues to be true. To solve the equation $2x = 8$, divide both sides of the equation by 2 using the division property of equality.

Domain The numbers that can be input into a function. When the function is defined as a set of ordered pairs or as a graph, the domain is the set of x-coordinates.

Dot plot A graphical way of representing a data set where each piece of data is represented with a dot.

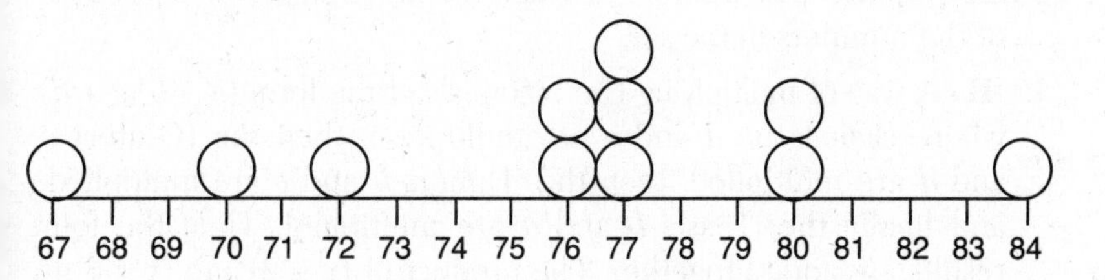

Elimination method A way of solving a system of linear equations by combining the two equations in such a way as to eliminate one of the variables.

In the set of equations,

$$x + 2y = 12$$
$$3x - 2y = -4$$

the y variable is eliminated by adding the two equations together.

Equation Two mathematical expressions with an equals sign between them. $3x + 2 = 8$ is an equation.

Exponential equation An equation in which the variable is an exponent. $2 \cdot 3^x = 18$ is an exponential equation.

Exponential function A function in which the variable is an exponent. $f(x) = 2 \cdot 3^x$ is an exponential function.

Expression Numbers and variables that are combined with the operations from math.

Factoring a polynomial Finding two polynomials that can be multiplied to become another polynomial. The polynomial $x^2 + 5x + 6$ can be factored into $(x + 2)(x + 3)$.

Factors The polynomials that evenly divide into a polynomial. The factors of $x^2 + 5x + 6$ are $(x + 2)$ and $(x + 3)$.

First quartile The number in a data set that is bigger than just 25% of the numbers in the set.

FOIL A way of multiplying two binomials of the form $(a + b)(c + d)$ where (F)irst the a and c are multiplied, then the (O)uters a and d are multiplied, then the (I)nners b and c are multiplied, and finally the (L)asts b and d are multiplied. Then the four results are added together. The product of $(x + 2)$ and $(x + 3)$ is $x^2 + 3x + 2x + 6 = x^2 + 5x + 6$ by this method.

Function Something that takes numbers as inputs and outputs numbers. Functions are often labeled with the letters f or g. The notation $f(2) = 7$ means that when the number 2 is input into function f, it outputs the number 7.

Geometric sequence A number sequence in which the ratio between two consecutive terms (what you get when you divide one term by the term before it) is a constant. The sequence 2, 6, 18, 54, 162, ... is a geometric sequence because the ratio between consecutive terms is always 3.

Graph A visual way to describe the solution set to an equation. Each solution to the equation corresponds to an ordered pair that is graphed as a point on the coordinate plane. Each of the ordered pairs that satisfy an equation produce a point on the coordinate plane and the collection of all the points is the graph of the equation.

Greatest common factor The largest expression that divides evenly into two or more monomials. The greatest common factor of $6x^2$ and $8x^3$ is $2x^2$.

Growth rate In an exponential expression $a \cdot b^x$, the b is the growth rate. For example, in the equation $y = 500 \cdot 1.05^x$, the growth rate is 1.05.

Histogram A way of representing data with repeated values. Each value is represented by a bar whose height corresponds to the number of times that value is repeated. There are no spaces between the bars.

Increasing A function is increasing on an interval if making the input value larger also makes the y output larger. On a graph, an increasing function "goes up" from left to right.

Inequality Like an equation, but there is a $<$, $>$, $\leq$, or $\geq$ sign between the two expressions. $x + 2 > 5$ in a one-variable inequality. $y \leq 2x + 6$ is a two-variable inequality.

Interquartile range The difference between the number that is the third quartile of a data set and the number that is the first quartile of a data set.

Isolating a variable A variable is isolated when it is by itself on one side of an equation. In the equation $x + 2 = 5$, the x is not yet isolated. Subtracting 2 from both sides of the equation transforms the original equation into $x = 3$ with the x now isolated.

Like terms Terms that have the same variable part. They can be combined by adding or subtracting. $2x^2$ and $3x^2$ are like terms. $2x^2$ and $3x^3$ are not like terms.

Line of best fit A line that comes closest to the set of points in a scatter plot.

Linear equation An equation in which the greatest exponent is a 1. The equation $2x + 3 = 7$ is a linear equation.

Linear function A function in which the greatest exponent is a 1. The function $f(x) = 2x + 3$ is a linear function.

Mean The average of the numbers in a data set. Calculate the mean by adding all the numbers and dividing the total by the number of numbers in the set.

Median The middle number in a data set after it has been arranged from least to greatest. If there are an even number of numbers in the data set, the median is found by adding the two middle numbers and dividing by 2.

Mode The most frequent number in a data set.

Monomial A mathematical expression that has a coefficient and/or a variable part. $3x^2$ is a monomial. $3 + x^2$ is not a monomial.

Multiplication property of equality A property of algebra that states that when equal values are multiplied by both sides of a true equation, the equation continues to be true. To solve the equation $(1/2)x = 5$, multiply both sides of the equation by 2, using the multiplication property of equality.

Ordered pair Two numbers written in the form (x, y). An ordered pair can be a solution to a two-variable equation. For example, (2, 5) is one solution to the equation $y = 2x + 1$. Ordered pairs can be graphed on the coordinate axes by locating the point with the x-coordinate equal to the x value and the y-coordinate equal to the y value.

Parabola A "U"-shaped curve that is the graph of the solution set of a quadratic equation.

Perfect square trinomial A quadratic trinomial of the form $x^2 + bx + (b/2)^2$, which can be factored into $(x + b/2)^2$. For example, $x^2 + 6x + 9 = (x + 3)^2$.

Piecewise function A function that has multiple rules for determining output values from input values, depending on what the input values are. If the function $f(x)$ is defined as

$$f(x) = \begin{cases} 2x + 1 & \text{if } x < 0 \\ x^2 & \text{if } x \geq 0 \end{cases}$$

then $f(-3) = 2(-3) + 1 = -1$ and $f(5) = 5^2 = 25$.

Polynomial The sum of one or more monomials. Each term of the polynomial has the form ax^n. $3x^4 + 2x^3 - 3x^2 + 6x - 1$ is a polynomial.

Quadratic equation An equation in which the highest power on a variable is a 2. $x^2 + 5x + 6 = 0$ is a quadratic equation.

Quadratic formula The formula $x = \left(-b \pm \sqrt{(b^2 - 4ac)}\right)/2a$. This formula can be used to find the two solutions to the quadratic equation $ax^2 + bx + c = 0$.

Quadratic function A function in which the highest power on a variable is 2. $f(x) = x^2 + 5x + 6$ is a quadratic function.

Quadratic polynomial A polynomial in which the highest power on a variable is 2. $x^2 + 5x + 6$ is a quadratic polynomial.

Range The set of values that can be output from a function is the range of that function.

Recursively defined sequence A way to define a sequence in which the first term or terms of the sequence are given and a formula is given for calculating the next term based on the previous term or terms.

$$a_1 = 5$$
$$a_n = 3 + a_{n-1} \text{ for } n > 1$$

is a recursive definition for the sequence 5, 8, 11, 14, ….

Regression Finding a curve that best fits a scatter plot. Three types of regression are linear, quadratic, and exponential.

Residual plot A set of points that represent how far points on a graph deviate from a curve of best fit.

Roots The roots of an equation are the values that solve that equation. The roots of $x^2 + 5x + 6 = 0$ are –3 and –2.

Sequence A list of numbers that usually have some kind of pattern.

Slope The slope of a line is a measure of the line's steepness. The equation for the slope of a line that passes through the two points (x_1, y_1) and (x_2, y_2) is $m = \frac{y_2 - y_1}{x_2 - x_1}$.

Slope-intercept form An equation in the form $y = mx + b$ where m and b are numbers in slope intercept form. $y = 2x - 1$ is in slope intercept form. When a two-variable equation is in slope intercept form, the graph of the equation has a y-intercept of $(0, b)$ and a slope of m.

Solution set The set of numbers or ordered pairs that satisfies an equation. The solution set of $x + 2 = 5$ is $\{3\}$. The solution set of $x + y = 10$ has an infinite number of ordered pairs in its solution set, including (2, 8), (3, 7), and (4, 6).

Substitution method A method for solving a system of equations in which one variable is isolated in one of the equations, and the expression equal to that variable is substituted for it in the other equation.

Subtraction property of equality A property of algebra that states that when equal values are subtracted from both sides of a true equation, the equation continues to be true. To solve the equation $x + 2 = 5$, subtract 2 from both sides of the equation by using the subtraction property of equality.

System of equations Two or more equations with two or more unknowns to solve for. An example of a system of equations with a solution of (8, 2) is

$$x + y = 10$$
$$x - y = 6$$

Third quartile The number in a data set that is greater than just 75% of the numbers in the set.

Translation A translation of a graph is when the points on it are each shifted the same amount in the same direction. Examples of translations are vertical translations, horizontal translations, and combinations of vertical and horizontal translations.

Trinomial A polynomial with three terms. The polynomial $x^2 + 5x + 6$ is a trinomial.

Variable A letter, often an x, y, or z, that represents a value in a mathematical expression. In an algebraic equation, the variable is often the unknown that needs to be solved for.

Vertex The turning point of a parabola is its vertex. If the parabola opens upward, the vertex is the minimum point. If the parabola opens downward, the vertex is the maximum point.

Vertical line test A way of testing to see if a graph can represent a function. If at least one vertical line can pass through at least two points on the graph, the graph fails the vertical line test and cannot represent a function. If there are no vertical lines that can pass through at least two points, then the graph can be the graph of a function.

x-intercept The location where a curve crosses the x-axis. The y-coordinate of the x-intercept is 0.

y-intercept The location where a curve crosses the y-axis. The x-coordinate of the y-intercept is 0. In slope-intercept form, $y = mx + b$, the y-intercept is located at $(0, b)$.

Zeros The zeros of a function f are the numbers that can be input into the function so that 0 is output from the function. For example, the function $f(x) = 2x - 6$ has the number 3 as its only zero since $f(3) = 2(3) - 6 = 6 - 6 = 0$.

Regents Examinations, Answers, and Self-Analysis Charts

Examination June 2014

Algebra I

HIGH SCHOOL MATH REFERENCE SHEET

Conversions

1 inch = 2.54 centimeters	1 cup = 8 fluid ounces
1 meter = 39.37 inches	1 pint = 2 cups
1 mile = 5280 feet	1 quart = 2 pints
1 mile = 1760 yards	1 gallon = 4 quarts
1 mile = 1.609 kilometers	1 gallon = 3.785 liters
	1 liter = 0.264 gallon
1 kilometer = 0.62 mile	1 liter = 1000 cubic centimeters
1 pound = 16 ounces	
1 pound = 0.454 kilogram	
1 kilogram = 2.2 pounds	
1 ton = 2000 pounds	

Formulas

Triangle	$A = \frac{1}{2}bh$
Parallelogram	$A = bh$
Circle	$A = \pi r^2$
Circle	$C = \pi d$ or $C = 2\pi r$

Formulas (continued)

General Prisms	$V = Bh$
Cylinder	$V = \pi r^2 h$
Sphere	$V = \frac{4}{3}\pi r^3$
Cone	$V = \frac{1}{3}\pi r^2 h$
Pyramid	$V = \frac{1}{3}Bh$
Pythagorean Theorem	$a^2 + b^2 = c^2$
Quadratic Formula	$x = \frac{-b \pm \sqrt{b^2 - 4ac}}{2a}$
Arithmetic Sequence	$a_n = a_1 + (n - 1)d$
Geometric Sequence	$a_n = a_1 r^{n-1}$
Geometric Series	$S_n = \frac{a_1 - a_1 r^n}{1 - r}$ where $r \neq 1$
Radians	1 radian = $\frac{180}{\pi}$ degrees
Degrees	1 degree = $\frac{\pi}{180}$ radians
Exponential Growth/Decay	$A = A_0 e^{k(t - t_0)} + B_0$

PART I

Answer all 24 questions in this part. Each correct answer will receive 2 credits. No partial credit will be allowed. For each statement or question, write in the space provided the numeral preceding the word or expression that best completes the statement or answers the question. [48 credits]

1 When solving the equation $4(3x^2 + 2) - 9 = 8x^2 + 7$, Emily wrote $4(3x^2 + 2) = 8x^2 + 16$ as her first step. Which property justifies Emily's first step?

(1) addition property of equality
(2) commutative property of addition
(3) multiplication property of equality
(4) distributive property of multiplication over addition

1 _____

2 Officials in a town use a function, C, to analyze traffic patterns. $C(n)$ represents the rate of traffic through an intersection where n is the number of observed vehicles in a specified time interval. What would be the most appropriate domain for the function?

(1) $\{\ldots -2, -1, 0, 1, 2, 3, \ldots\}$
(2) $\{-2, -1, 0, 1, 2, 3\}$
(3) $\left\{0, \frac{1}{2}, 1, 1\frac{1}{2}, 2, 2\frac{1}{2}\right\}$
(4) $\{0, 1, 2, 3, \ldots\}$

2 _____

3 If $A = 3x^2 + 5x - 6$ and $B = -2x^2 - 6x + 7$, then $A - B$ equals

(1) $-5x^2 - 11x + 13$
(2) $5x^2 + 11x - 13$
(3) $-5x^2 - x + 1$
(4) $5x^2 - x + 1$

3 _____

4 Given: $y + x > 2$
$y \leq 3x - 2$

Which graph shows the solution of the given set of inequalities?

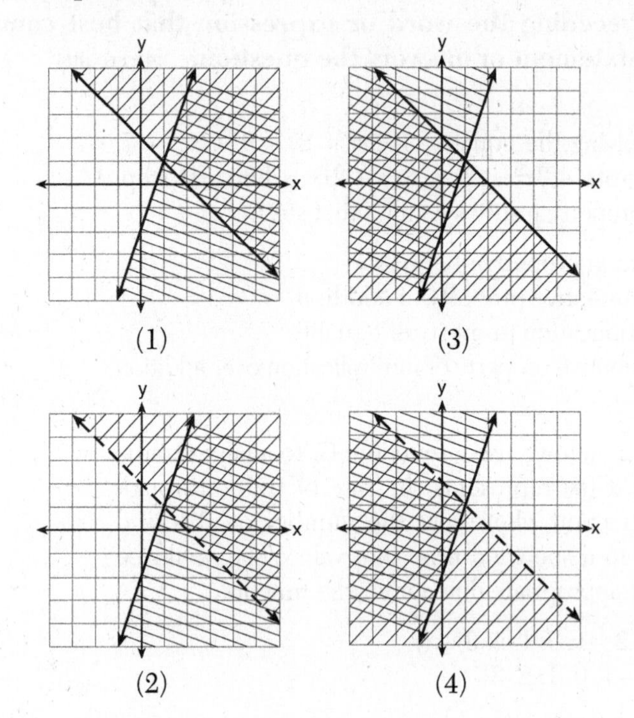

4 _____

5 Which value of x satisfies the equation $\frac{7}{3}\left(x + \frac{9}{28}\right) = 20$?

(1) 8.25
(2) 8.89
(3) 19.25
(4) 44.92

5 _____

6 The table below shows the average yearly balance in a savings account where interest is compounded annually. No money is deposited or withdrawn after the initial amount is deposited.

Year	Balance, in Dollars
0	380.00
10	562.49
20	832.63
30	1232.49
40	1824.39
50	2700.54

Which type of function best models the given data?

(1) linear function with a negative rate of change
(2) linear function with a positive rate of change
(3) exponential decay function
(4) exponential growth function

6 _____

7 A company that manufactures radios first pays a start-up cost, and then spends a certain amount of money to manufacture each radio. If the cost of manufacturing r radios is given by the function $c(r) = 5.25r + 125$, then the value 5.25 best represents

(1) the start-up cost
(2) the profit earned from the sale of one radio
(3) the amount spent to manufacture each radio
(4) the average number of radios manufactured

7 _____

8 Which equation has the same solution as $x^2 - 6x - 12 = 0$?

(1) $(x + 3)^2 = 21$
(2) $(x - 3)^2 = 21$
(3) $(x + 3)^2 = 3$
(4) $(x - 3)^2 = 3$

8 _____

9 A ball is thrown into the air from the edge of a 48-foot-high cliff so that it eventually lands on the ground. The graph below shows the height, y, of the ball from the ground after x seconds.

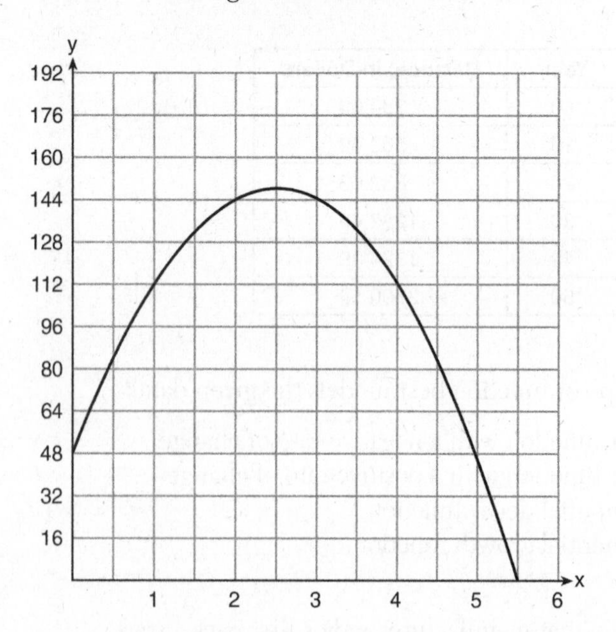

For which interval is the ball's height always *decreasing*?

(1) $0 \leq x \leq 2.5$ (3) $2.5 < x < 5.5$
(2) $0 < x < 5.5$ (4) $x \geq 2$ 9 _____

10 What are the roots of the equation $x^2 + 4x - 16 = 0$?

(1) $2 \pm 2\sqrt{5}$ (3) $2 \pm 4\sqrt{5}$
(2) $-2 \pm 2\sqrt{5}$ (4) $-2 \pm 4\sqrt{5}$ 10 _____

11 What is the correlation coefficient of the linear fit of the data shown below, to the *nearest hundredth*?

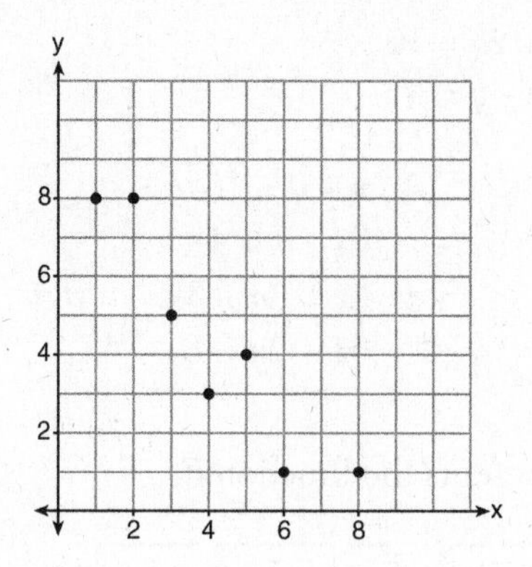

(1) 1.00
(2) 0.93
(3) –0.93
(4) –1.00

11 ____

12 Keith determines the zeros of the function $f(x)$ to be –6 and 5. What could be Keith's function?

(1) $f(x) = (x + 5)(x + 6)$
(2) $f(x) = (x + 5)(x - 6)$
(3) $f(x) = (x - 5)(x + 6)$
(4) $f(x) = (x - 5)(x - 6)$

12 ____

13 Given:

$L = \sqrt{2}$

$M = 3\sqrt{3}$

$N = \sqrt{16}$

$P = \sqrt{9}$

Which expression results in a rational number?

(1) $L + M$
(2) $M + N$
(3) $N + P$
(4) $P + L$

13 ____

14 Which system of equations has the same solution as the system below?

$$2x + 2y = 16$$
$$3x - y = 4$$

(1) $2x + 2y = 16$
$6x - 2y = 4$

(2) $2x + 2y = 16$
$6x - 2y = 8$

(3) $x + y = 16$
$3x - y = 4$

(4) $6x + 6y = 48$
$6x + 2y = 8$

14 _____

15 The table below represents the function F.

x	3	4	6	7	8
F(x)	9	17	65	129	257

The equation that represents this function is

(1) $F(x) = 3^x$
(2) $F(x) = 3x$
(3) $F(x) = 2^x + 1$
(4) $F(x) = 2x + 3$

15 _____

16 John has four more nickels than dimes in his pocket, for a total of $1.25. Which equation could be used to determine the number of dimes, x, in his pocket?

(1) $0.10(x + 4) + 0.05(x) = \$1.25$
(2) $0.05(x + 4) + 0.10(x) = \$1.25$
(3) $0.10(4x) + 0.05(x) = \$1.25$
(4) $0.05(4x) + 0.10(x) = \$1.25$

16 _____

17 If $f(x) = \frac{1}{3}x + 9$, which statement is always true?

(1) $f(x) < 0$
(2) $f(x) > 0$
(3) If $x < 0$, then $f(x) < 0$.
(4) If $x > 0$, then $f(x) > 0$.

17 _____

18 The Jamison family kept a log of the distance they traveled during a trip, as represented by the graph below.

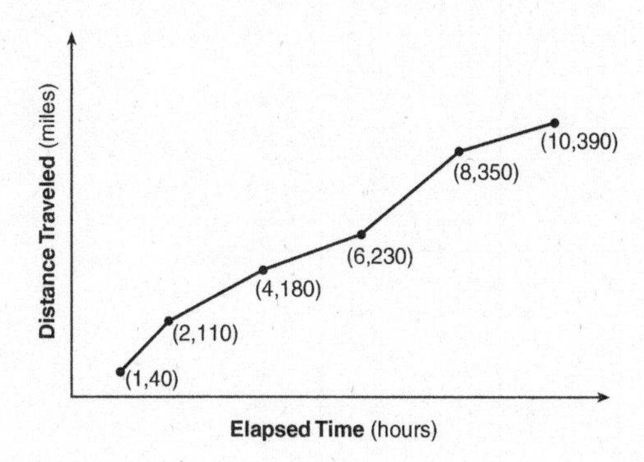

During which interval was their average speed the greatest?

(1) the first hour to the second hour
(2) the second hour to the fourth hour
(3) the sixth hour to the eighth hour
(4) the eighth hour to the tenth hour

18 ____

19 Christopher looked at his quiz scores shown below for the first and second semesters of his Algebra class.

Semester 1: 78, 91, 88, 83, 94
Semester 2: 91, 96, 80, 77, 88, 85, 92

Which statement about Christopher's performance is correct?

(1) The interquartile range for semester 1 is greater than the interquartile range for semester 2.
(2) The median score for semester 1 is greater than the median score for semester 2.
(3) The mean score for semester 2 is greater than the mean score for semester 1.
(4) The third quartile for semester 2 is greater than the third quartile for semester 1.

19 ____

20 The graph of $y = f(x)$ is shown below.

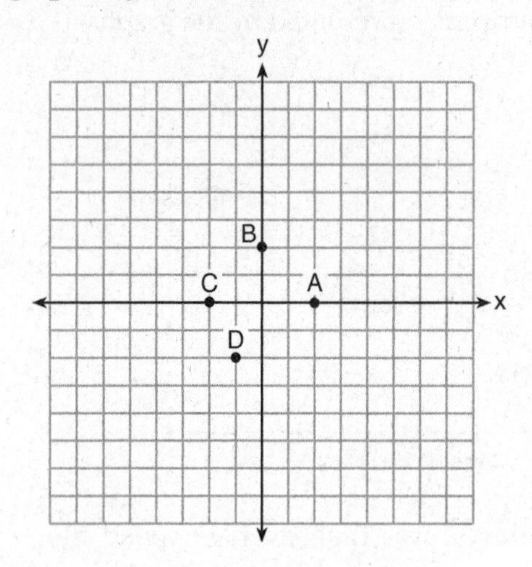

Which point could be used to find $f(2)$?

(1) *A*
(2) *B*
(3) *C*
(4) *D*

20 _____

21 A sunflower is 3 inches tall at week 0 and grows 2 inches each week. Which function(s) shown below can be used to determine the height, $f(n)$, of the sunflower in n weeks?

I. $f(n) = 2n + 3$
II. $f(n) = 2n + 3(n - 1)$
III. $f(n) = f(n - 1) + 2$, where $f(0) = 3$

(1) I and II
(2) II, only
(3) III, only
(4) I and III

21 _____

22 A cell phone company charges \$60.00 a month for up to 1 gigabyte of data. The cost of additional data is \$0.05 per megabyte. If d represents the number of additional megabytes used and c represents the total charges at the end of the month, which linear equation can be used to determine a user's monthly bill?

(1) $c = 60 - 0.05d$
(2) $c = 60.05d$
(3) $c = 60d - 0.05$
(4) $c = 60 + 0.05d$

22 _____

23 The formula for the volume of a cone is $V = \frac{1}{3}\pi r^2 h$.

The radius, r, of the cone may be expressed as

(1) $\sqrt{\frac{3V}{\pi h}}$
(2) $\sqrt{\frac{V}{3\pi h}}$
(3) $3\sqrt{\frac{V}{\pi h}}$
(4) $\frac{1}{3}\sqrt{\frac{V}{\pi h}}$

23 _____

24 The diagrams below represent the first three terms of a sequence.

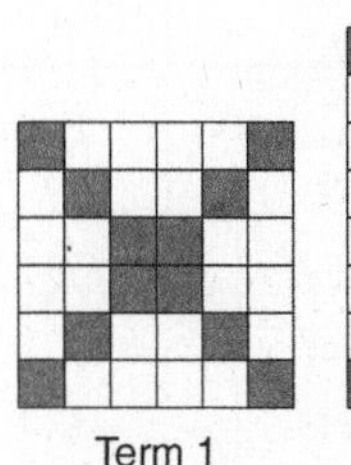
Term 1

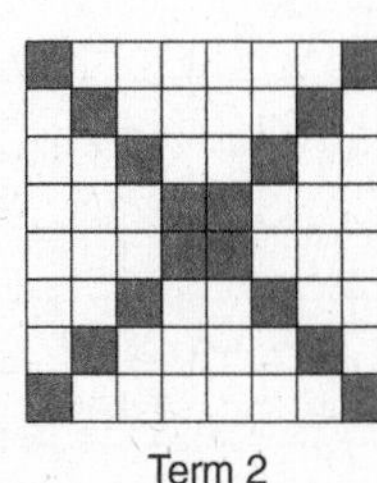
Term 2

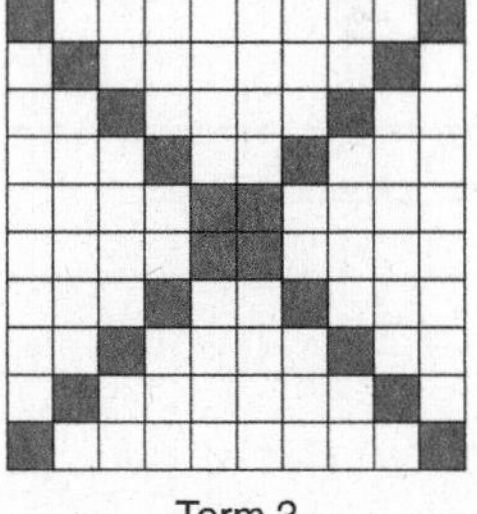
Term 3

Assuming the pattern continues, which formula determines a_n, the number of shaded squares in the nth term?

(1) $a_n = 4n + 12$
(2) $a_n = 4n + 8$
(3) $a_n = 4n + 4$
(4) $a_n = 4n + 2$

24 _____

PART II

Answer all 8 questions in this part. Each correct answer will receive 2 credits. Clearly indicate the necessary steps, including appropriate formula substitutions, diagrams, graphs, charts, etc. For all questions in this part, a correct numerical answer with no work shown will receive only 1 credit. [16 credits]

25 Draw the graph of $y = \sqrt{x} - 1$ on the set of axes below.

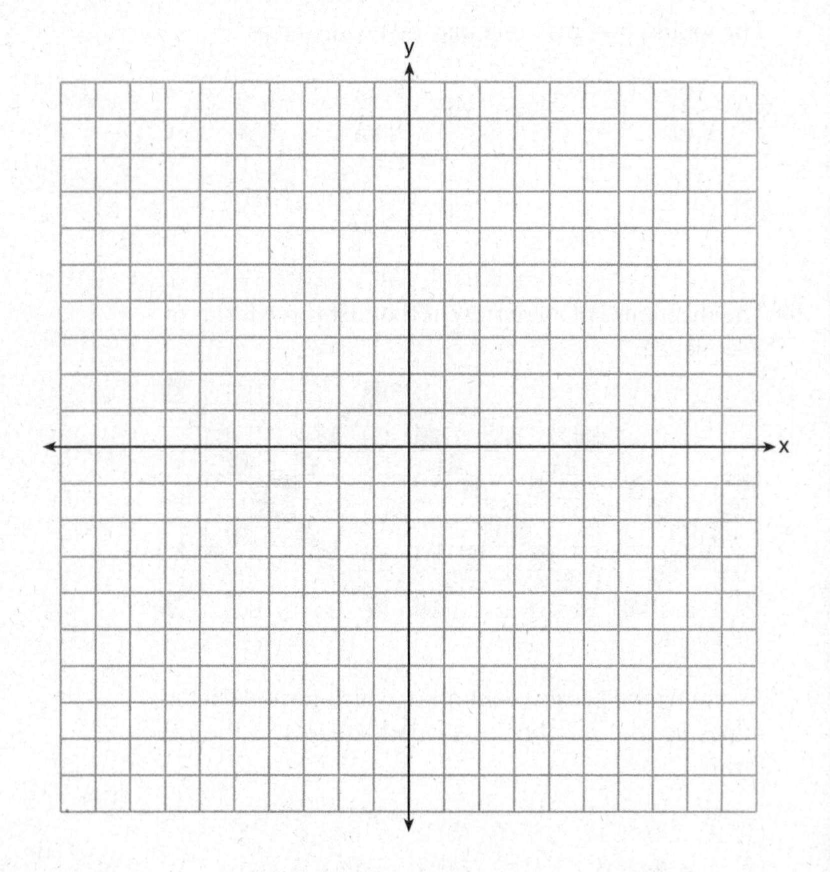

26 The breakdown of a sample of a chemical compound is represented by the function $p(t) = 300(0.5)^t$, where $p(t)$ represents the number of milligrams of the substance and t represents the time, in years. In the function $p(t)$, explain what 0.5 and 300 represent.

27 Given $2x + ax - 7 > -12$, determine the largest integer value of a when $x = -1$.

28 The vertex of the parabola represented by $f(x) = x^2 - 4x + 3$ has coordinates (2, –1). Find the coordinates of the vertex of the parabola defined by $g(x) = f(x - 2)$. Explain how you arrived at your answer.

[The use of the set of axes below is optional.]

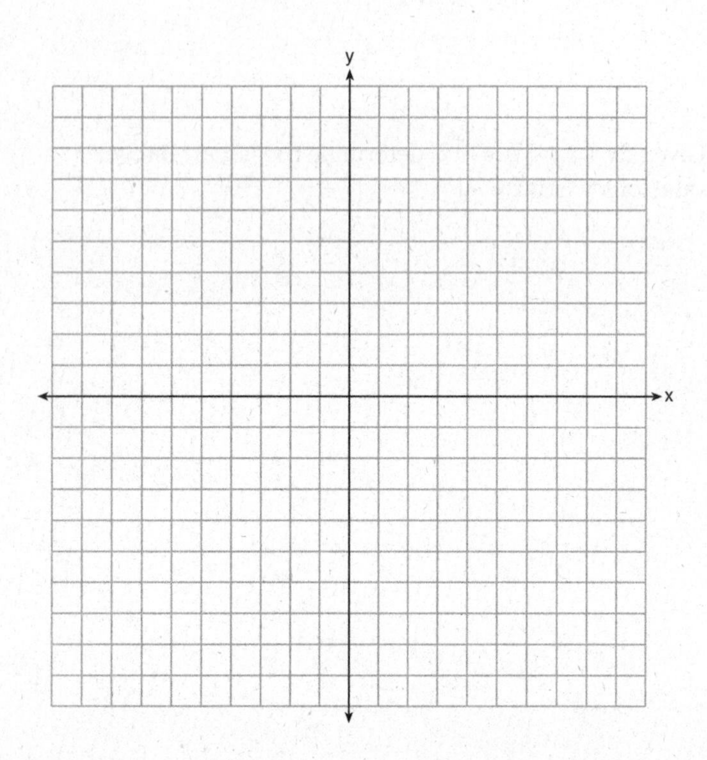

29 On the set of axes below, draw the graph of the equation $y = -\frac{3}{4}x + 3$.

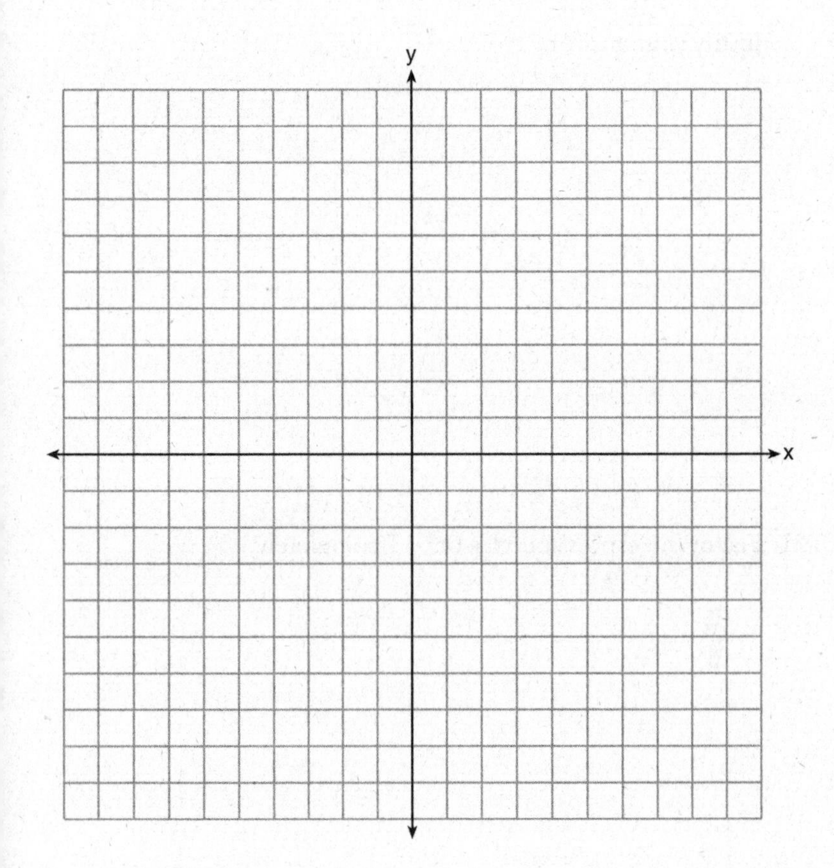

Is the point (3,2) a solution to the equation? Explain your answer based on the graph drawn.

30 The function f has a domain of {1, 3, 5, 7} and a range of {2, 4, 6}.

Could f be represented by {(1,2), (3,4), (5,6), (7,2)}?

Justify your answer.

31 Factor the expression $x^4 + 6x^2 - 7$ completely.

32 Robin collected data on the number of hours she watched television on Sunday through Thursday nights for a period of 3 weeks. The data are shown in the table below.

	Sun	Mon	Tues	Wed	Thurs
Week 1	4	3	3.5	2	2
Week 2	4.5	5	2.5	3	1.5
Week 3	4	3	1	1.5	2.5

Using an appropriate scale on the number line below, construct a box plot for the 15 values.

PART III

Answer all 4 questions in this part. Each correct answer will receive 4 credits. Clearly indicate the necessary steps, including appropriate formula substitutions, diagrams, graphs, charts, etc. For all questions in this part, a correct numerical answer with no work shown will receive only 1 credit. [16 credits]

33 Write an equation that defines $m(x)$ as a trinomial where $m(x) = (3x - 1)(3 - x) + 4x^2 + 19$.

Solve for x when $m(x) = 0$.

34 A rectangular garden measuring 12 meters by 16 meters is to have a walkway installed around it with a width of x meters, as shown in the diagram below. Together, the walkway and the garden have an area of 396 square meters.

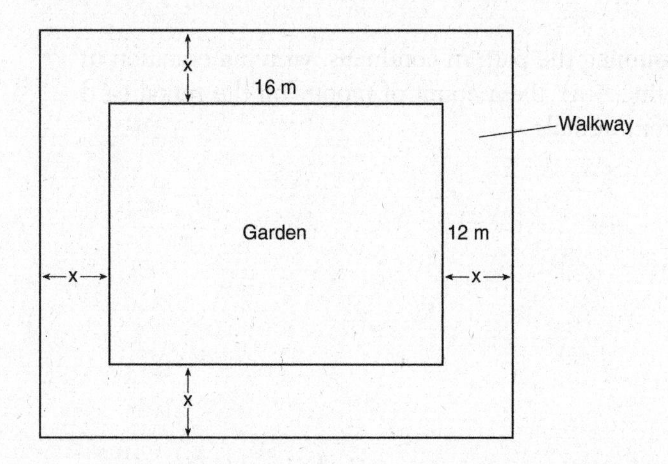

Write an equation that can be used to find x, the width of the walkway.

Describe how your equation models the situation.

Determine and state the width of the walkway, in meters.

35 Caitlin has a movie rental card worth $175. After she rents the first movie, the card's value is $172.25. After she rents the second movie, its value is $169.50. After she rents the third movie, the card is worth $166.75.

Assuming the pattern continues, write an equation to define $A(n)$, the amount of money on the rental card after n rentals.

Caitlin rents a movie every Friday night. How many weeks in a row can she afford to rent a movie, using her rental card only? Explain how you arrived at your answer.

36 An animal shelter spends $2.35 per day to care for each cat and $5.50 per day to care for each dog. Pat noticed that the shelter spent $89.50 caring for cats and dogs on Wednesday.

Write an equation to represent the possible numbers of cats and dogs that could have been at the shelter on Wednesday.

Pat said that there might have been 8 cats and 14 dogs at the shelter on Wednesday. Are Pat's numbers possible? Use your equation to justify your answer.

Later, Pat found a record showing that there were a total of 22 cats and dogs at the shelter on Wednesday. How many cats were at the shelter on Wednesday?

PART IV

Answer the question in this part. A correct answer will receive 6 credits. Clearly indicate the necessary steps, including appropriate formula substitutions, diagrams, graphs, charts, etc. A correct numerical answer with no work shown will receive only 1 credit. [6 credits]

37 A company is considering building a manufacturing plant. They determine the weekly production cost at site A to be $A(x) = 3x^2$, while the production cost at site B is $B(x) = 8x + 3$, where x represents the number of products, *in hundreds*, and $A(x)$ and $B(x)$ are the production costs, *in hundreds of dollars*.

Graph the production cost functions on the set of axes below and label them site A and site B.

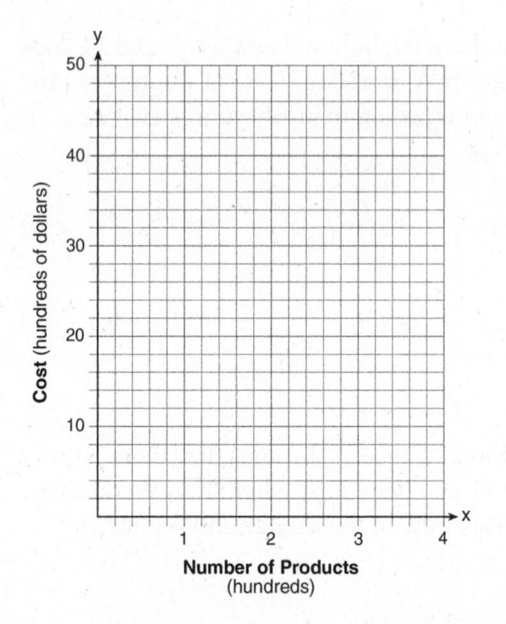

Question 37 is continued on the next page.

Question 37 continued

State the positive value(s) of x for which the production costs at the two sites are equal. Explain how you determined your answer.

If the company plans on manufacturing 200 products per week, which site should they use? Justify your answer.

Answers June 2014

Algebra I

Answer Key

PART I

1. (1)	**5.** (1)	**9.** (3)	**13.** (3)	**17.** (4)	**21.** (4)
2. (4)	**6.** (4)	**10.** (2)	**14.** (2)	**18.** (1)	**22.** (4)
3. (2)	**7.** (3)	**11.** (3)	**15.** (3)	**19.** (3)	**23.** (1)
4. (2)	**8.** (2)	**12.** (3)	**16.** (2)	**20.** (1)	**24.** (2)

PART II

25. A graph should be created with points (0, –1), (1, 0), (4, 1), and (9, 2)

26. 300 is the initial value and .5 is the rate of decay.

27. $a = 2$

28. (4, –1)

29. (3, 2) is not a solution

30. Yes, f can be represented by the domain and range given.

31. $(x^2 + 7)(x + 1)(x - 1)$

32. Plot the relevant numbers on the line: 1 the first quartile, 2 the median, 3 the third quartile, 4 and the maximum, 5.

PART III

33. $x^2 + 10x + 16$, $x = -2$ or $x = -8$

34. $(12 + 2x)(16 + 2x) = 396$, 3 meters

35. $A(n) = 175 - 2.75n$, 63 weeks

36. $2.35c + 5.50d = 89.50$, 10 cats

PART IV

37. Since $A(2) < B(2)$, it would be less expensive to build at site A.

In Parts II–IV, you are required to show how you arrived at your answers. For sample methods of solutions, see the *Answers Explained* section.

Answers Explained

PART I

1. The two differences between the equations $4(3x^2 + 2) - 9 = 8x^2 + 7$ and $4(3x^2 + 2) = 8x^2 + 16$ are that the -9 is gone from the left side of the equals sign and there is a $+16$ on the right side of the equals sign instead of a $+7$. The way to make both of these changes happen is to add 9 to both sides of the first equation. When you add to both sides of an equation, it is called the addition property of equality.

   The correct choice is **(1)**.

2. The number of cars is most appropriately measured in whole numbers. Choices 1 and 2 have negative numbers so they can be eliminated. Choice 3 can be eliminated because cars generally come in whole units.

   The correct choice is **(4)**.

3. To subtract the two polynomials, first write them as
   $(3x^2 + 5x - 6) - (-2x^2 - 6x + 7)$

   Now, treat the minus sign between the two polynomials as a negative one and distribute it to eliminate the parentheses.
   $(3x^2 + 5x - 6) - 1(-2x^2 - 6x + 7)$
   $3x^2 + 5x - 6 + 2x^2 + 6x - 7$

   Now, combine like terms.
   $3x^2 + 2x^2 + 5x + 6x - 6 - 7$
   $5x^2 + 11x - 13$

   Note: A common wrong answer is choice 4, which is what you get if you do not put parentheses around the two polynomials.

   $$3x^2 + 5x - 6 - -2x^2 - 6x + 7$$

   The correct choice is **(2)**.

4. Write each inequality in slope-intercept form:
   $y > -x + 2$
   $y \leq 3x - 2$

   The line that passes through $(0, 2)$ is the boundary of $y > -x + 2$ and the line that passes through $(0, -2)$ is $y \leq 3x - 2$.

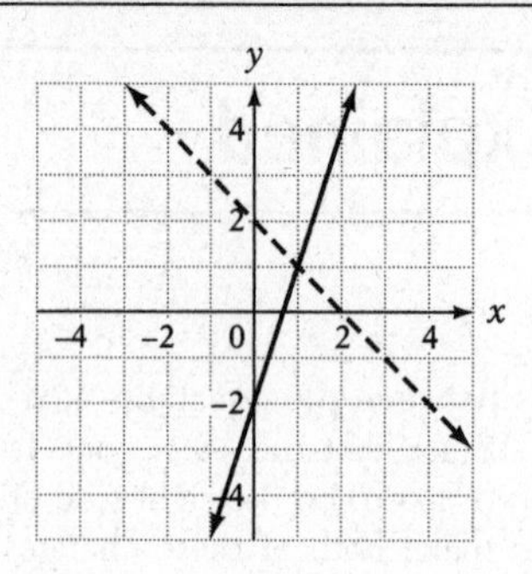

Notice that each of the four choices has the same two lines.

Because $y > -x + 2$ has a $>$ sign rather than a $\geq$ sign, that line requires a dotted line. This eliminates choices 1 and 3.

When you plug (0, 0) into the inequality $y > -x + 2$, it becomes $0 > 0 + 2$, which is false so the region that does not contain (0, 0) gets shaded.

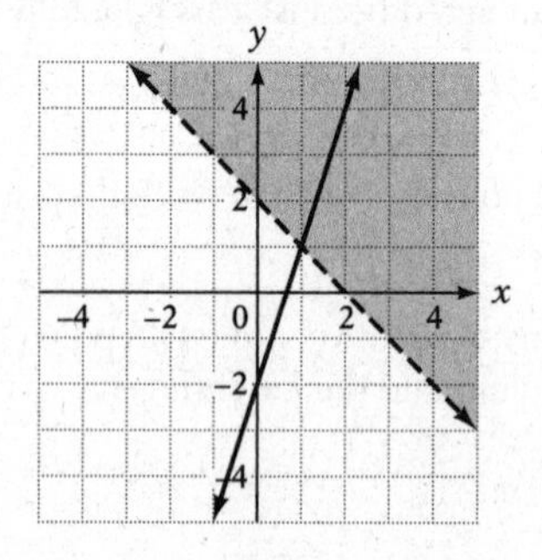

When you plug (0, 0) into the inequality $y \leq 3x - 2$, it becomes $0 \leq 0 - 2$, which is false so the region that does not contain (0, 0) gets shaded.

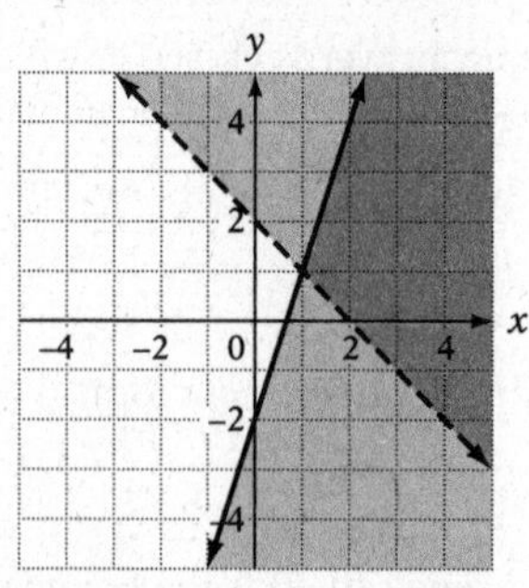

The correct choice is **(2)**.

5. First distribute the $\frac{7}{3}$ through on the left-hand side of the equals sign.

$$\frac{7}{3}x+\frac{63}{84}=20$$
$$-\frac{63}{84}=-\frac{63}{84}$$
$$\frac{7}{3}x=20-\frac{63}{84}$$
$$\frac{7}{3}x=\frac{1{,}680}{84}-\frac{63}{84}$$
$$\frac{7}{3}x=\frac{1{,}617}{84}$$

Multiply both sides of the equation by $\frac{3}{7}$ to isolate the x.

$$\frac{3}{7}\cdot\frac{7}{3}x=\frac{3}{7}\cdot\frac{1{,}617}{84}$$
$$x=\frac{4{,}851}{588}=8.25$$

The correct choice is **(1)**.

6. The balance is increasing so choices 1 and 3 can be eliminated, since each of those would represent decreasing values. If the function were linear, then for each ten-year increase there would be approximately the same amount of change in the balance. But between years 0 and 10 there was an increase of about $180, while between years 10 and 20 there was an increase of about $270. For exponential growth, the percent increase for each ten-year interval will be approximately the same. The growth rate from year 0 to 10 is $\frac{562.49}{380}=1.48$, and the growth rate from year 10 to 20 is $\frac{832.63}{562.49}=1.48$.

An alternate way to do this question is to make a graph of the 6 points.

For the TI-84:

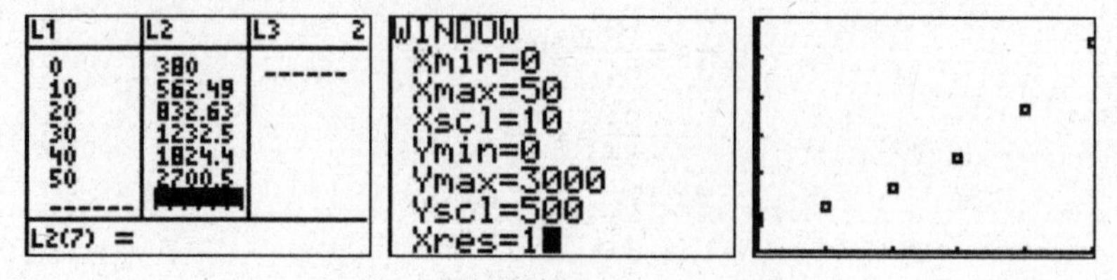

For the TI-Nspire:

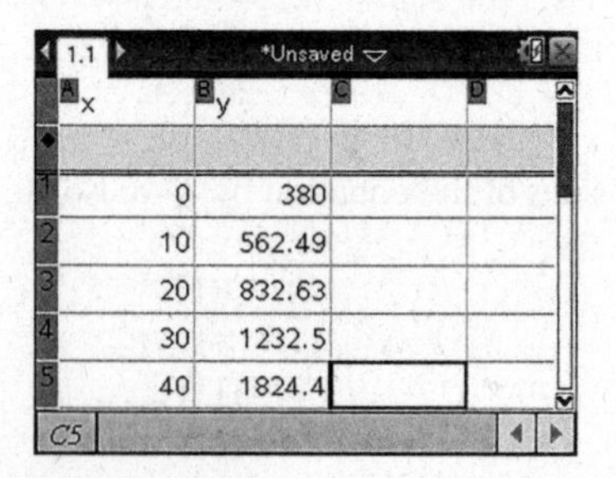

x	y
0	380
10	562.49
20	832.63
30	1232.5
40	1824.4

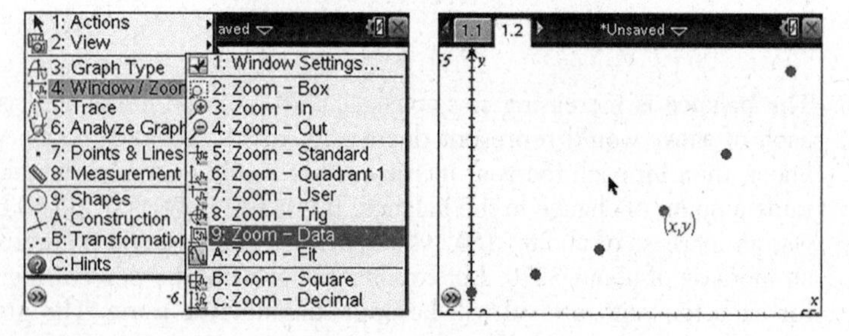

This is the shape of a graph of an exponential growth function.

The correct choice is **(4)**.

7. When no radios are manufactured, r is 0, which makes the cost $c(0) = 5.25(0) + 125 = 125$. This \$125, then, is the start-up cost, so choice 1 can be eliminated. When r is increased to 1, the cost becomes $c(1) = 5.25(1) + 125 = 130.25$. This is \$5.25 more than it cost to manufacture zero radios so \$5.25 must be the amount spent to manufacture each radio.

   Another way to approach this question is to know that when an equation has the form $y = mx + b$, the m is the amount y increases for each time x increases by 1 and that b is the constant term. In a real world scenario, the initial value will be the constant, b, and the expression mx will be the part that changes when x changes.

   The correct choice is **(3)**.

8. To make an equation like $x^2 - 6x - 12 = 0$ resemble the answer choices, you have to do a process called completing the square.

   First eliminate the constant from the left-hand side of the equation by adding 12 to each side.

$$\begin{aligned} x^2 - 6x - 12 &= 0 \\ +12 &= +12 \\ x^2 - 6x &= 12 \end{aligned}$$

   Now to make the left-hand side into a perfect square trinomial, multiply half of the coefficient of the x by itself, $\left(\frac{-6}{2}\right)^2 = (-3)^2 = 9$ and add that to both sides of the equation.

$$\begin{aligned} x^2 - 6x + 9 &= 12 + 9 \\ x^2 - 6x + 9 &= 21 \end{aligned}$$

   The left-hand side of the equation can now be factored.

$$(x - 3)^2 = 21$$

   The correct choice is **(2)**.

9. Between 0 and 2.5 seconds the ball's height increases from 48 feet to more than 144 feet. After the ball reaches its maximum height at 2.5 seconds, the height decreases until it lands at 5.5 seconds.

   The correct choice is **(3)**.

10. Use the quadratic formula with $a = 1$, $b = 4$, and $c = -16$.

$$x = \frac{-b \pm \sqrt{b^2 - 4ac}}{2a}$$

$$x = \frac{-4 \pm \sqrt{4^2 - 4(1)(-16)}}{2(1)}$$

$$x = \frac{-4 \pm \sqrt{16+64}}{2}$$

$$x = \frac{-4 \pm \sqrt{80}}{2}$$

$$x = \frac{-4 + \sqrt{16 \cdot 5}}{2}$$

$$x = \frac{-4 \pm 4\sqrt{5}}{2}$$

$$x = -2 \pm 2\sqrt{5}$$

The correct choice is **(2)**.

11. The correlation coefficient of the linear fit is a number between –1 and 1, which measures how well the points in a scatter plot look like a straight line. If the slope of that line is positive, the correlation coefficient is positive. If the slope of that line is negative, the correlation coefficient is negative. In this case, the correlation coefficient must be negative, so choices 1 and 2 can be eliminated.

Since the points on this graph do not all lie perfectly on the same line, the answer is not –1, which leaves –0.93. A negative number close to –1 is a correlation coefficient for a collection of points that approximately lie on the same downward sloping line.

Had there been another choice, also close to –1, like –0.75, you would be required to use a calculator to calculate the exact value of the correlation coefficient.

For the TI-84:

Put the x values of all the points into list 1 and all the corresponding y values into list 2. In the stat menu, go to CALC and select the LinReg option. If the r value does not display, go to the catalog and enable Diagnostics On.

L1	L2	L3
1	8	
2	8	
3	5	
4	3	
5	4	
6	1	
8	1	

L3(1)=

EDIT CALC TESTS
1:1-Var Stats
2:2-Var Stats
3:Med-Med
4:LinReg(ax+b)
5:QuadReg
6:CubicReg
7↓QuartReg

LinReg
y=ax+b
a=-1.12704918
b=8.954918033
r²=.8609403461
r=-.9278687117

For the TI-Nspire:

Select Add Lists & Spreadsheet and enter the x values into column 1 and the y values into column 2. At the top of column 1, put the label x. At the top of column 2, put the label y. Move to cell B2 and press [menu], [4], [1], [3], [enter]. Put x into the X List and y into the Y List and press the OK button.

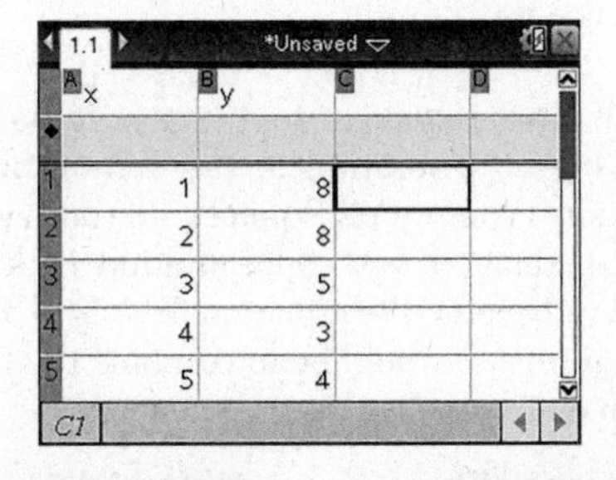

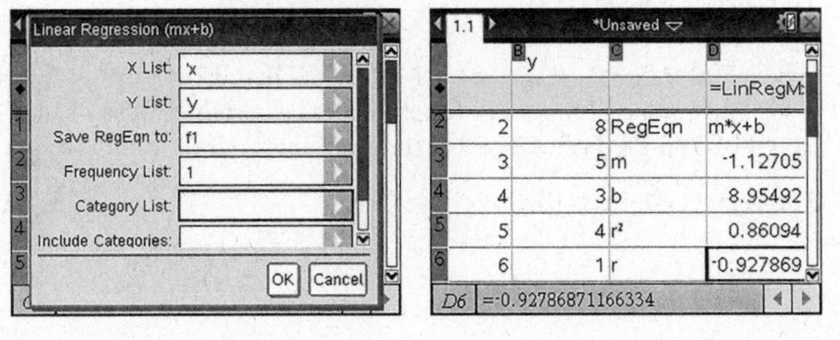

The "r" represents the correlation coefficent, which is approximately –.93.

The correct choice is **(3)**.

12. The zeros of a function are the values that when substituted into the function result in a value of zero. The zeros of the function described in choice 3 are –6 and 5 since

$f(-6) = (-6 - 5)(-6 + 6) = (-11)(0) = 0$ and
$f(5) = (5 - 5)(5 + 6) = (0)(11) = 0$.

In general, the zeros of a function have the opposite signs of the constants in the factors of the function.

The correct choice is **(3)**.

13. The sum of two rational numbers is always rational. L and M are irrational numbers while N and P are actually rational since 16 and 9 are both perfect squares.

$N + P = \sqrt{16} + \sqrt{9} = 4 + 3 = 7$, which is rational.

The sum of a rational and an irrational is always irrational, as in choices 2 and 4, whereas the sum of two irrationals is usually irrational. (An example of the sum of two irrationals being rational is $(\sqrt{2}) + (5 - \sqrt{2}) = 5$.)

The correct choice is **(3)**.

14. When you multiply an equation in a system of equations by a constant, it does not change the solution to the system because you are using the Multiplication Property of Equality. If you had to solve the system described in the equation, you could multiply both sides of the equation $3x - y = 4$ by 2 to become the equation $6x - 2y = 8$. If you were asked to solve for x, the next step would be to combine the two equations, making the y term drop out.

The correct choice is **(2)**.

15. When $F(x) = 2^x + 1$, $F(3) = 2^3 + 1 = 8 + 1 = 9$ and $F(4) = 2^4 + 1 = 16 + 1 = 17$. Since none of the other function choices produce 9 and 17 for input values of 3 and 4, $F(x) = 2^x + 1$ is the answer.

The correct choice is **(3)**.

16. If the number of dimes is represented by x, then the number of nickels is $x + 4$.

Since each dime is worth 10 cents, the value of x dimes is $0.10(x)$ dollars. Since each nickel is worth 5 cents, the value of $x + 4$ nickels is $0.05\ (x + 4)$ dollars.

The combined value is $0.05\ (x + 4) + 0.10(x)$.

To calculate x, the number of dimes, you would set up the equation

$0.05\ (x + 4) + 0.10(x) = \$1.25$

The correct choice is **(2)**.

17. The function value can be negative, for example $f(-30) = \frac{1}{3}(-30) + 9 = -10 + 9 = -1$. It can also be positive, for example $f(6) = \frac{1}{3}(6) + 9 = 2 + 9 = 11$. So it isn't *always* negative or *always* positive.

If x is a small enough negative number, then $f(x)$ can be positive, such as $f(-3) = \frac{1}{3}(-3) + 9 = -1 + 9 = 8$, so having a negative x doesn't *always* mean that $f(x)$ is negative.

But for any positive x value, $f(x)$ will be greater than 9 so it will *always* be greater than zero.

This question can also be solved by creating the graph of $f(x)$.

On the TI-84:

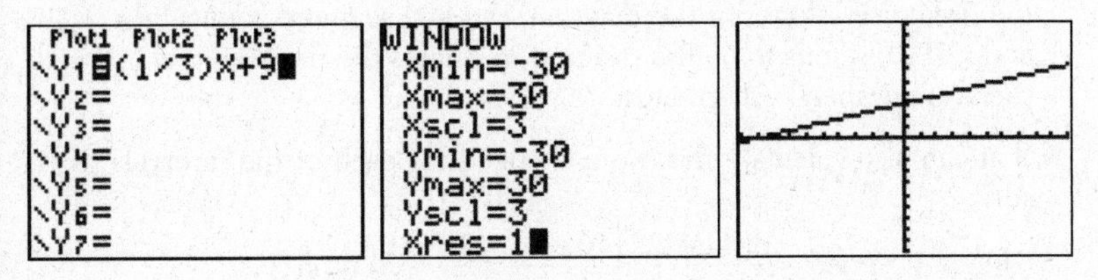

On the TI-Nspire:

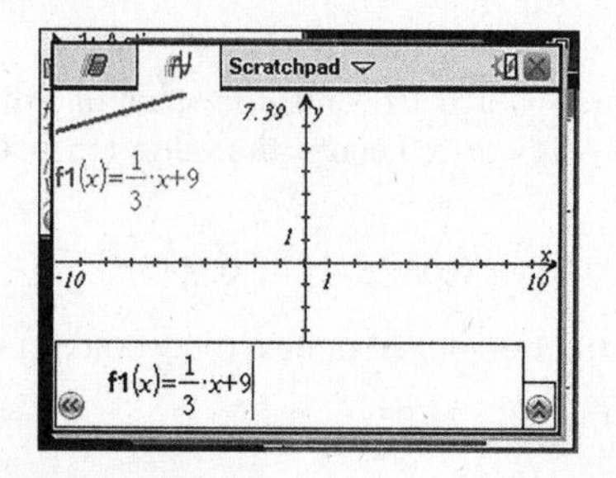

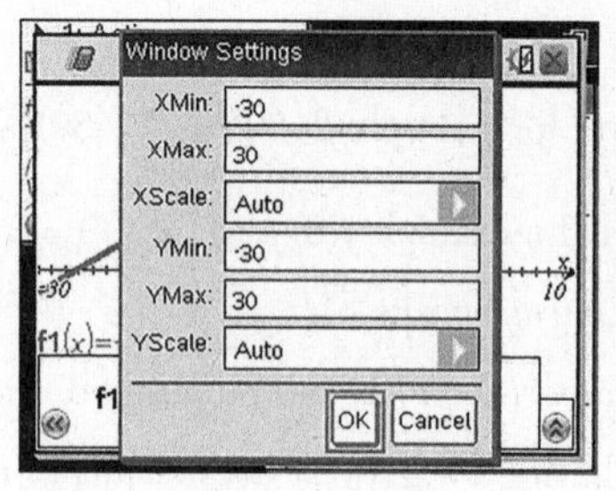

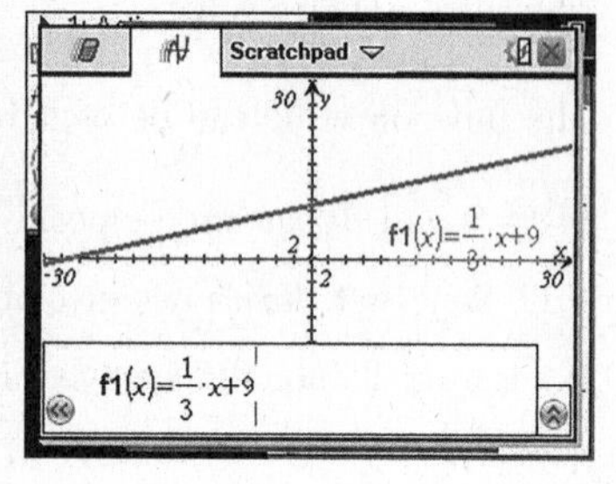

Notice that for all points on the graph to the right of the y-axis, the y-value, which is the $f(x)$ value, is positive.

The correct choice is **(4)**.

18. The speed for each time interval is equivalent to the *slope* of the line segment representing that time interval. The steeper the line segment, the greater its slope. In the diagram, the line segment connecting (1,40) and (2,110) seems to be the steepest, so that is the time interval in which the average speed was greatest.

    You can also calculate the average speed for each of the intervals to be sure.

    Choice 1: $\text{speed} = \dfrac{\text{distance}}{\text{time}} = \dfrac{110-40}{2-1} = \dfrac{70}{1} = 70 \text{ mph}$

    Choice 2: $\text{speed} = \dfrac{\text{distance}}{\text{time}} = \dfrac{180-110}{4-2} = \dfrac{70}{2} = 35 \text{ mph}$

    Choice 3: $\text{speed} = \dfrac{\text{distance}}{\text{time}} = \dfrac{350-230}{8-6} = \dfrac{120}{2} = 60 \text{ mph}$

    Choice 4: $\text{speed} = \dfrac{\text{distance}}{\text{time}} = \dfrac{390-350}{10-8} = \dfrac{40}{2} = 20 \text{ mph}$

    The correct choice is **(1)**.

19. You have to calculate the mean, median, third quartile, and interquartile range for both data sets.

For the median, arrange the numbers in each data set from least to greatest, and pick the middle number (or the average of the two middle numbers if the set has an even number of elements). For semester 1, the scores are 78, 83, 88, 91, 94, so the median is 88. For semester 2, the scores are 77, 80, 85, 88, 91, 92, 96, so the median is also 88. Choice 2 can be eliminated since the two medians are equal.

For the third quartile, create a set of all the numbers greater than the median, and find the median of that set. For semester 1, the scores greater than 88 are just 91 and 94. Since that is two numbers, there is no middle number so you take the average of the two middle numbers, which are just 91 and 94. So the third quartile is 92.5. For semester 2, the numbers greater than the mean are 91, 92, and 96. With three numbers, the median of that is the middle number, which is 92. Choice 4 can be eliminated since the third quartile for semester 2 is less than the third quartile for set 1.

To calculate the interquartile range, you find the difference between the third quartile, calculated above, and the first quartile. To calculate the first quartile, you find the median of all the numbers less than the median of the set. For semester 1, this is the median of 78 and 83. Since there is no number between them, you average the two numbers to get a first quartile of 80.5. For semester 2, the median of the three numbers 77, 80, and 85 is 80. Using the values calculated for the third quartile, you find that the interquartile range for semester 1 is $92.5 - 80.5 = 12$ while the interquartile range for semester 2 is $92 - 80$, which is also 12. Choice 1 can be eliminated.

The mean is also called the average, and you calculate it by adding all the numbers and then dividing that sum by the number of numbers in the set. For semester 1 this is $\frac{78+91+88+83+94}{5} = 86.8$. For semester 2 this is $\frac{91+96+80+77+88+85+92}{7} = 87$, so the mean score for semester 2 is greater than the mean score for semester 1.

For the TI-84:

Press [STAT], [1] to edit. Enter the numbers for semester 1 into column L1. Press [STAT], [right arrow], [1], [ENTER] for 1-Var Stats

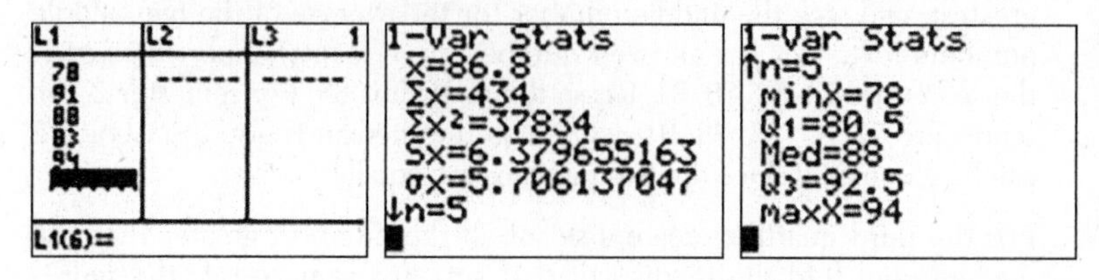

For the TI-Nspire:

In a spreadsheet label column 1 "s1" and enter the data for semester 1 into that column. Press [menu], [4], [1], [1] for One-Variable Statistics. Select 1 list and set the X1 List to s1 and press the OK button. Do the same for semester 2.

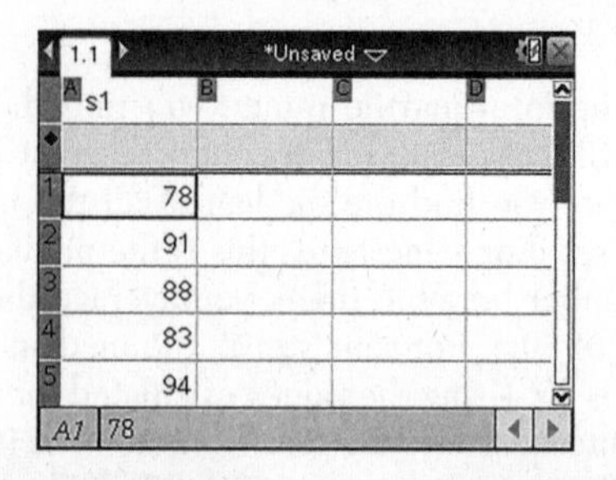

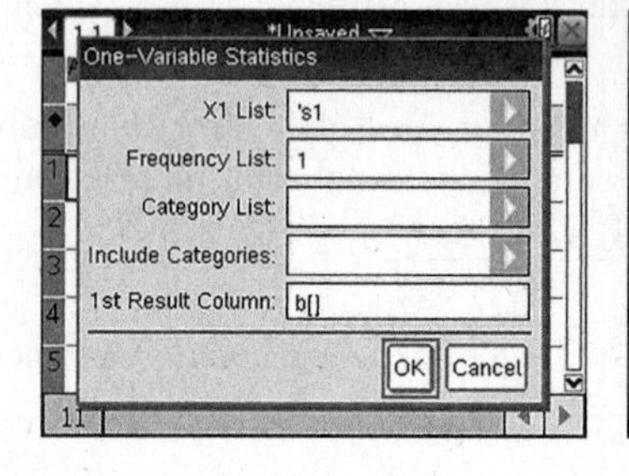

1.1 *Unsaved

A s1	B	C	D
		=OneVar('s	
91	x̄	86.8	
88	Σx	434.	
83	Σx²	37834.	
94	sx := sn-1...	6.37966	
	σx := σnx...	5.70614	

C2 =86.8

The correct choice is **(3)**.

20. When a function is represented with a graph, each (x, y) point can be interpreted as $f(x) = y$. Since point A is located at $(2, 0)$, $f(2) = 0$ so point A can be used to find $f(2)$.

Point B can be used to find that $f(0) = 2$, point C can be used to find that $f(-2) = 0$, and point D can be used to find that $f(-1) = -2$.

The correct choice is **(1)**.

21. First, read the question carefully and create a list of the heights of the sunflower for a few weeks. The list would start 3, 5, 7, 9.

    For the function described in I, if you substitute 0 for n, you get $f(0) = 2(0) + 3 = 3$ and that $f(1) = 2(1) + 3 = 5$, which agrees with the list.

    For the function described in II, if you substitute 0 for n, you get $f(0) = 2(0) + 3(0 - 1) = -3$, so function II does not agree with the list.

    The function described in III is a *recursive* function, which is more involved to calculate.

    According to the function description, to get $f(0)$, no calculation is needed since it says that $f(0) = 3$, which agrees with the list. To get $f(1)$, you substitute both occurrences of n in the function with a 1. This will become $f(1) = f(1 - 1) + 2 = f(0) + 2 = 3 + 2 = 5$. This also agrees with the list.

    Another way to think about option III is that if $f(n)$ is the nth term, then $f(n - 1)$ is the term right before the nth term on the list. So the definition $f(n) = f(n - 1)+2$ says that each term is 2 more than the term right before it, which is what is happening in this scenario.

    The correct choice is **(4)**.

22. For a specific value, like $d = 10$ additional megabytes, the total charges would be calculated as $c = \$60.00 + 0.05(10)$. Since 10 is the value of d, the general formula will be

    $c = 60.00 + 0.05d$.

    In equations of the form $y = mx + b$ or, in this case, $y = b + mx$, the b is the constant term and the m is the amount that the expression changes by each time x is increased by 1. Since the \$60.00 is always the initial cost regardless of the value of d, it is the constant term. Since the cost of the additional data is 0.05 for each of the d gigabytes, the cost of all the additional data is $.05d$.

    The correct choice is **(4)**.

23. It takes three steps to isolate the r in an equation like this. First multiply both sides of the equation by 3.

$$V = \frac{1}{3}\pi r^2 h$$
$$3V = 3\left(\frac{1}{3}\right)\pi r^2 h = \pi r^2 h$$

Then divide both sides of the equation by □*h*.

$$\frac{3V}{\pi h} = \frac{\pi r^2 h}{\pi h}$$
$$\frac{3V}{\pi h} = r^2$$

Finally, take the square root of both sides of the equation, keeping only the positive square root since the radius can't be negative.

$$\sqrt{\frac{3V}{\pi h}} = \sqrt{r^2}$$
$$\sqrt{\frac{3V}{\pi h}} = r$$

The correct choice is **(1)**.

24. Count the number of shaded squares in each picture. Term 1 has 12 shaded squares, term 2 has 16 shaded squares, and term 3 has 20 shaded squares. These numbers can be written in a sequence so that $a_1 = 12$, $a_2 = 16$, and $a_3 = 20$. If you substitute $n = 1$ into each of the equations, only choice 2 will give you $a_1 = 12$.

The correct choice is **(2)**.

PART II

25. Since $\sqrt{x}-1$ is undefined for negative values of x, make a chart for x values between 0 and 9.

x	0	1	2	3	4	5	6	7	8	9
$\sqrt{x}-1$	–1	0	$\sqrt{2}-1$	$\sqrt{3}-1$	1	$\sqrt{5}-1$	$\sqrt{6}-1$	$\sqrt{7}-1$	$\sqrt{8}-1$	2

Graph the points (0, –1), (1, 0), (4, 1), and (9, 2), and be sure to put an arrow at the end of the curve.

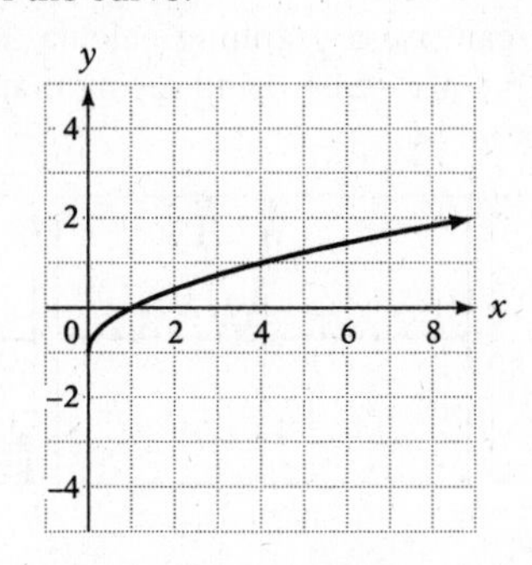

26. If you substitute 0 for t, you find that $p(0) = 300(0.5)^0 = 300(1) = 300$, which means that the 300 represents the number of milligrams of the substance in the beginning of the process. If you substitute 1 for t, you get $p(1) = 300(0.5)^1 = 300(0.5) = 150$, so the 0.5 is the decay rate.

In general, when you have an exponential expression $a \cdot b^x$, a is the initial value and b is the growth rate, when $b > 1$ and b is the decay rate when b is between 0 and –1.

27. When $x = -1$, the equation becomes

$$
\begin{aligned}
2(-1)+a(-1)-7 &> -12 \\
-2-a-7 &> -12 \\
-a-9 &> -12 \\
+9 &= +9 \\
-a &> -3 \\
\frac{-a}{-1} &< \frac{-3}{-1}
\end{aligned}
$$

In an inequality, whenever you divide by a negative, the direction of the inequality sign switches, so this becomes

$$a < 3$$

The largest integer value that makes this inequality true is $a = 2$.

28. The graph of $f(x - 2)$ is the same as the graph of $f(x)$ with each point translated two units to the right. So if the vertex of the parabola represented by $f(x)$ is $(2, -1)$, the vertex of the parabola represented by $f(x - 2)$ will be $(2, -1)$ translated two units to the right, which is $(4, -1)$.

Alternatively, you can use a graphing calculator to graph $f(x - 2)$, by entering $y = (x - 2)^2 - 4(x - 2) + 3$ and seeing that the vertex of the graph of this function is at $(4, -1)$.

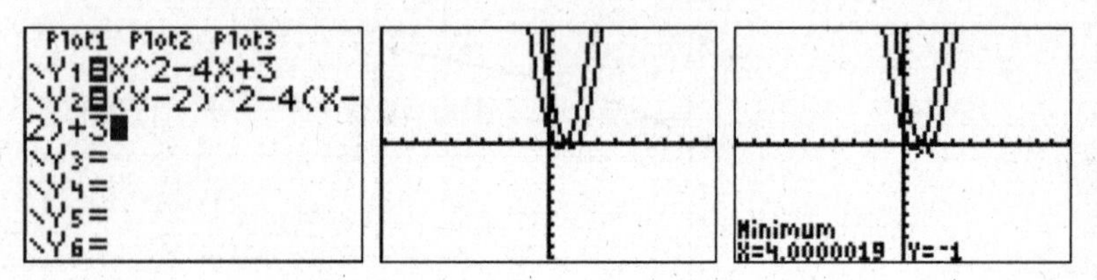

29. To graph the solution set of $y = -\frac{3}{4}x + 3$, plot the y-intercept $(0, 3)$ and then, because the slope is $-\frac{3}{4}$, plot the next point four units to the right and three down, at $(4, 0)$.

Connect the points $(0, 3)$ and $(4, 0)$ and continue this segment in both directions to graph the line. Each ordered pair that is a solution to the equation $y = -\frac{3}{4}x + 3$ corresponds to a point on the line. Since $(3, 2)$ is not on the line, $(3, 2)$ is not a solution to the equation.

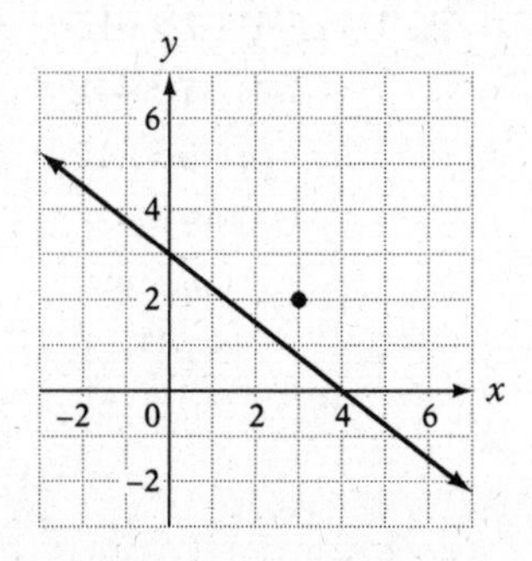

Though plugging a 3 in for x and a 2 in for y will verify that (3, 2) is not a solution, you would not get credit for that method since the question says your explanation must be "based on the graph drawn" and not on the equation.

30. For the function to be represented by this set of ordered pairs, three conditions must be met:

(a) Each of the x-coordinates must be part of the domain. Since 1, 3, 5, and 7 are the x-coordinates and they are all in the domain, this condition is met.

(b) Each of the y-coordinates must be part of the range. Since 2, 4, and 6 are all in the range, this condition is also met.

(c) Each x-coordinate must correspond to only one y-value. This condition is also met. It is OK that two different x-values correspond to the same y-value (1, 2) and (7, 2). Since all three conditions are met, f could be represented by this set of ordered pairs.

Had there been an ordered pair like (1, 4) in addition to the four ordered pairs listed, that would violate the condition.

31. $x^4 + 6x^2 - 7$ can be rewritten as $(x^2)^2 + 6(x^2) - 7$ so it has the same structure of a quadratic trinomial. Just as something like $y^2 + 6y - 7$ would factor into $(y + 7)(y - 1)$, $(x^2)^2 + 6(x^2) - 7$ can factor into $(x^2 + 7)(x^2 - 1)$. Then, since $x^2 - 1$ is the difference of perfect squares, it can be further factored into $(x - 1)(x + 1)$. The complete factorization, then, is $(x^2 + 7)(x + 1)(x - 1)$.

32. To make a box plot (also known as a box-and-whisker plot), first list all the data in order from least to greatest. There were 15 numbers in the data set:

1, 1.5, 1.5, 2, 2, 2.5, 2.5, 3, 3, 3, 3.5, 4, 4, 4.5, 5

For the box plot, five things need to be calculated:

(a) The minimum element, which for this example is 1.

(b) The maximum element, which for this example is 5.

(c) The median number is determined by locating the middle element on the list after the elements are arranged from least to greatest. Since there are 15 elements, the eighth element is the median.

1, 1.5, 1.5, 2, 2, 2.5, 2.5, (3) 3, 3, 3.5, 4, 4, 4.5, 5.

(d) The first quartile is the median of the numbers 1, 1.5, 1.5, 2, 2, 2.5, and 2.5, all the numbers less than the median of the entire data set. The median of these seven numbers is the fourth number, which is 2.

(e) The third quartile is the median of the numbers 3, 3, 3.5, 4, 4, 4.5, and 5, all the numbers greater than the median of the entire data set. The median of these seven numbers is the fourth number, which is 4.

The five relevant numbers, then, are the minimum, 1, the first quartile, 2, the median, 3, the third quartile, 4, and the maximum, 5. These must be plotted on the number line.

It is reasonable to put 1 on the first mark, a 2 on the third, a 3 on the fifth, a 4 on the seventh, and a 5 on the ninth.

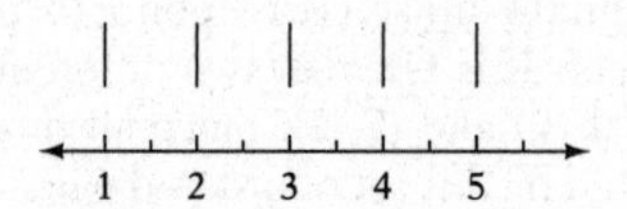

To complete the box plot, make the box between the first quartile and the third quartile. Then make the whiskers between the minimum and the first quartile and between the third quartile and the maximum.

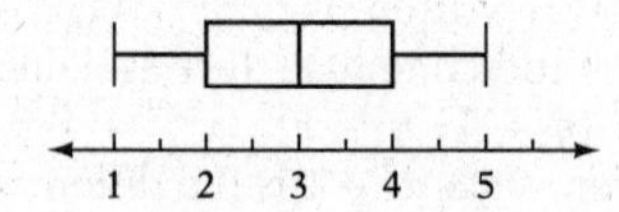

PART III

33. A trinomial is a polynomial with three terms. In order to write $m(x)$ as a trinomial, you have to multiply the expression $(3x - 1)(3 - x)$ and then combine like terms.

$$m(x) = (3x - 1)(3 - x) + 4x^2 + 19$$
$$m(x) = 9x - 3x^2 - 3 + x + 4x^2 + 19$$
$$m(x) = -3x^2 + 4x^2 + 9x + x - 3 + 19$$
$$m(x) = 1x^2 + 10x + 16$$

If $m(x) = 0$,

$$0 = x^2 + 10x + 16$$
$$0 = (x + 2)(x + 8)$$
$$(x + 2) = 0 \text{ or } (x + 8) = 0$$
$$x = -2 \text{ or } x = -8$$

34. The walkway and garden form a rectangle with a base of $16 + 2x$ and a height of $12 + 2x$.

Since the area is known to be 396 square meters, and the area of a rectangle is base times height, the equation is $(16 + 2x)(12 + 2x) = 396$.

To calculate the width of the walkway, solve the equation:

$$(16 + 2x)(12 + 2x) = 396$$
$$192 + 32x + 24x + 4x^2 = 396$$
$$4x^2 + 56x + 192 = 396$$
$$-396 = -396$$

$$4x^2 + 56x - 204 = 0$$

This quadratic equation can be solved several ways. One way is through factoring:

$$4(x^2 + 14x - 51) = 0$$
$$4(x + 17)(x - 3) = 0$$

$$(x + 17) = 0 \text{ or } (x - 3) = 0$$
$$x = -17 \text{ or } x = 3$$

Another way is with the quadratic formula:

$$\begin{aligned} x &= \frac{-56 \pm \sqrt{56^2 - 4(4)(-204)}}{2(4)} \\ &= \frac{-56 \pm \sqrt{3136 + 3264}}{8} \\ &= \frac{-56 \pm \sqrt{6400}}{8} \\ &= \frac{-56 \pm 80}{8} \\ x &= -17 \text{ or } x = 3 \end{aligned}$$

But since the width of the walkway must be greater than zero, it must have a width of 3 meters.

35. In order for the value of the card to decrease from \$175 to \$172.25, the first movie must have cost \$2.75. To decrease from \$172.25 to \$169.50 is also a decrease of \$2.75. So each movie rented costs \$2.75. If you rent one movie, the value left on the card is \$175 – \$2.75 = \$172.25. If you rent two movies, the value left on the card is \$175 – 2(\$2.75) = \$169.50. For n movies, the amount left on the card will be \$175 – n(\$2.75).

So the equation can be written as $A(n) = 175 - 2.75n$.

Caitlin can rent movies until another rental would cause the value on her card to be below \$0. To calculate the number of weeks she can rent, you can set up the inequality:

$$175 - 2.75n \geq 0.$$

This inequality is solved much like an equality.

$$\begin{gathered} 175 - 2.75n \geq 0 \\ -175 = -175 \\ -2.75n \geq -175 \\ \frac{-2.75n}{-2.75} \leq \frac{-175}{-2.75} \end{gathered}$$

In an inequality, when you divide both sides by a negative, the direction of the inequality sign changes.

$$n \leq 63.\overline{63}$$

So the maximum number of weeks is 63.

36. If c is the number of cats and d is the number of dogs, then the cost for the cats is $2.35c$, and the cost for the dogs is $5.50d$. The total cost is $2.35c + 5.50d$.

    Since you are told that the total cost is $89.50, the equation for the possible numbers of cats and dogs is $2.35c + 5.50d = 89.50$.

    To check if Pat is correct about there being 8 cats and 14 dogs, test to see if $c = 8$ and $d = 14$ is a valid solution of the equation $2.35c + 5.50d = 89.50$.

$$\begin{aligned} 2.35(8) + 5.50(14) &\stackrel{?}{=} 89.50 \\ 18.80 + 77 &\stackrel{?}{=} 89.50 \\ 95.80 &\neq 89.50 \end{aligned}$$

    So 8 cats and 14 dogs is not a valid solution.

    If the total number of animals is 22, there is a second equation $c + d = 22$. This equation, together with the other equation $2.35a + 5.50d = 89.50$ form a system of equations.

$$\begin{aligned} 2.35c + 5.50d &= 89.50 \\ c + d &= 22 \end{aligned}$$

    The most convenient way to solve this system of equations is to isolate one of the variables in the second equation and use the substitution method.

$$c = 22 - d$$

    Replace the c in the first equation with $22 - d$.

$$\begin{aligned} 2.35(22 - d) + 5.50d &= 89.50 \\ 51.7 - 2.35d + 5.50d &= 89.50 \\ 51.7 + 3.15d &= 89.50 \\ -51.7 &= -51.7 \\ 3.15d &= 37.8 \\ /3.15 &/ 3.15 \\ d &= 12 \end{aligned}$$

    To get the value of c, substitute 12 for d in the bottom equation.

$$\begin{aligned} c + d &= 22 \\ c + 12 &= 22 \\ -12 &= -12 \\ c &= 10 \end{aligned}$$

PART IV

37. The graph of the weekly production cost at site A is half of a parabola.

    The graph of the weekly production cost at site B is a line.

    When the two functions are graphed on the same set of axes, it looks like this:

    On the TI-84:

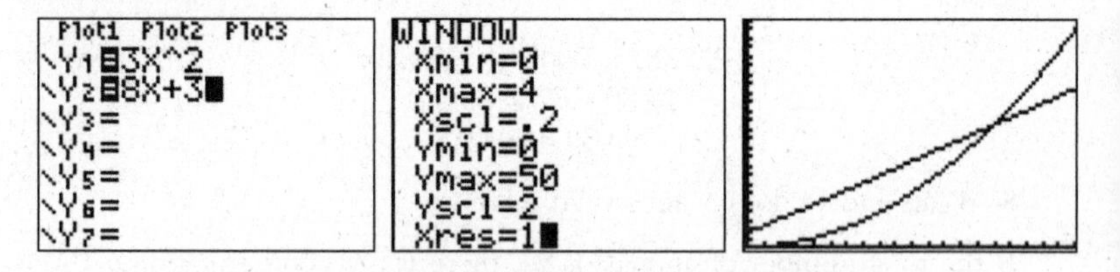

On the TI-Nspire:

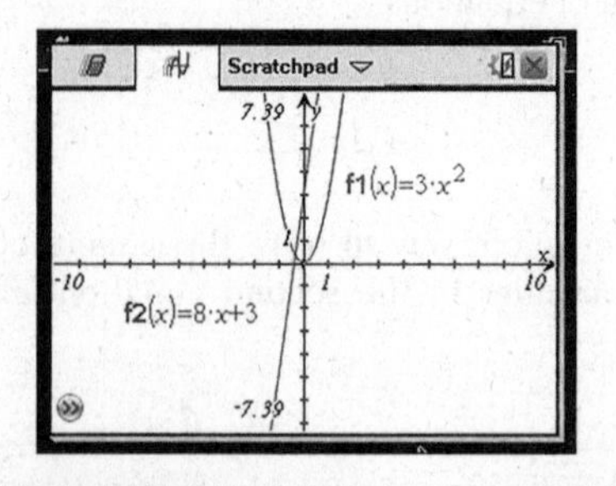

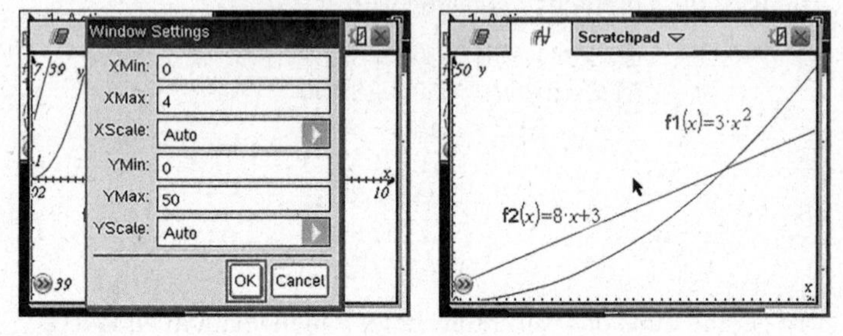

The x-coordinates of the intersection points of the line and the parabola represent the values of x when the production costs at the two sites are equal.

This can be done on a graphing calculator with the intersect feature.

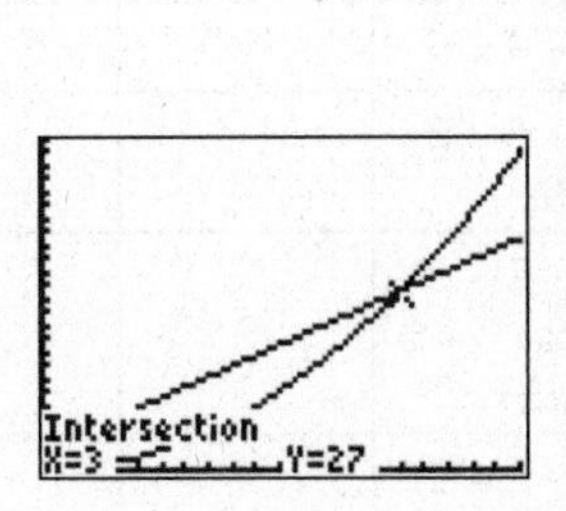

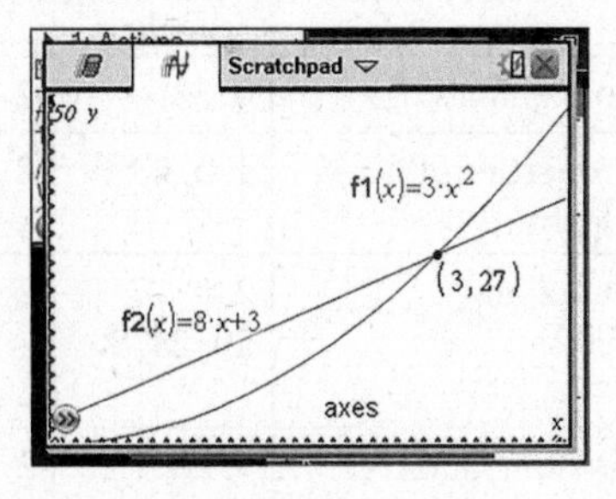

This can also be done with algebra by solving the equation

$$3x^2 = 8x + 3$$
$$3x^2 - 8x - 3 = 0$$
$$(3x+1)(x-3) = 0$$
$$3x + 1 = 0 \text{ or } x - 3 = 0$$
$$x = -\frac{1}{3} \text{ or } x = 3$$

Since they ask for the positive solution for x, the answer is 3.

On the graph it can be seen that at $x = 2$, the y-coordinate of the point on $A(x)$ is less than the y-coordinate of the point on $B(x)$. So to save money, they would choose site A if they were only planning to manufacture 200 products per week because the cost is less at that point on the graph.

This can also be determined numerically by calculating the value of $A(2)$ and $B(2)$.

$$A(2) = 3 \cdot 2^2 = 12$$
$$B(2) = 8 \cdot 2 + 3 = 19$$

Since $A(2) < B(2)$, it would be less expensive to build at site A instead of site B for the planned production.

On the graphing calculator, a table of values can be constructed in order to compare the values of $A(x)$ and $B(x)$ for various values of x.

X	Y1	Y2
0	3	0
1	11	3
2	19	12
3	27	27
4	35	48
5	43	75
6	51	108

Y2=108

1.1 1.2 *Unsaved

x	f1(x):= 8*x+3	f2(x):... 3*x^2
0.	3.	0.
1.	11.	3.
2.	19.	12.
3.	27.	27.
4.	35.	48.

0.

Topic	Question Numbers	Number of Points	Your Points	Your Percentage
1. Polynomials	3, 33	2 + 4 = 6		
2. Properties of Algebra	1, 5, 23	2 + 2 + 2 = 6		
3. Functions	2, 6, 17, 20, 28, 30	2 + 2 + 2 + 2+ 2 + 2 = 12		
4. Creating and Interpreting Equations	7, 16, 18, 22	2 + 2 + 2 + 2 = 8		
5. Inequalities	4, 27	2 + 2 = 4		
6. Sequences and Series	21, 24, 35	2 + 6 = 8		
7. Systems of Equations	14, 36	2 + 4 = 6		
8. Quadratic Equations and Factoring	8, 9, 10, 12, 31, 34	2 + 2 + 2 + 2 + 6 = 14		
9. Regression	11	2		
10. Exponential Equations	15, 26	2 + 2 = 4		
11. Graphing	25, 29, 37	2 + 4 + 4 = 10		
12. Statistics	19, 32	2 + 2 = 4		
13. Number Properties	13	2		

HOW TO CONVERT YOUR RAW SCORE TO YOUR ALGEBRA I REGENTS EXAMINATION SCORE

The accompanying conversion chart must be used to determine your final score on the June 2014 Regents Examination in Algebra I. To find your final exam score, locate in the column labeled "Raw Score" the total number of points you scored out of a possible 86 points. Since partial credit is allowed in Parts II, III, and IV of the test, you may need to approximate the credit you would receive for a solution that is not completely correct. Then locate in the adjacent column to the right the scale score that corresponds to your raw score. The scale score is your final Algebra I Regents Examination score.

Regents Examination in Algebra I—June 2014
Chart for Converting Total Test Raw Scores to Final Examination Scores (Scaled Scores)

Raw Score	Scale Score	Performance Level	Raw Score	Scale Score	Performance Level	Raw Score	Scale Score	Performance Level
86	100	5	57	75	4	28	64	2
85	99	5	56	74	4	27	63	2
84	97	5	55	74	4	26	62	2
83	96	5	54	74	4	25	61	2
82	95	5	53	73	3	24	60	2
81	94	5	52	73	3	23	59	2
80	92	5	51	73	3	22	58	2
79	91	5	50	72	3	21	56	2
78	90	5	49	72	3	20	55	2
77	89	5	48	72	3	19	54	1
76	88	5	47	72	3	18	52	1
75	87	5	46	71	3	17	50	1
74	86	5	45	71	3	16	49	1
73	85	5	44	71	3	15	47	1
72	84	4	43	70	3	14	45	1
71	83	4	42	70	3	13	42	1
70	82	4	41	70	3	12	40	1
69	82	4	40	70	3	11	38	1
68	81	4	39	69	3	10	35	1
67	80	4	38	69	3	9	32	1
66	79	4	37	69	3	8	30	1
65	79	4	36	68	3	7	26	1
64	78	4	35	68	3	6	23	1
63	78	4	34	67	3	5	20	1
62	77	4	33	67	3	4	16	1
61	77	4	32	66	3	3	12	1
60	76	4	31	66	3	2	9	1
59	76	4	30	65	3	1	4	1
58	75	4	29	64	2	0	0	1

Examination August 2014

Algebra I

HIGH SCHOOL MATH REFERENCE SHEET

Conversions

1 inch = 2.54 centimeters	1 cup = 8 fluid ounces
1 meter = 39.37 inches	1 pint = 2 cups
1 mile = 5280 feet	1 quart = 2 pints
1 mile = 1760 yards	1 gallon = 4 quarts
1 mile = 1.609 kilometers	1 gallon = 3.785 liters
	1 liter = 0.264 gallon
1 kilometer = 0.62 mile	1 liter = 1000 cubic centimeters
1 pound = 16 ounces	
1 pound = 0.454 kilogram	
1 kilogram = 2.2 pounds	
1 ton = 2000 pounds	

Formulas

Triangle	$A = \frac{1}{2}bh$
Parallelogram	$A = bh$
Circle	$A = \pi r^2$
Circle	$C = \pi d$ or $C = 2\pi r$

Formulas (continued)

General Prisms	$V = Bh$
Cylinder	$V = \pi r^2 h$
Sphere	$V = \frac{4}{3}\pi r^3$
Cone	$V = \frac{1}{3}\pi r^2 h$
Pyramid	$V = \frac{1}{3}Bh$
Pythagorean Theorem	$a^2 + b^2 = c^2$
Quadratic Formula	$x = \frac{-b \pm \sqrt{b^2 - 4ac}}{2a}$
Arithmetic Sequence	$a_n = a_1 + (n-1)d$
Geometric Sequence	$a_n = a_1 r^{n-1}$
Geometric Series	$S_n = \frac{a_1 - a_1 r^n}{1-r}$ where $r \neq 1$
Radians	1 radian = $\frac{180}{\pi}$ degrees
Degrees	1 degree = $\frac{\pi}{180}$ radians
Exponential Growth/Decay	$A = A_0 e^{k(t - t_0)} + B_0$

PART I

Answer all 24 questions in this part. Each correct answer will receive 2 credits. No partial credit will be allowed. For each statement or question, write in the space provided the numeral preceding the word or expression that best completes the statement or answers the question. [48 credits]

1 Which statement is *not* always true?

(1) The product of two irrational numbers is irrational.
(2) The product of two rational numbers is rational.
(3) The sum of two rational numbers is rational.
(4) The sum of a rational number and an irrational number is irrational. 1 ____

2 A satellite television company charges a one-time installation fee and a monthly service charge. The total cost is modeled by the function $y = 40 + 90x$. Which statement represents the meaning of each part of the function?

(1) y is the total cost, x is the number of months of service, \$90 is the installation fee, and \$40 is the service charge per month.
(2) y is the total cost, x is the number of months of service, \$40 is the installation fee, and \$90 is the service charge per month.
(3) x is the total cost, y is the number of months of service, \$40 is the installation fee, and \$90 is the service charge per month.
(4) x is the total cost, y is the number of months of service, \$90 is the installation fee, and \$40 is the service charge per month. 2 ____

3 If $4x^2 - 100 = 0$, the roots of the equation are

(1) –25 and 25 (3) –5 and 5
(2) –25, only (4) –5, only 3 ____

4 Isaiah collects data from two different companies, each with four employees. The results of the study, based on each worker's age and salary, are listed in the tables below.

Company 1

Worker's Age in Years	Salary in Dollars
25	30,000
27	32,000
28	35,000
33	38,000

Company 2

Worker's Age in Years	Salary in Dollars
25	29,000
28	35,500
29	37,000
31	65,000

Which statement is true about these data?

(1) The median salaries in both companies are greater than $37,000.
(2) The mean salary in company 1 is greater than the mean salary in company 2.
(3) The salary range in company 2 is greater than the salary range in company 1.
(4) The mean age of workers at company 1 is greater than the mean age of workers at company 2. 4 ____

5 Which point is *not* on the graph represented by $y = x^2 + 3x - 6$?

(1) (–6,12) (3) (2,4)
(2) (–4,–2) (4) (3,–6) 5 ____

6 A company produces x units of a product per month, where $C(x)$ represents the total cost and $R(x)$ represents the total revenue for the month. The functions are modeled by $C(x) = 300x + 250$ and $R(x) = -0.5x^2 + 800x - 100$. The profit is the difference between revenue and cost where $P(x) = R(x) - C(x)$. What is the total profit, $P(x)$, for the month?

(1) $P(x) = -0.5x^2 + 500x - 150$
(2) $P(x) = -0.5x^2 + 500x - 350$
(3) $P(x) = -0.5x^2 - 500x + 350$
(4) $P(x) = -0.5x^2 + 500x + 350$

6 ____

7 What is one point that lies in the solution set of the system of inequalities graphed below?

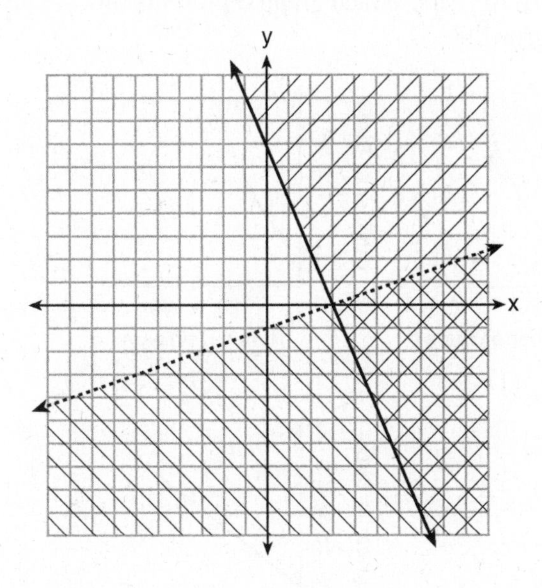

(1) (7,0)
(2) (3,0)
(3) (0,7)
(4) (–3,5)

7 ____

8 The value of the x-intercept for the graph of $4x - 5y = 40$ is

(1) 10 (3) $-\frac{4}{5}$

(2) $\frac{4}{5}$ (4) -8 8 ____

9 Sam and Jeremy have ages that are consecutive odd integers. The product of their ages is 783. Which equation could be used to find Jeremy's age, j, if he is the younger man?

(1) $j^2 + 2 = 783$ (3) $j^2 + 2j = 783$

(2) $j^2 - 2 = 783$ (4) $j^2 - 2j = 783$ 9 ____

10 A population that initially has 20 birds approximately doubles every 10 years. Which graph represents this population growth?

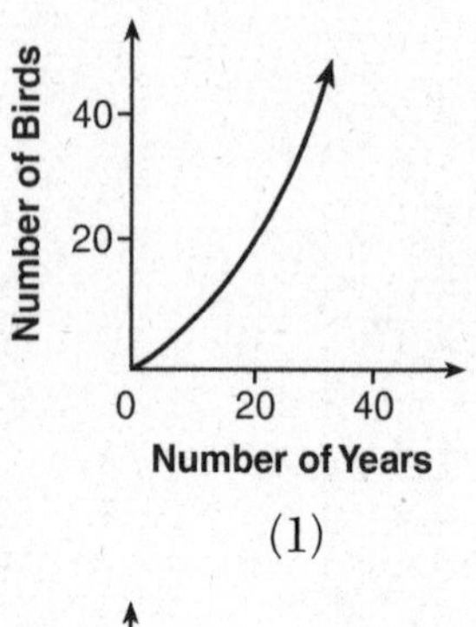

(1)

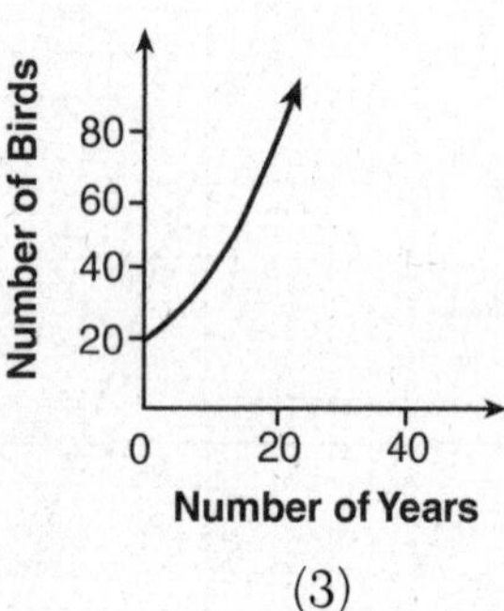

(3)

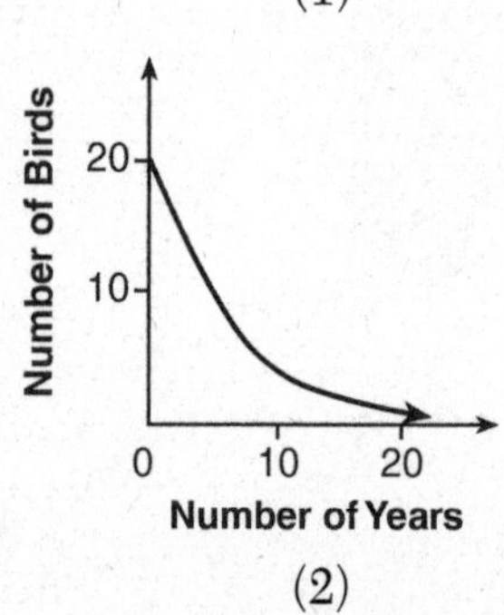

(2)

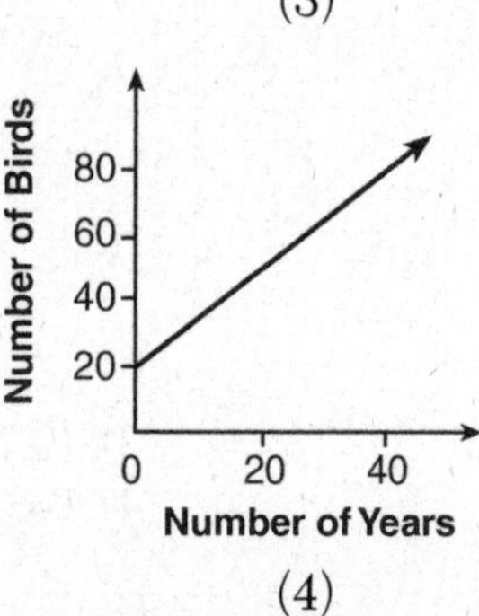

(4)

10 ____

11 Let f be a function such that $f(x) = 2x - 4$ is defined on the domain $2 \leq x \leq 6$. The range of this function is

(1) $0 \leq y \leq 8$
(2) $0 \leq y \leq \infty$
(3) $2 \leq y \leq 6$
(4) $-\infty < y < \infty$

11 ____

12 Which situation could be modeled by using a linear function?

(1) a bank account balance that grows at a rate of 5% per year, compounded annually
(2) a population of bacteria that doubles every 4.5 hours
(3) the cost of cell phone service that charges a base amount plus 20 cents per minute
(4) the concentration of medicine in a person's body that decays by a factor of one-third every hour

12 ____

13 Which graph shows a line where each value of y is three more than half of x?

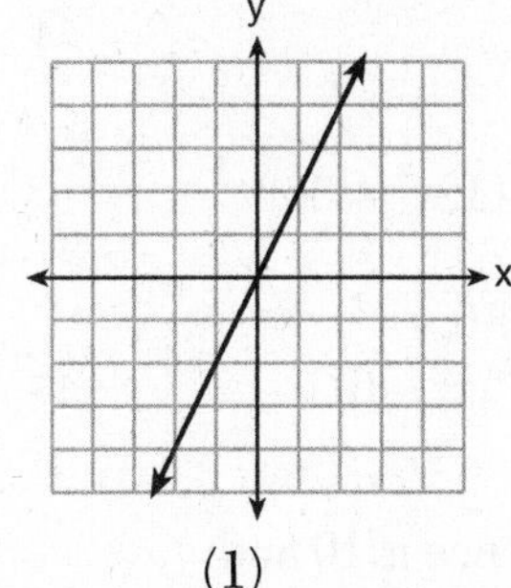

(1)

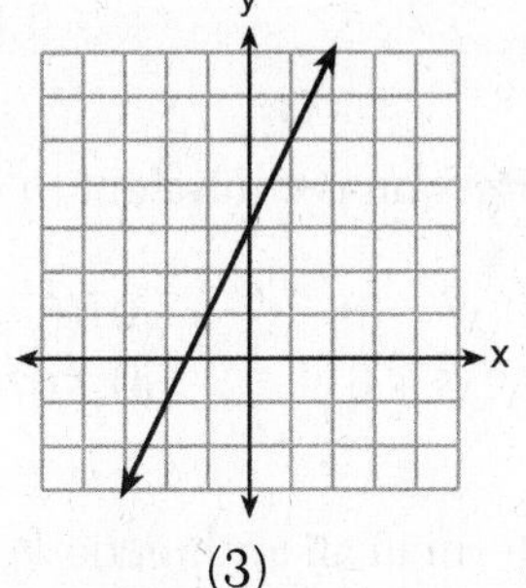

(3)

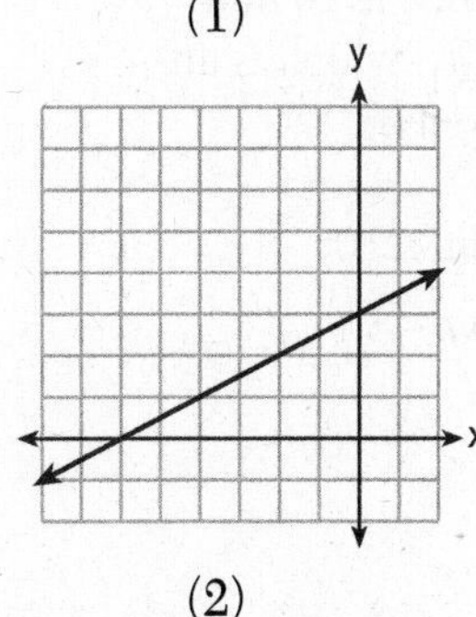

(2)

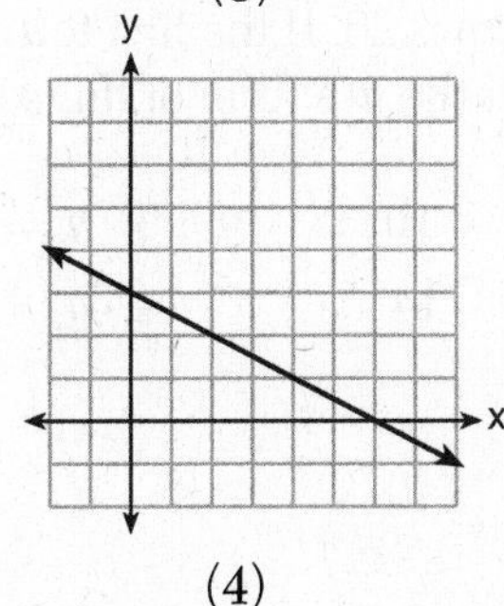

(4)

13 ____

14 The table below shows the average diameter of a pupil in a person's eye as he or she grows older.

Age (years)	Average Pupil Diameter (mm)
20	4.7
30	4.3
40	3.9
50	3.5
60	3.1
70	2.7
80	2.3

What is the average rate of change, in millimeters per year, of a person's pupil diameter from age 20 to age 80?

(1) 2.4 (3) –2.4
(2) 0.04 (4) –0.04

14 _____

15 Which expression is equivalent to $x^4 - 12x^2 + 36$?

(1) $(x^2 - 6)(x^2 - 6)$ (3) $(6 - x^2)(6 + x^2)$
(2) $(x^2 + 6)(x^2 + 6)$ (4) $(x^2 + 6)(x^2 - 6)$

15 _____

16 The third term in an arithmetic sequence is 10 and the fifth term is 26. If the first term is a_1, which is an equation for the nth term of this sequence?

(1) $a_n = 8n + 10$ (3) $a_n = 16n + 10$
(2) $a_n = 8n - 14$ (4) $a_n = 16n - 38$

16 _____

17 The graph of the equation $y = ax^2$ is shown below.

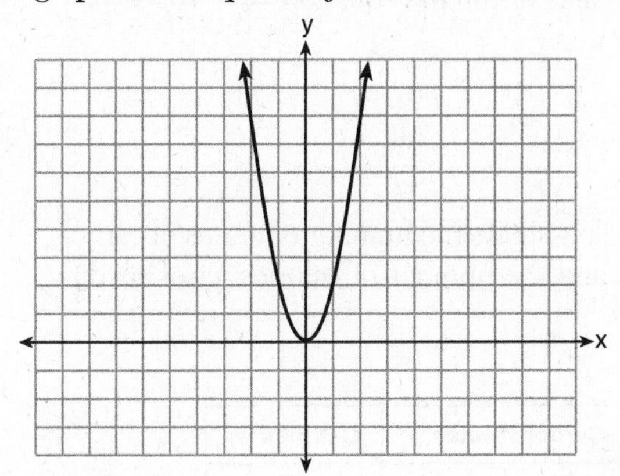

If a is multiplied by $-\frac{1}{2}$, the graph of the new equation is

(1) wider and opens downward
(2) wider and opens upward
(3) narrower and opens downward
(4) narrower and opens upward

17 _____

18 The zeros of the function $f(x) = (x + 2)^2 - 25$ are

(1) –2 and 5
(2) –3 and 7
(3) –5 and 2
(4) –7 and 3

18 _____

19 During the 2010 season, football player McGee's earnings, m, were 0.005 million dollars more than those of his teammate Fitzpatrick's earnings, f. The two players earned a total of 3.95 million dollars. Which system of equations could be used to determine the amount each player earned, in millions of dollars?

(1) $m + f = 3.95$
$m + 0.005 = f$

(2) $m - 3.95 = f$
$f + 0.005 = m$

(3) $f - 3.95 = m$
$m + 0.005 = f$

(4) $m + f = 3.95$
$f + 0.005 = m$

19 _____

20 What is the value of x in the equation $\frac{x-2}{3} + \frac{1}{6} = \frac{5}{6}$?

(1) 4 (3) 8
(2) 6 (4) 11

20 ____

21 The table below shows the number of grams of carbohydrates, x, and the number of calories, y, of six different foods.

Carbohydrates (x)	Calories (y)
8	120
9.5	138
10	147
6	88
7	108
4	62

Which equation best represents the line of best fit for this set of data?

(1) $y = 15x$ (3) $y = 0.1x - 0.4$
(2) $y = 0.07x$ (4) $y = 14.1x + 5.8$

21 ____

22 A function is graphed on the set of axes below.

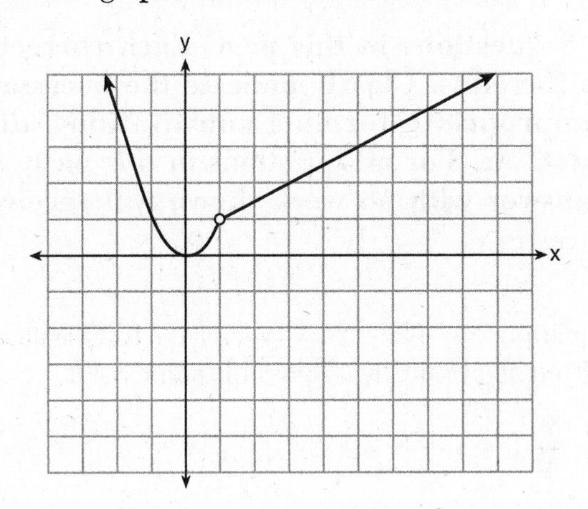

Which function is related to the graph?

(1) $f(x) = \begin{cases} x^2, & x<1 \\ x-2, & x>1 \end{cases}$

(2) $f(x) = \begin{cases} x^2, & x<1 \\ \frac{1}{2}x+\frac{1}{2}, & x>1 \end{cases}$

(3) $f(x) = \begin{cases} x^2, & x<1 \\ 2x-7, & x>1 \end{cases}$

(4) $f(x) = \begin{cases} x^2, & x<1 \\ \frac{3}{2}x-\frac{9}{2}, & x>1 \end{cases}$

22 _____

23 The function $h(t) = -16t^2 + 144$ represents the height, $h(t)$, in feet, of an object from the ground at t seconds after it is dropped. A realistic domain for this function is

(1) $-3 \le t \le 3$
(2) $0 \le t \le 3$
(3) $0 \le h(t) \le 144$
(4) all real numbers

23 _____

24 If $f(1) = 3$ and $f(n) = -2f(n-1) + 1$, then $f(5) =$

(1) –5
(2) 11
(3) 21
(4) 43

24 _____

PART II

Answer all 8 questions in this part. Each correct answer will receive 2 credits. Clearly indicate the necessary steps, including appropriate formula substitutions, diagrams, graphs, charts, etc. For all questions in this part, a correct numerical answer with no work shown will receive only 1 credit. [16 credits]

25 In the equation $x^2 + 10x + 24 = (x + a)(x + b)$, b is an integer. Find algebraically *all* possible values of b.

26 Rhonda deposited $3000 in an account in the Merrick National Bank, earning 4.2% interest, compounded annually. She made no deposits or withdrawals. Write an equation that can be used to find B, her account balance after t years.

27 Guy and Jim work at a furniture store. Guy is paid \$185 per week plus 3% of his total sales in dollars, x, which can be represented by $g(x) = 185 + 0.03x$. Jim is paid \$275 per week plus 2.5% of his total sales in dollars, x, which can be represented by $f(x) = 275 + 0.025x$. Determine the value of x, in dollars, that will make their weekly pay the same.

28 Express the product of $2x^2 + 7x - 10$ and $x + 5$ in standard form.

29 Let f be the function represented by the graph below.

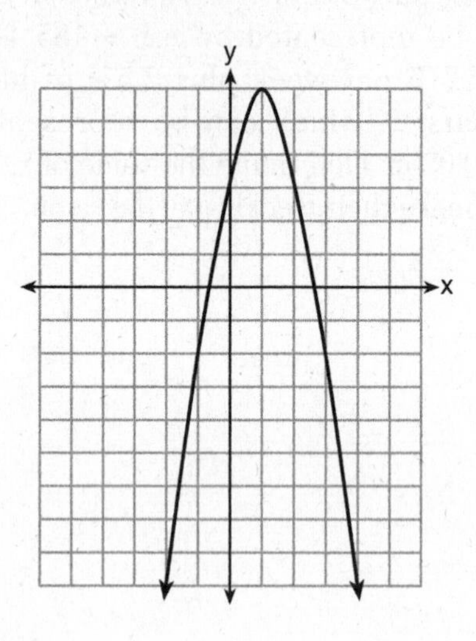

Let g be a function such that $g(x) = -\frac{1}{2}x^2 + 4x + 3$.

Determine which function has the larger maximum value. Justify your answer.

30 Solve the inequality below to determine and state the smallest possible value for x in the solution set.

$$3(x + 3) \leq 5x - 3$$

31 The table below represents the residuals for a line of best fit.

x	2	3	3	4	6	7	8	9	9	10
Residual	2	1	−1	−2	−3	−2	−1	2	0	3

Plot these residuals on the set of axes below.

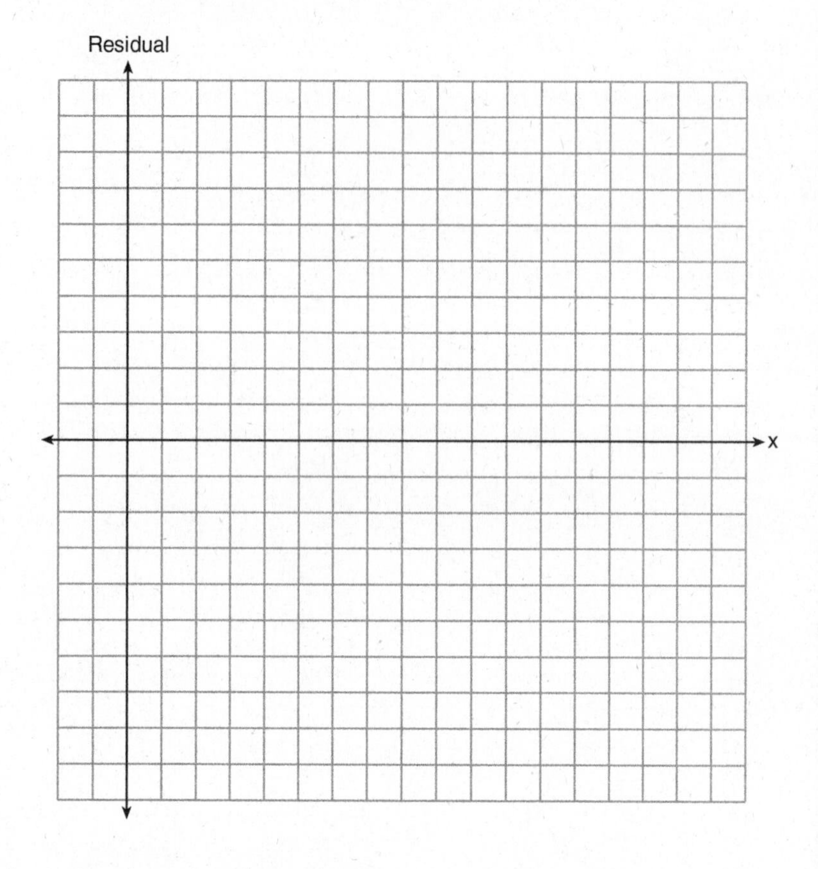

Using the plot, assess the fit of the line for these residuals and justify your answer.

32 A student was given the equation $x^2 + 6x - 13 = 0$ to solve by completing the square. The first step that was written is shown below.

$$x^2 + 6x = 13$$

The next step in the student's process was $x^2 + 6x + c = 13 + c$.

State the value of c that creates a perfect square trinomial.

Explain how the value of c is determined.

PART III

Answer all 4 questions in this part. Each correct answer will receive 4 credits. Clearly indicate the necessary steps, including appropriate formula substitutions, diagrams, graphs, charts, etc. For all questions in this part, a correct numerical answer with no work shown will receive only 1 credit. [16 credits]

33 On the axes below, graph $f(x) = |3x|$.

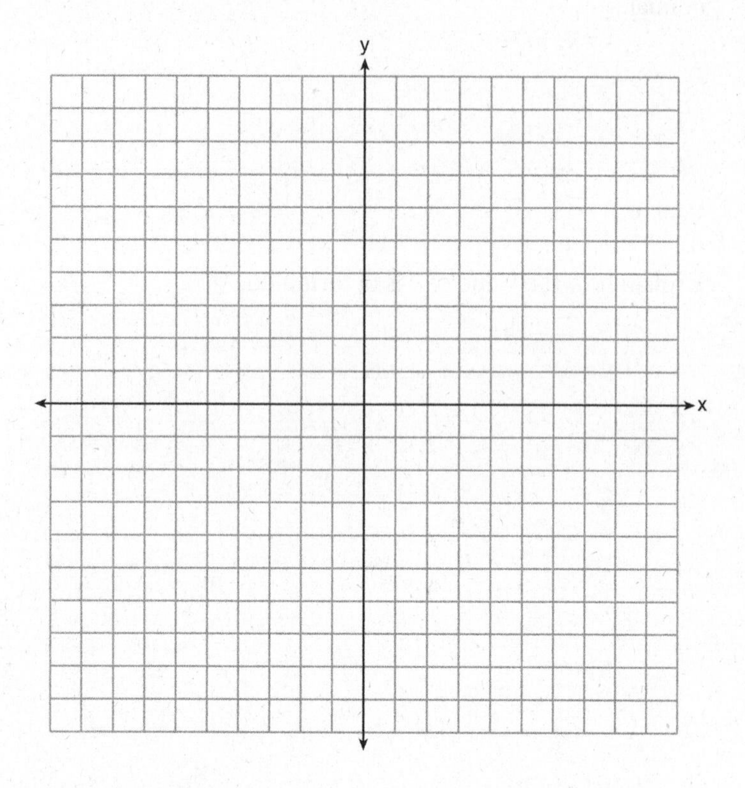

If $g(x) = f(x) - 2$, how is the graph of $f(x)$ translated to form the graph of $g(x)$?

If $h(x) = f(x - 4)$, how is the graph of $f(x)$ translated to form the graph of $h(x)$?

34 The formula for the area of a trapezoid is

$$A = \frac{1}{2}h(b_1 + b_2).$$

Express b_1 in terms of A, h, and b_2.

The area of a trapezoid is 60 square feet, its height is 6 ft, and one base is 12 ft. Find the number of feet in the other base.

35 Let $f(x) = -2x^2$ and $g(x) = 2x - 4$. On the set of axes below, draw the graphs of $y = f(x)$ and $y = g(x)$.

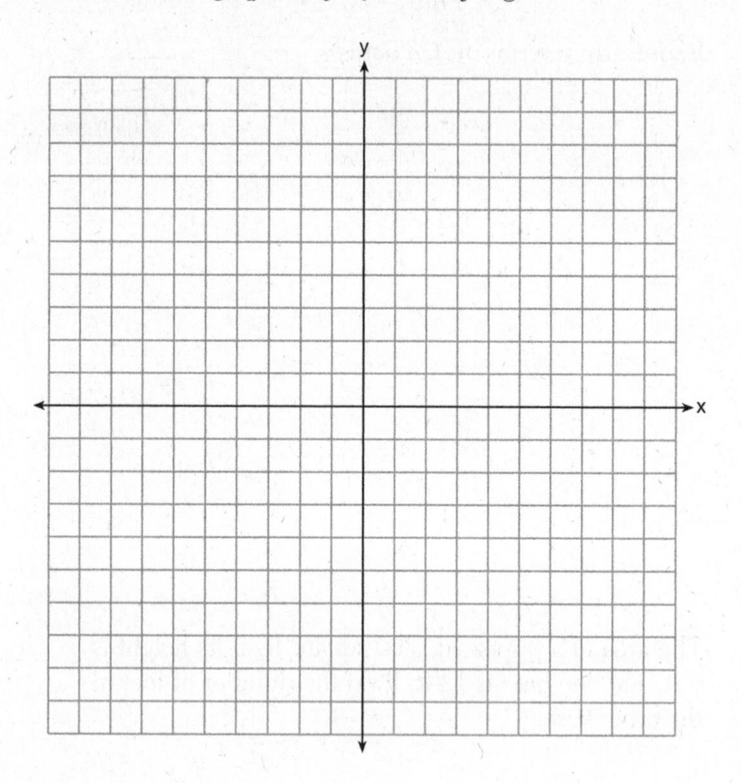

Using this graph, determine and state *all* values of x for which $f(x) = g(x)$.

36 A school is building a rectangular soccer field that has an area of 6000 square yards. The soccer field must be 40 yards longer than its width. Determine algebraically the dimensions of the soccer field, in yards.

PART IV

Answer the question in this part. A correct answer will receive 6 credits. Clearly indicate the necessary steps, including appropriate formula substitutions, diagrams, graphs, charts, etc. A correct numerical answer with no work shown will receive only 1 credit. [6 credits]

37 Edith babysits for x hours a week after school at a job that pays \$4 an hour. She has accepted a job that pays \$8 an hour as a library assistant working y hours a week. She will work both jobs. She is able to work *no more than* 15 hours a week, due to school commitments. Edith wants to earn *at least* \$80 a week, working a combination of both jobs.

Write a system of inequalities that can be used to represent the situation.

Graph these inequalities on the set of axes below.

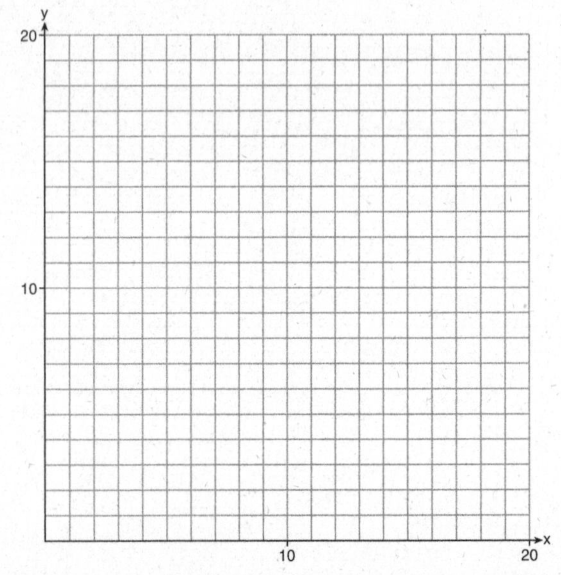

Question 37 is continued on the next page.

Question 37 continued

Determine and state one combination of hours that will allow Edith to earn *at least* $80 per week while working *no more than* 15 hours.

Answers August 2014

Algebra I

Answer Key

PART I

1. (1)	**5.** (4)	**9.** (3)	**13.** (2)	**17.** (1)	**21.** (4)
2. (2)	**6.** (2)	**10.** (3)	**14.** (4)	**18.** (4)	**22.** (2)
3. (3)	**7.** (1)	**11.** (1)	**15.** (1)	**19.** (4)	**23.** (2)
4. (3)	**8.** (1)	**12.** (3)	**16.** (2)	**20.** (1)	**24.** (4)

PART II

25. The two possible values of b are 4 and 6.

26. $B = 3000\,(1.042)^t$

27. $x = 18{,}000$

28. $2x^3 + 17x^2 + 25x - 50$

29. Maximum of the parabola for $g(x)$ is higher

30. 6

31. The line of best fit does not accurately model the data. Another type of curve is more appropriate.

32. 9 makes $x^2 + 6x + 9$ a perfect square trinomial.

PART III

33. Two units down to form the graph $g(x)$. Four units to the right to form graph $h(x)$.

34. $b_1 = \dfrac{2A}{h} - b_2$. $b_1 = 8$.

35. $x = -2$ and $x = 1$

36. Width is 60 yards. Length is 100 yards.

PART IV

37. The system of equalities is $4x + 8y \geq 80$, $x + y \leq 15$; one possible answer is 2 hours babysitting and 10 hours in the library.

In Parts II–IV, you are required to show how you arrived at your answers. For sample methods of solutions, see the *Answers Explained* section.

Answers Explained

PART I

1. The product of two irrational numbers does not have to be irrational. For example, $\sqrt{5} \cdot \sqrt{5} = 5$, which is a rational number. Choices 2, 3, and 4 are always true.

   The correct choice is **(1)**.

2. Since in the equation y is the sum of the number 40 and the expression $90x$, the y represents the total cost. This eliminates choices 3 and 4. In an equation of the form $y = mx + b$ or $y = b + mx$, the b is the constant, and the m is the amount that y increases for each time the x increases by 1.

   To check the equation $y = 40 + 90x$, if 0 is substituted for x, it becomes $y = 40 + 90(0) = \$40$, which should be the total cost for 0 months, or just the cost of installation.

   The correct choice is **(2)**.

3. The simplest way to solve this equation is to first add 100 to both sides of the equation.

$$\begin{aligned} 4x^2 - 100 &= 0 \\ +100 &= +100 \\ \frac{4x^2}{4} &= \frac{100}{4} \\ x^2 &= 25 \\ x &= \pm\sqrt{25} \\ x &= \pm 5 \end{aligned}$$

   Another way is to recognize that the left-hand side of the equation is an example of the difference between two squares and can be factored.

$$\begin{aligned} 4x^2 - 100 &= 0 \\ (2x - 10)(2x + 10) &= 0 \\ 2x - 10 = 0 &\text{ or } 2x + 10 = 0 \\ +10 = +10 &\quad -10 = -10 \\ \frac{2x}{2} = \frac{10}{2} &\text{ or } \frac{2x}{2} = -\frac{10}{2} \\ x = 5 &\text{ or } x = -5 \end{aligned}$$

   The correct choice is **(3)**.

4. To calculate the mean of four values, add them and divide by four. The mean for company 1 is $\frac{30{,}000+32{,}000+35{,}000+38{,}000}{4}$ = 33,750. The mean for company 2 is $\frac{29{,}000+35{,}500+37{,}000+65{,}000}{4}$ = 41,625. This eliminates choice 2. The mean age of the workers at company 1 is $\frac{25+27+28+33}{4}$ = 28.25 and the mean age of the workers at company 2 is $\frac{25+28+29+31}{4}$ = 28.25. This eliminates choice 4.

   To calculate the median of four values, arrange them from smallest to largest. Since there is an even number of numbers, there is no one middle number. Add the two middle numbers and divide by two. For company 1 the median is $\frac{32{,}000+35{,}000}{2}$ = 33,500. For company 2 the median is 36,250. This eliminates choice 1.

   The range is calculated by subtracting the smallest value from the largest. For company 1 the range is 38,000 – 30,000 = 8,000. For company 2 the range is 65,000 – 29,000 = 36,000. The salary range of company 2 is greater than the salary range of company 1.

   The correct choice is **(3)**.

5. The coordinates of a point on the graph represented by $y = x^2 + 3x - 6$ will satisfy the equation when the x-coordinate is substituted for x and the y-coordinate is substituted for y. Test each ordered pair until you find the one that does not yield a true equation.

   Testing choice 1:

   $12 \stackrel{?}{=} (-6)^2 + 3(-6) - 6$
   $12 \stackrel{?}{=} 36 - 18 - 6$
   $12 \stackrel{\checkmark}{=} 12$

   Testing choice 2:

   $-2 \stackrel{?}{=} (-4)^2 + 3(-4) - 6$
   $-2 \stackrel{?}{=} 16 - 12 - 6$
   $-2 \stackrel{\checkmark}{=} -2$

   Testing choice 3:

   $4 \stackrel{?}{=} 2^2 + 3(2) - 6$
   $4 \stackrel{?}{=} 4 + 6 - 6$
   $4 \stackrel{\checkmark}{=} 4$

Testing choice 4:

$-6 \stackrel{?}{=} 3^2 + 3(3) - 6$
$-6 \stackrel{?}{=} 9 + 9 - 6$
$-6 \neq 12$

The correct choice is **(4)**.

6. $R(x) - C(x) = (-0.5x^2 + 800x - 100) - 1(300x + 250)$
Distribute the negative one through the parentheses on the right.

$-0.5x^2 + 800x - 100 - 300x - 250$
$-0.5x^2 + 800x - 300x - 100 - 250$
$-0.5x^2 + 500x - 350$

The correct choice is **(2)**.

7. The solution set of the system of inequalities is the portion that has both types of shading. Plot the points from the four choices on the graph to see which lies in the double shaded region.

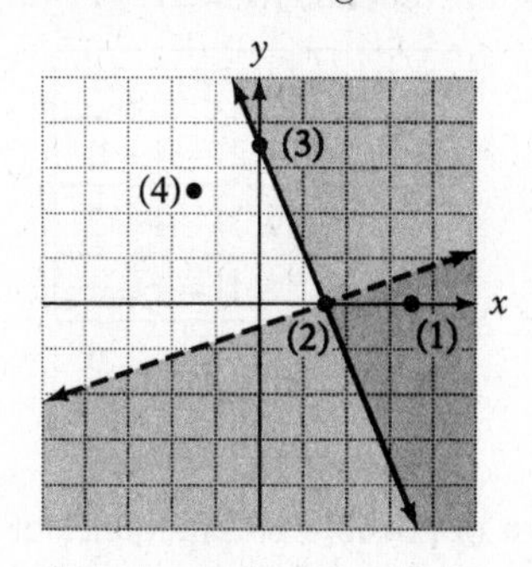

Though choice 2 is close to the double shaded region, choice 1 is clearly in the double shaded region.

The correct choice is **(1)**.

8. An x-intercept of a graph is a point on the x-axis with a y-coordinate of 0. Substitute 0 into the equation for y and solve for x:

$$4x - 5(0) = 40$$

$$\frac{4x}{4} = \frac{40}{4}$$

$$x = 10$$

The correct choice is **(1)**.

9. Consecutive odd integers are two numbers apart. For example, 11 and 13 are consecutive odd integers. Consecutive even integers are also two numbers apart. If Jeremy's age is j, Sam's age is $j + 2$. The product of their ages is $j(j + 2)$. If the product of their ages is 783, the equation is

$$j(j + 2) = 783$$

Using the distributive property,

$$j^2 + 2j = 783$$

The correct choice is **(3)**.

10. Since the initial population is 20, the graph must have the point (0,20). This eliminates choice 1. The population doubles every 10 years, so the graph must also have the points (10,40) and (20,80).

Of the choices, only choice 3 has those three points.

The correct choice is **(3)**.

11. A table of values for the function $f(x)= 2x - 4$ with domain $2 \le x \le 6$ is

x	$f(x)$
2	$2(2) - 4 = 0$
3	$2(3) - 4 = 2$
4	$2(4) - 4 = 4$
5	$2(5) - 4 = 6$
6	$2(6) - 4 = 8$

The range is the list of possible output values of the function. Since the smallest output value is 0 and the largest output value is 8, the range is $0 \le y \le 8$.

The graph of $f(x) = 2x - 4$ with the domain $2 \le x \le 6$ is a line segment with endpoints (2,0) and (6,8). The y-values on this segment can be seen to be $0 \le y \le 8$.

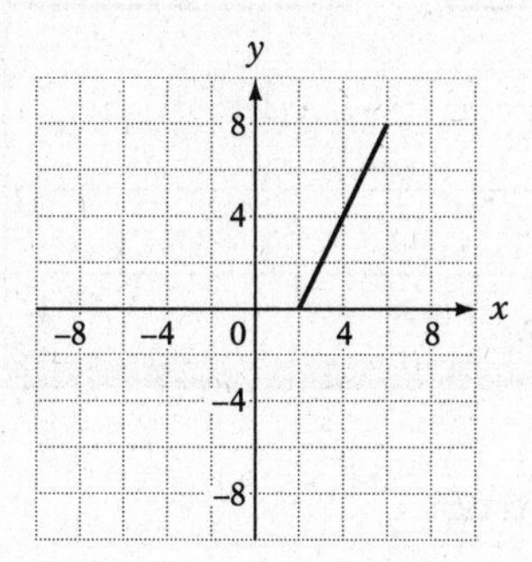

The correct choice is **(1)**.

12. A cell phone service that charges a base amount plus 20 cents per minute would have an equation like $C = 0.20M + B$, which is a linear equation since there are no exponents involved.

    Choices 1, 2, and 4 are all exponential equations. Choice 1 is $A = P(1.05)t$, choice 2 is $P = A \cdot 2^{\frac{t}{4.5}}$, and choice 4 is $Q = q(\frac{1}{3})^t$.

    The correct choice is **(3)**.

13. The sentence describes the equation $y = \frac{1}{2}x + 3$. The graph will pass through the y-intercept (0,3) and also increase as it goes from left to right because the slope of $\frac{1}{2}$ is positive.

    On the TI-84:

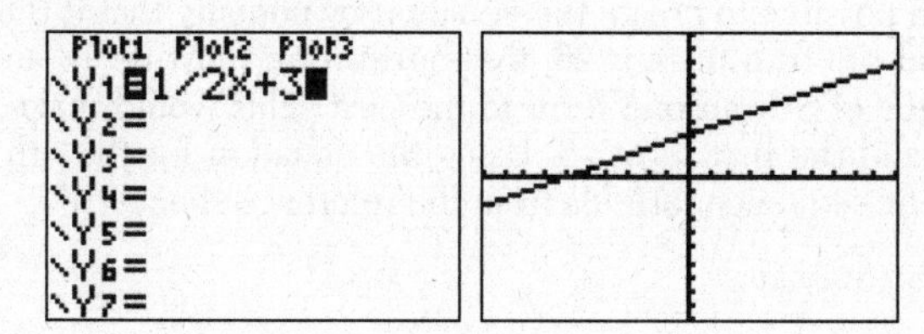

On the TI-Nspire:

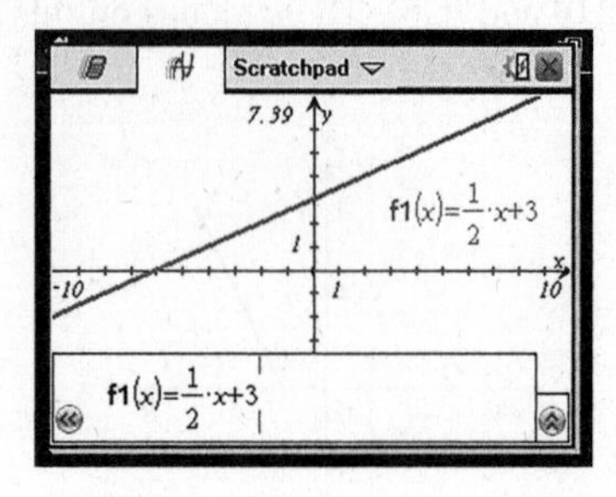

The correct choice is **(2)**.

14. To calculate the average rate of change, divide the change in the average pupil diameter by the change in the age. From 20 to 80, the change in the average pupil diameter is 2.3 – 4.7 = –2.4. The change in the age is 80 – 20 = 60. Divide –2.4 by 60 to get –0.04.

    The correct choice is **(4)**.

15. The given expression has the structure as a quadratic trinomial. It can be rewritten as $(x^2)^2 - 12(x^2) + 36$. Just as $u^2 - 12u + 36$ could be factored into $(u - 6)(u - 6)$, $(x^2)^2 - 12(x^2) + 36$ can be factored into $(x^2 - 6)(x^2 - 6)$.

    The correct choice is **(1)**.

16. Check the four choices to see which make $a_3 = 10$ and $a_5 = 26$:

    Testing choice (1): $a_3 = 8 \cdot 3 + 10 = 24 + 10 = 34$. No.

    Testing choice (2): $a_3 = 8 \cdot 3 - 14 = 24 - 14 = 10$. Also need to check for $n = 5$. $a_5 = 8 \cdot 5 - 14 = 40 - 14 = 26$. It works for both values.

    It is also possible to create the equation by noticing that if the third term is 10 and the fifth term is 26, the fourth term must be 18 and there is a difference of 8 from one term to the next. This would make the second term 2 and the first term –6. Using the equation for the nth term of an arithmetic sequence formula from the reference sheet:

    $a_n = a_1 + (n - 1)d$

    In this case $a_1 = -6$ and $d = 8$.

    $a_n = -6 + (n - 1)8 = -6 + 8n - 8 = 8n - 14$.

    The correct choice is **(2)**.

17. The given parabola opens upward so the value of a must be positive. Multiplying it by $-\frac{1}{2}$ would create an equation whose graph is a parabola that opens downward. This eliminates choices 2 and 4. Multiplying the a value by a fraction between –1 and 1 also makes the parabola wider.

On the graphing calculator, here is a comparison between $y = 2x^2$ and $y = -\frac{1}{2} \cdot 2x^2$.

On the TI-84:

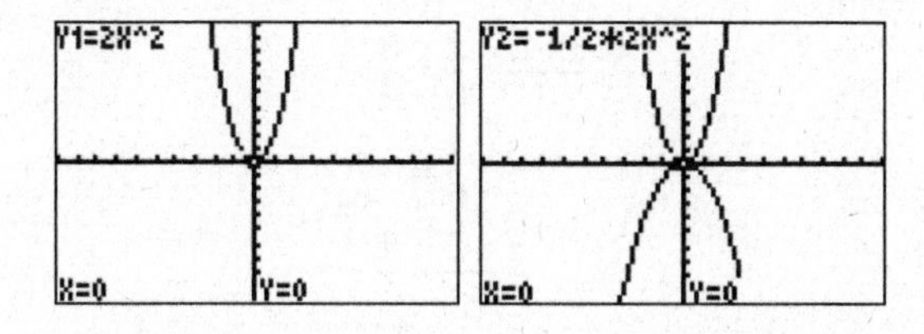

On the TI-Nspire:

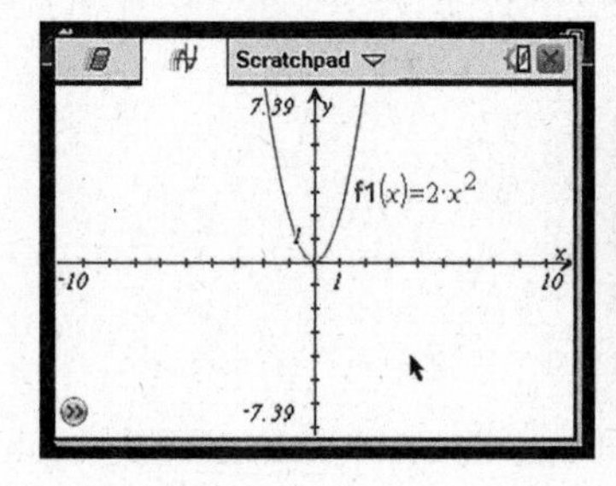

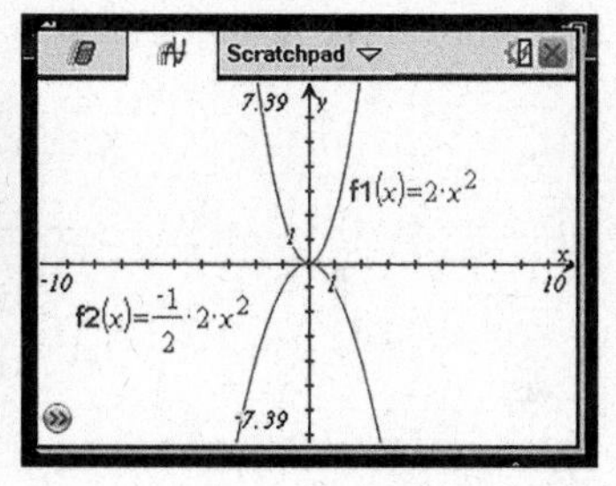

The correct choice is **(1)**.

18. The zeros of a function are the values for x that make $f(x) = 0$.

Solve the equation $0 = (x + 2)^2 - 25$.

$$\begin{aligned} 0 &= (x + 2)^2 - 25 \\ +25 &= +25 \\ 25 &= (x + 2)^2 \\ \pm\sqrt{25} &= x + 2 \\ \pm 5 &= x + 2 \\ -2 \pm 5 &= x \end{aligned}$$

$x = -2 + 5 = 3$ or $x = -2 - 5 = -7$

The correct choice is **(4)**.

19. The 0.005 must be added to the smaller variable to make it equal to the larger variable. Since m is 0.005 greater than f, then one equation is $f + 0.005 = m$. Since their combined salaries are 3.95 million, the other equation is $m + f = 3.95$. You do not have to solve for m and f. You only need the system of equations.

    The correct choice is **(4)**.

20. Use the subtraction property of equality to subtract $\frac{1}{6}$ from both sides of the equation.

$$\frac{x-2}{3}+\frac{1}{6}=\frac{5}{6}$$
$$-\frac{1}{6}=-\frac{1}{6}$$
$$\frac{x-2}{3}=\frac{4}{6}$$

Use the multiplication property of equality to multiply both sides of the equation by 3. Then use the addition property of equality to add 2 to both sides of the equation.

$$3\cdot\frac{x-2}{3}=3\cdot\frac{4}{6}$$
$$x-2=2$$
$$+2=+2$$
$$x=4$$

This question could also have been done by substituting each of the answer choices into the equation to see which one makes a true statement.

To test choice (1):

$$\frac{4-2}{3}+\frac{1}{6}=\frac{5}{6}$$
$$\frac{2}{3}+\frac{1}{6}=\frac{5}{6}$$
$$\frac{4}{6}+\frac{1}{6}=\frac{5}{6}$$
$$\frac{5}{6}=\frac{5}{6}$$

The correct choice is **(1)**.

21. This question requires the linear regression feature of the graphing calculator.

For the TI-84:

Press [STAT] and [1]; then enter the carbohydrates into L1 and the calories into L2. Press [STAT], [Right Arrow], [4], and [ENTER] to see the equation for the line of best fit.

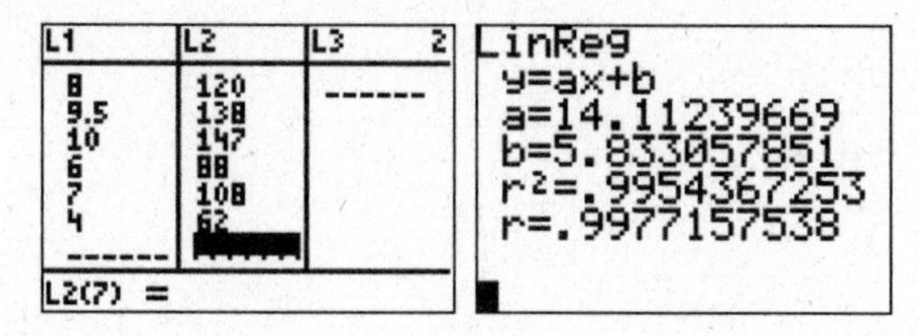

For the TI-Nspire:

Press [home] and select Add Lists and Spreadsheet. Label column A "car" for carbohydrates, and label column B "cal" for calories. Enter the data into the two columns. Press [menu], [4], [1], and [3]. Make X List "car" and Y List "cal" and press the OK button.

1.1 1.2 1.3 *Unsaved

	A car	B cal	C	D
1	8	120		
2	9.5	138		
3	10	147		
4	6	88		
5	7	108		

B1 120

1.1 1.2 1.3 *Unsaved

	ir	B cal	C	D
◆				=LinRegM
1	8	120	Title	Linear Re..
2	9.5	138	RegEqn	m*x+b
3	10	147	m	14.1124
4	6	88	b	5.83306
5	7	108	r²	0.995437

D3 =14.112396694215

The correct choice is **(4)**.

22. This piecewise function is composed of a parabola for x values less than 1 and a line for x values greater than 1. All four choices have the parabola as $y = x^2$ so this must be the first equation.

To get the slope of the line, locate two points that have integer coordinates such as (3,2) and (5,3).

Use the slope formula

$$m = \frac{3-2}{5-3} = \frac{1}{2}$$

To get the y-intercept of the line, substitute the coordinates of either of the two points and the slope into the equation $y = mx + b$.

Using the point (3,2), it becomes

$$2 = \left(\frac{1}{2}\right)(3) + b$$

$$2 = \frac{3}{2} + b$$

$$-\frac{3}{2} = -\frac{3}{2}$$

$$\frac{1}{2} = b$$

The equation of the line is $y = \left(\frac{1}{2}\right)x + \frac{1}{2}$.

The correct choice is **(2)**.

23. The domain is the set of input values. In this case, the variable name of the input values is t, so choice 3 is eliminated. The object is dropped at $t = 0$. When 3 is substituted into the function, it becomes $h(3) = -16(3^2) + 144 = 0$, which means that the object is on the ground at 3 seconds so it is not realistic to use the function for t values greater than 3.

 The correct choice is **(2)**.

24. According to the definition of f, $f(1) = 3$.

 To find $f(2)$, substitute 2 for n into the equation $f(n) = -2f(n - 1) + 1$. It becomes

$$f(2) = -2f(2 - 1) + 1 = -2f(1) + 1$$

 Then since $f(1) = 3$, $f(2) = -2(3) + 1 = -6 + 1 = -5$, so $f(2) = -5$

 To find $f(3)$, substitute 3 for n into the second equation:

$$f(3) = -2f(3 - 1) + 1 = -2f(2) + 1 = -2(-5) + 1 = 10 + 1 = 11, \text{ so } f(3) = 11.$$

 Continue with $n = 4$ and $n = 5$.

$$f(4) = -2f(4 - 1) + 1 = -2f(3) + 1 = -2(11) + 1 = -22 + 1 = -21, \text{ so } f(4) = -21$$

$$f(5) = -2f(5 - 1) + 1 = -2f(4) + 1 = -2(-21) + 1 = 42 + 1 = 43, \text{ so } f(5) = 43$$

 The correct choice is **(4)**.

PART II

25. To factor the expression $x^2 + 10x + 24$ into $(x + a)(x + b)$, the numbers for a and b must have a sum of 10 and a product of 24. The two numbers are 4 and 6. So $x^2 + 10x + 24 = (x + 4)(x + 6)$, which can also be written as $(x + 6)(x + 4)$. The two possible values of b are 4 and 6.

26. The compound interest formula is $B = P(1 + r)^t$, where P is the initial deposit, r is the percent interest, and t is the number of years. For this example, $P = 3{,}000$ and $r = 0.042$.

    The equation is $B = 3000(1.042)^t$.

    Note, if you use $r = 4.2$ instead and get $B = 3000(5.2)^2$, you lose half credit.

27. The quickest way to solve this question is to set the functions equal to each other and to solve for x.

$$\begin{aligned} 185 + 0.03x &= 275 + 0.025x \\ -0.025x &= -0.025x \\ 185 + 0.005x &= 275 \\ -185 &= -185 \\ \frac{0.005x}{0.005} &= \frac{90}{0.005} \\ x &= 18{,}000 \end{aligned}$$

Another way to solve this is to graph both functions and to find the x-coordinate of the intersection point.

On the TI-84:

Press [Y=] and enter the two functions into Y1 and Y2. Set the window dimensions. Press [GRAPH]. To find the intersection, press [2ND], [TRACE], and [5]. Select the two lines and move the cursor near the intersection and press [ENTER].

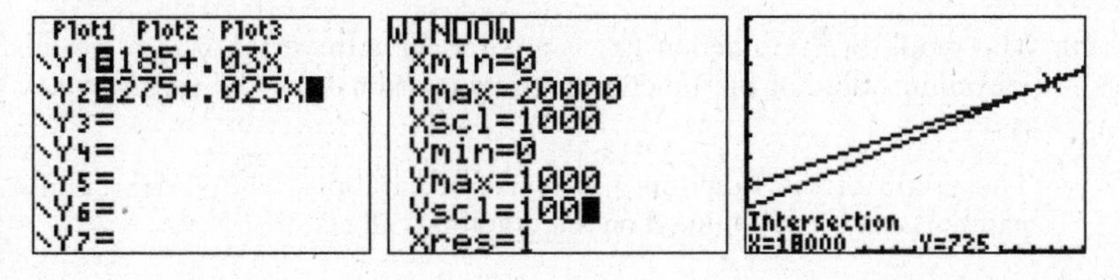

On the TI-Nspire:

Press [home] and [b] to get to the calculator Scratchpad. Graph the two equations, set the window dimensions, and press [menu], [6], and [4] to find the intersection point.

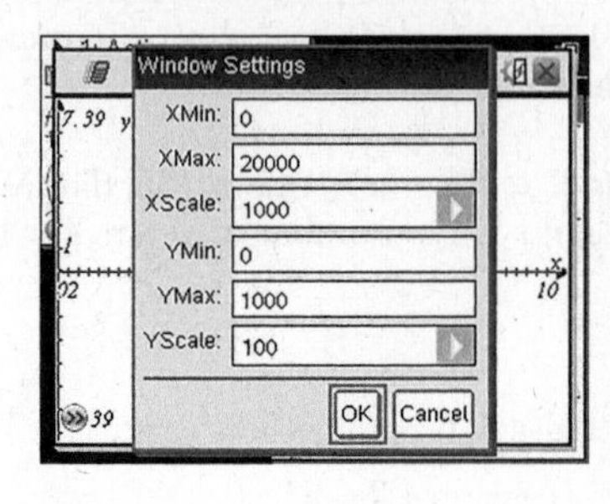

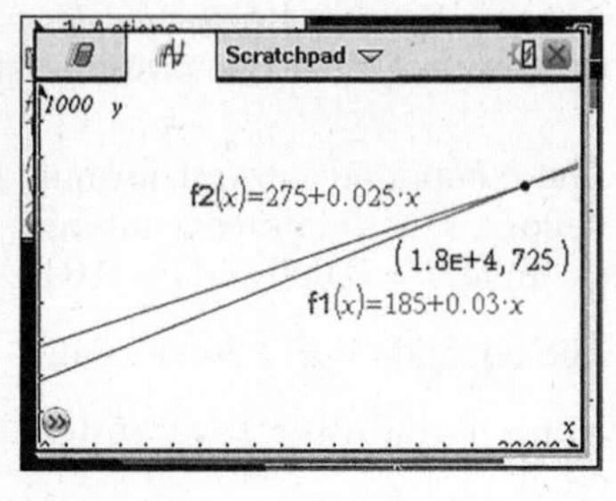

28. Polynomials like these can be multiplied in a way similar to the way a three digit number can be multiplied by a two digit number.

		$2x^2$	+	$7x$	−	10
×				x	+	5
		$10x^2$	+	$35x$	−	50
$2x^3$	+	$7x^2$	−	$10x$		
$2x^3$	+	$17x^2$	+	$25x$	−	50

This is in standard form since the exponents are in decreasing order.

This question can also be done with the distributive property, multiplying $2x^2 + 7x - 10$ by each term in the $(x + 5)$ factor.

$$(2x^2 + 7x - 10)(x + 5) = (2x^2 + 7x - 10)(x) + (2x^2 + 7x - 10)(5)$$

Then use the distributive property twice more, multiplying the x by each of the terms in the $(2x^2 + 7x - 10)$ factor and multiplying 5 by each of the terms in the other $(2x^2 + 7x - 10)$ factor.

$$2x^3 + 7x^2 - 10x + 10x^2 + 35x - 50 = 2x^3 + 17x^2 + 25x - 50$$

29. The graph of the function $f(x)$ is a parabola with vertex at (1,6) so the maximum value of the function is the y-coordinate of the vertex, which is 6.

The graph of the function $g(x)$ is also a parabola. The vertex of this parabola can be determined on the graphing calculator.

For the TI-84:

Press the [Y=] button and enter the expression into Y1. Since the standard graphing window will not show the vertex for this parabola, change the Ymax to 15. Press [GRAPH], [2ND], [TRACE], and [4] to find the maximum.

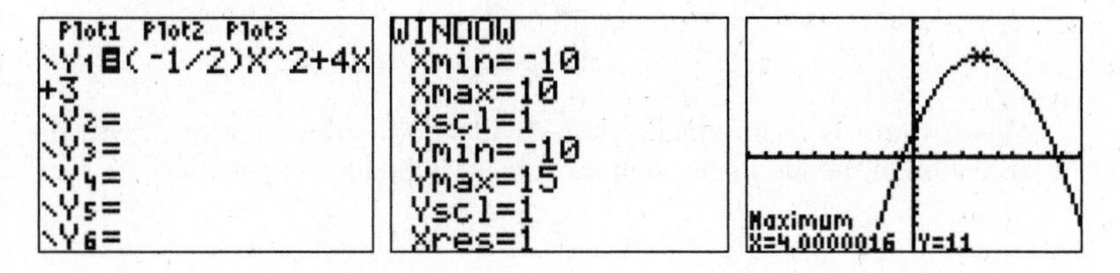

For the TI-Nspire:

Press [home] and [B] and enter the expression after $f1(x)$=. Press [menu], [4], and [1] to change the YMax to 15. Press [menu], [6], and [4] to find the maximum point.

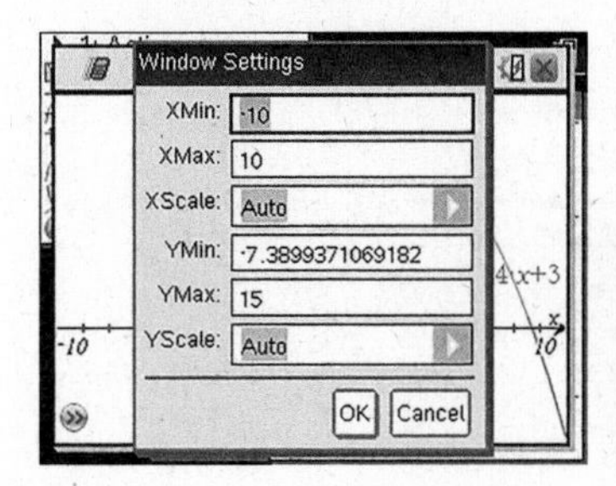

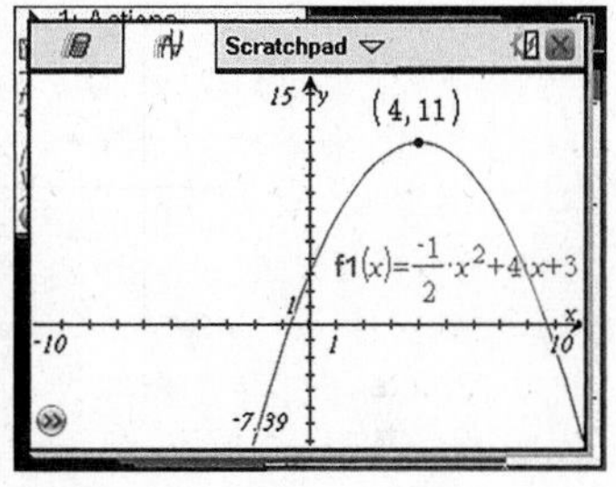

The maximum of the parabola defined by g(x), 11, is higher than the maximum of the parabola defined by g(x), 6.

As an alternate way of determining the y-coordinate of the vertex of the parabola defined by $g(x) = -\frac{1}{2}x^2 + 4x + 3$, the x-coordinate of the vertex can be found with the equation $x = -\frac{b}{2a}$, where b is the coefficient of the x term, 4 in this case, and a is the coefficient of the x^2 term, $-\frac{1}{2}$ in this case.

$x = \dfrac{-(4)}{2(-\frac{1}{2})} = \dfrac{-4}{-1} = 4$. 4 is the x-coordinate of the vertex.

To get the y-coordinate of the vertex, substitute 4 into the function.

$g(4) = -(\frac{1}{2})(4^2) + 4(4) + 3 = -(\frac{1}{2})(16) + 16 + 3 = -8 + 16 + 3 = 11.$

Hence, g(x) has the larger maximum value.

30. Solving inequalities is very similar to solving equations.

$$
\begin{aligned}
3(x+3) &\leq 5x-3 \\
3x+9 &\leq 5x-3 \\
-5x &= -5x \\
-2x+9 &\leq -3 \\
-9 &= -9 \\
\frac{-2x}{-2} &\geq \frac{-12}{-2}
\end{aligned}
$$

When there is an inequality you divide both sides by a negative; the direction of the inequality sign must be switched to keep it true.

$$x \geq 6$$

The smallest possible value that makes $x \geq 6$ true is 6.

31. The residual plot looks like this:

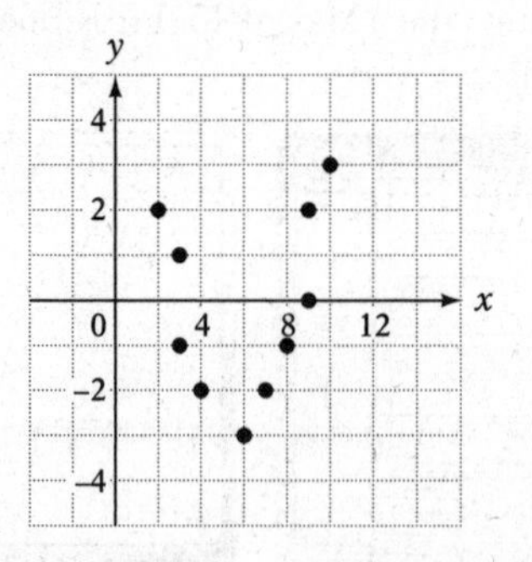

When a line is a good fit, the residual plot often looks like a random scattering of points. Since this residual plot does not look like a random scattering of points, the test makers want you to say that the line of best fit does not accurately model the data and that another type of curve is more appropriate.

32. In a perfect square trinomial $x^2 + bx + c$, the value of c is equal to $(\frac{b}{2})^2$.

In this example, since b is 6, c would have to be $(\frac{6}{2})^2 = 3^2 = 9$.

PART III

33. The graph of $f(x) = |3x|$ can be created with a chart.

x	$f(x)$
–3	$|3(-3)| = |-9| = 9$
–2	$|3(-2)| = |-6| = 6$
–1	$|3(-1)| = |-3| = 3$
0	$|3(0)| = |0| = 0$
1	$|3(1)| = |3| = 3$
2	$|3(2)| = |6| = 6$
3	$|3(3)| = |9| = 9$

Graph these values.

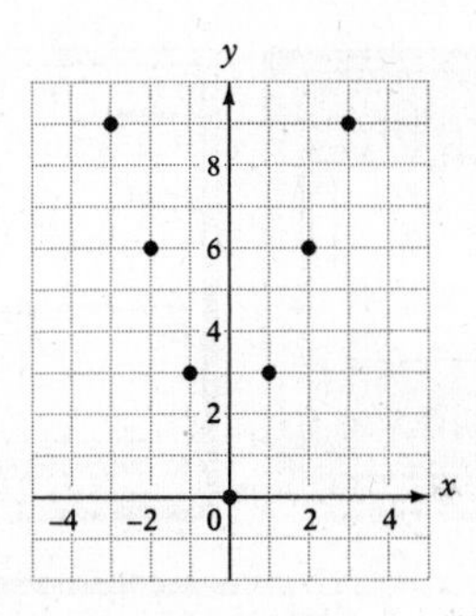

Connect the points to form the classic V shape of a linear absolute value graph.

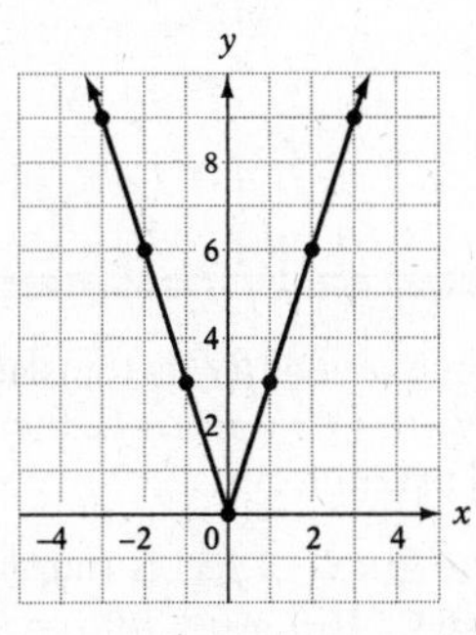

This graph can also be produced on the graphing calculator.

For the TI-84:

Press [Y=], [MATH], [right arrow], and [1] to access the absolute value function. Press [ZOOM] and [6] to see the graph.

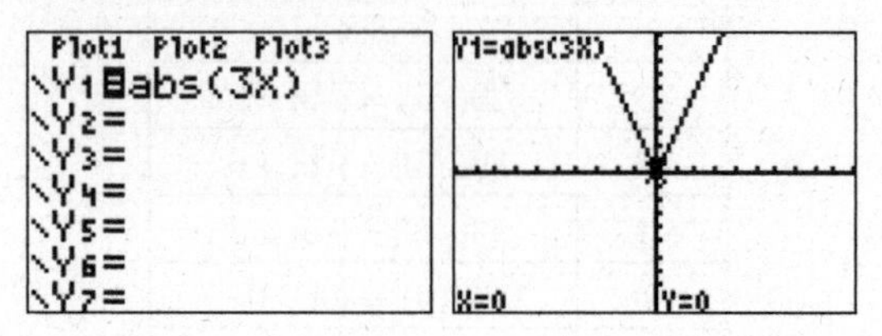

For the TI-Nspire:

Press [home] and [B] to get to the graphing Scratchpad. To enter the absolute value function press the [catalog (looks like a book)] button, which is two buttons to the right of the [9].

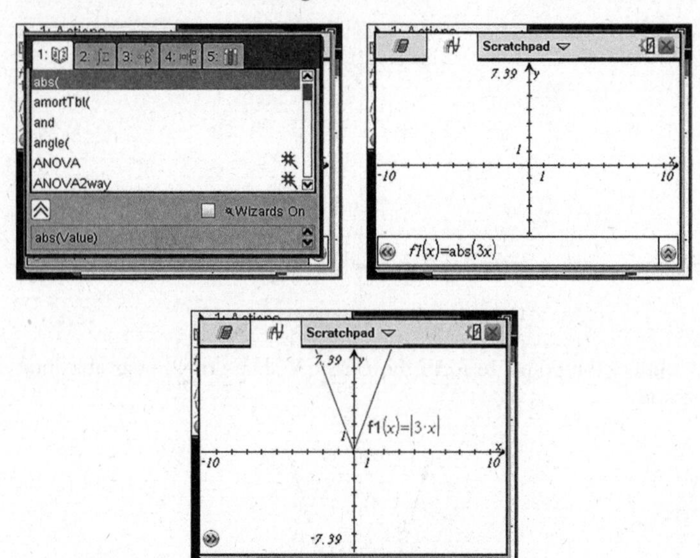

Since $g(x) = f(x) - 2$, the graph of $f(x)$ is translated two units down to form the graph of $g(x)$. If it were $g(x) = f(x) + 2$, the graph of $g(x)$ would be the graph of $f(x)$ translated two units up.

Since $h(x) = f(x - 4)$ the graph of $f(x)$ is translated four units to the right to form the graph of $h(x)$. If it were $h(x) = f(x + 4)$, the graph of $h(x)$ would be the graph of $f(x)$ translated four units to the left.

34. Using the properties of algebra, eliminate everything from the right-hand side of the equation except for the b_1.

Multiply both sides of the equation by 2 to eliminate the $\frac{1}{2}$ from the right-hand side.

$$A = \frac{1}{2}h(b_1 + b_2)$$

$$2 \cdot A = 2 \cdot \frac{1}{2}h(b_1 + b_2)$$

$$2A = h(b_1 + b_2)$$

Divide both sides of the equation by h to eliminate the h from the right-hand side.

$$\frac{2A}{h} = \frac{h(b_1 + b_2)}{h}$$

$$\frac{2A}{h} = b_1 + b_2$$

Subtract b_2 from both sides of the equation to eliminate b_2 from the right-hand side.

$$\frac{2A}{h} = b_1 + b_2$$

$$-b_2 = -b_2$$

$$\frac{2A}{h} - b_2 = b_1$$

Using this equation with $A = 60$, $h = 6$, and $b_2 = 12$, solve for b_1.

$$\frac{2A}{h} - b_2 = b_1$$

$$\frac{2 \cdot 60}{6} - 12 = b_1$$

$$\frac{120}{6} - 12 = b_1$$

$$20 - 12 = b_1$$

$$8 = b_1$$

35. To make the graphs without a graphing calculator, make a table of values.

x	$f(x)$	$g(x)$
–2	$-2(-2)^2 = -2(4) = -8$	$2(-2) - 4 = -4 - 4 = -8$
–1	$-2(-1)^2 = -2(1) = -2$	$2(-1) - 4 = -2 - 4 = -6$
0	$-2(0)^2 = -2(0) = 0$	$2(0) - 4 = 0 - 4 = -4$
1	$-2(1)^2 = -2(1) = -2$	$2(1) - 4 = 2 - 4 = -2$
2	$-2(2)^2 = -2(4) = -8$	$2(2) - 4 = 4 - 4 = 0$

For the graph of $f(x)$, plot the five points (–2,–8), (–1,–2), (0,0), (1,–2), (2,–8) and draw a parabola through them.

For the graph of $g(x)$, plot the five points (–2,–8), (–1,–6), (0,–4), (1,–2), (2,0) and draw a line through them.

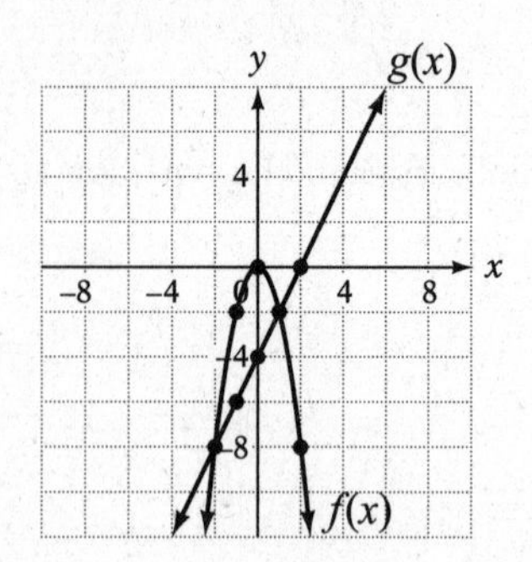

According to the graph, the x-coordinates of the two intersection points are $x = -2$ and $x = 1$. These are the two x values for which $f(x) = g(x)$.

The graphs and intersections can also be produced on the graphing calculator.

For the TI-84:

Press [Y=] and enter the equation for $f(x)$ into Y1 and $g(x)$ into Y2. Press [ZOOM] and [6] to see the graphs. Press [2ND], [TRACE], and [5] to find the first intersection points and then again to find the second intersection point.

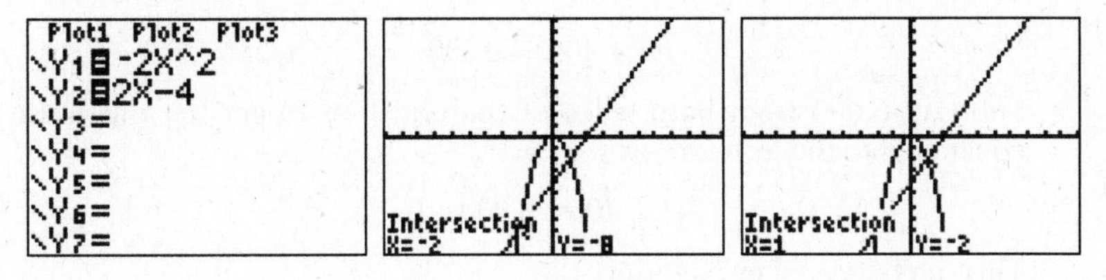

For the TI-Nspire:

Press [home] and [B] to get to the graphing Scratchpad. Enter the two equations into $f_1(x)$ and $f_2(x)$. Change the window so the YMin is –10. Press [menu], [6], and [4] to find the intersections.

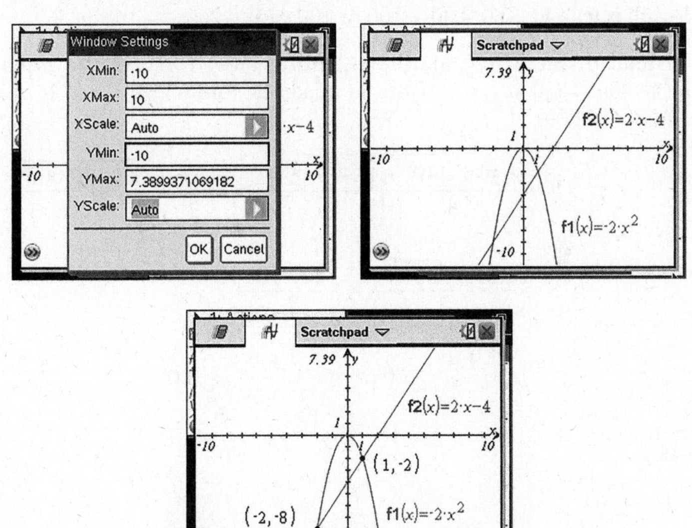

36. If the width is represented by w, the length can be expressed as $w + 40$. The area of a rectangle is length times width. For the area to be 6,000 square yards, solve the equation

$$(w + 40)w = 6{,}000$$

This becomes

$$w^2 + 40w = 6{,}000$$

Subtract 6,000 from both sides of the equation to get the quadratic equation into the form $ax^2 + bx + c = 0$

$$w^2 + 40w - 6{,}000 = 0$$

This can be solved by factoring

$$(w + 100)(w - 60) = 0$$
$$w + 100 = 0 \text{ or } w - 60 = 0$$
$$w = -100 \text{ or } w = 60$$

Since the width must be positive, the answer is $w = 60$ or 60 yards. The length is $w + 40 = 60 + 40 = 100$ or 100 yards.

Instead of factoring, another algebraic way to solve the equation $w^2 + 40w - 6{,}000 = 0$ is with the quadratic formula with $a = 1$, $b = 40$, and $c = -6{,}000$.

$$w = \frac{-40 \pm \sqrt{40^2 - 4(1)(-6000)}}{2(1)} = \frac{-40 \pm \sqrt{1600 + 24000}}{2}$$

$$= \frac{-40 \pm \sqrt{25600}}{2} = \frac{-40 \pm 160}{2}$$

$$w = \frac{120}{2} = 60 \text{ or } w = \frac{-200}{2} = -100$$

PART IV

37. x is the number of hours Edith babysits. y is the number of hours Edith works at the library.

    The total number of dollars Edith earns for x hours of babysitting and y hours at the library is $4x + 8y$.

    To earn at least $80 a week, the inequality is

$$4x + 8y \geq 80$$

    To work no more than 15 hours, the sum of x and y must be less than or equal to 15.

$$x + y \leq 15$$

    The system of inequalities is

$$\begin{aligned} 4x + 8y &\geq 80 \\ x + y &\leq 15 \end{aligned}$$

    To graph the inequality $4x + 8y \geq 80$ by hand, first graph the line defined by the equality $4x + 8y = 80$.

    Put $4x + 8y = 80$ into slope-intercept form by isolating the y variable.

$$\begin{aligned} 4x + 8y &= 80 \\ -4x &= -4x \end{aligned}$$

$$\frac{8y}{8} = \frac{-4x + 80}{8}$$

$$y = -\frac{4}{8}x + 10$$

$$y = -\frac{1}{2}x + 10$$

    The y-intercept is (0,10) and the slope is $-\frac{1}{2}$.

    Graph the line $y = -\frac{1}{2}x + 10$ with a solid line because the original inequality had a $\geq$ sign. If it had just a $>$ sign, it would be a dotted line.

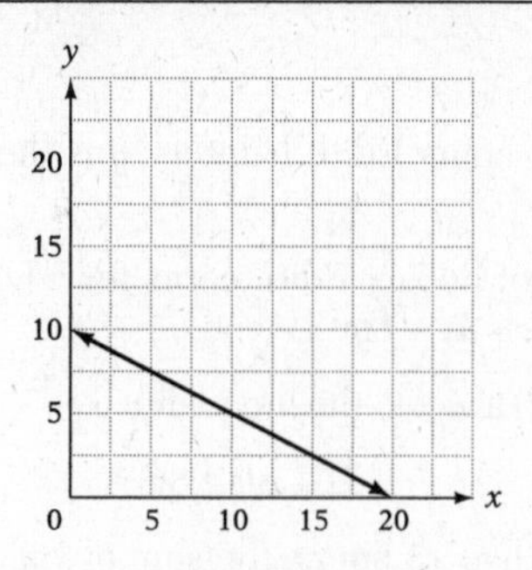

Substitute (0,0) into the inequality $4x + 8y \geq 80$. Since it becomes $0 \geq 80$, which is not true, shade the side of the line that does not contain (0,0).

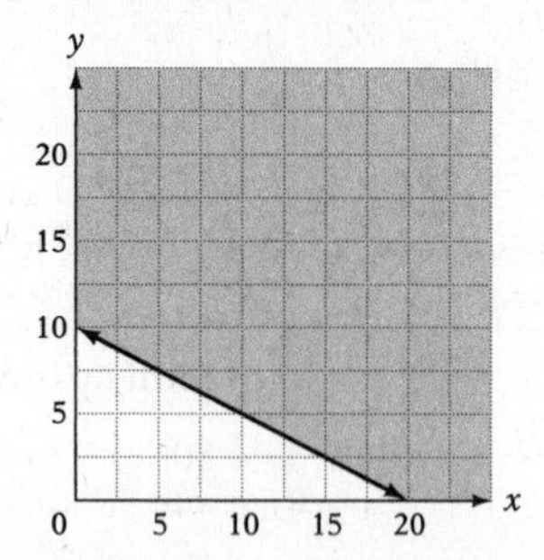

To graph $x + y \leq 15$, first graph the line determined by $x + y = 15$.

Subtract x from both sides of the equation to change to slope-intercept form.

$$\begin{aligned} x + y &= 15 \\ -x &= -x \\ y &= -x + 15 \end{aligned}$$

The y-intercept is (0,15), and the slope is -1.

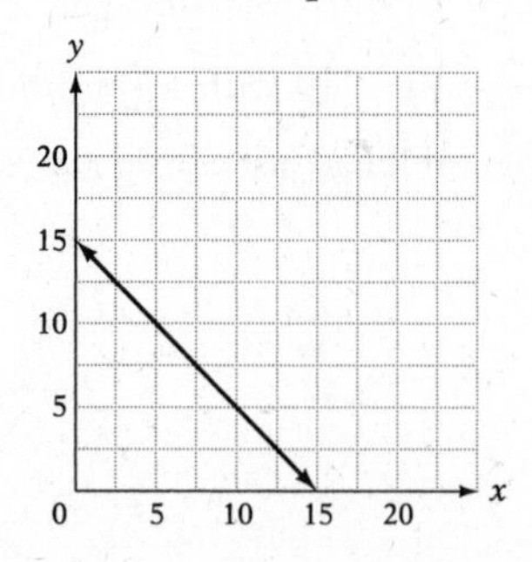

Substitute (0,0) into the inequality $x + y \leq 15$. Since it becomes $0 \leq 80$, which is true, shade the side of the line that does contain (0,0).

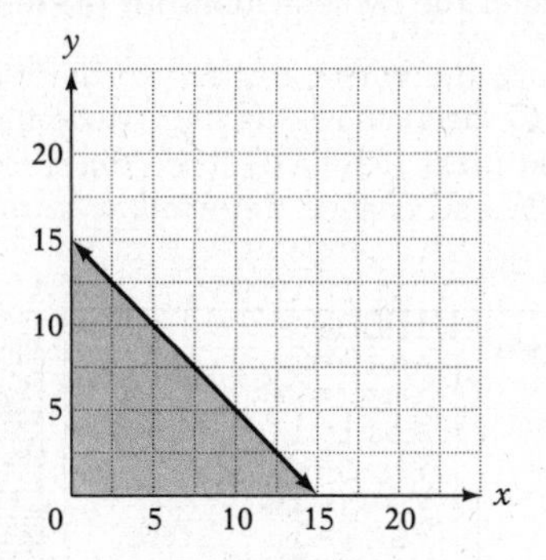

When both inequalities are graphed on the same set of axes, it looks like this:

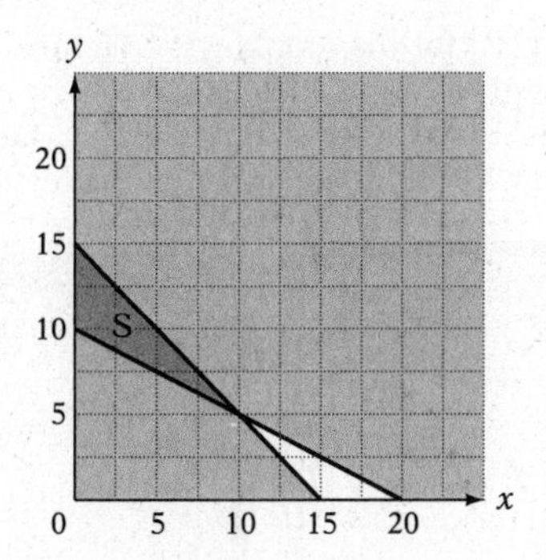

The double shaded triangular region labeled with an S is the graph of the solution set of the system of inequalities.

All points in the shaded region correspond to ordered pairs in the solution set. One example is (2,10). For the second part of the question, this means that if she works for 2 hours babysitting and 10 hours in the library, she will earn more than \$80 and work less than 15 hours. In this case, she will earn \$88 for 12 hours of work.

The system of inequalities can also be graphed on a graphing calculator.

For the TI-84:

Press [Y=] and enter the two equations for the lines $y = -(\frac{1}{2})x + 10$ and $y = -x + 15$. Move the cursor to the \ to the left of the Y1 and press [ENTER] twice to shade above the line. Move the cursor to the \ to the left of the Y2 and press [ENTER] three times to shade below the line. Press [WINDOW] and change the window to match the graph on the test booklet.

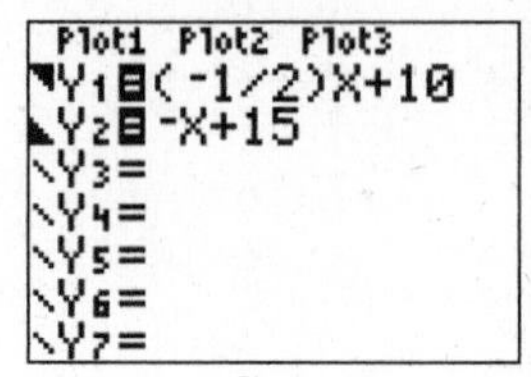

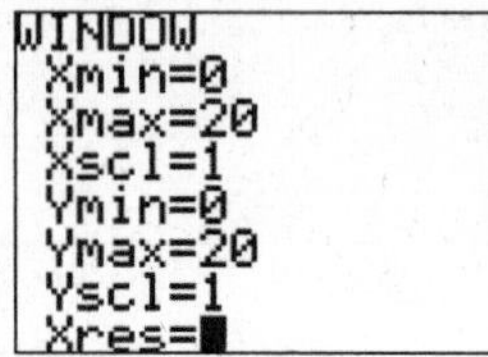

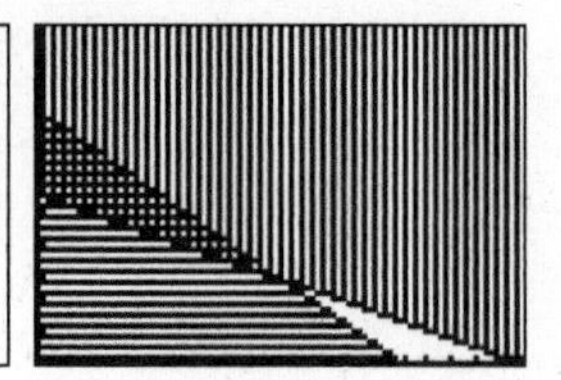

For the TI-Nspire:

Press [home] and [B] for the graphing Scratchpad. To change the = sign to one of the inequality signs, delete the = sign and a menu of choices will appear. Enter the two inequalities. Press [menu], [4], and [1], and change the window to match the graph on the test booklet.

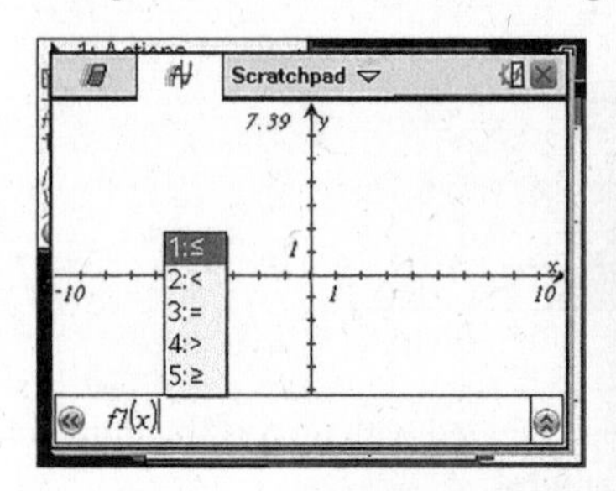

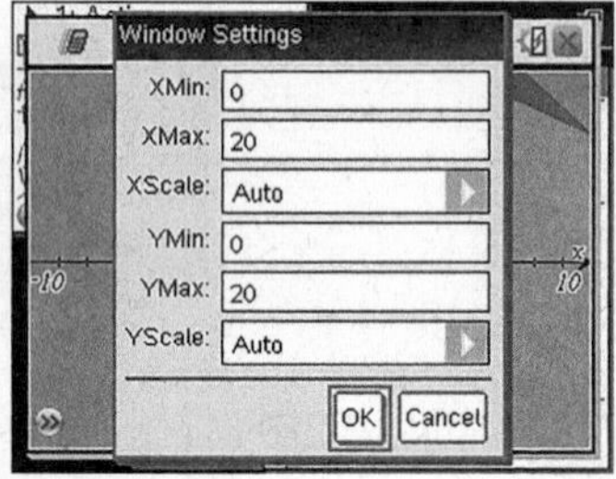

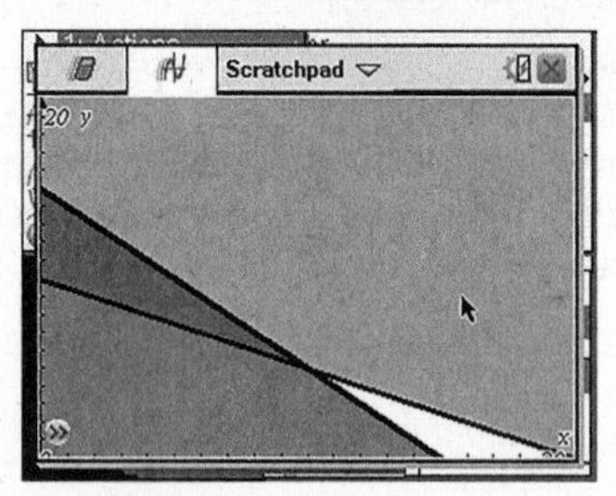

Topic	Question Numbers	Number of Points	Your Points	Your Percentage
1. Polynomials	6, 28	2 + 2 = 4		
2. Properties of Algebra	20, 34	2 + 4 = 6		
3. Functions	11, 22, 23, 33	2 + 2 + 2 + 4 = 10		
4. Creating and Interpreting Equations	2, 9, 12, 14	2 + 2 + 4 = 8		
5. Inequalities	7, 30, 37	2 + 2 + 2 + 4 = 10		
6. Sequences and Series	16, 24	2 + 2 = 4		
7. Systems of Equations	19, 27, 35	2 + 2 + 4 =8		
8. Quadratic Equations and Factoring	3, 15, 18, 25, 32, 36	2 + 2 + 2 + 2 + 2 + 4 = 14		
9. Regression	21, 31	2 + 2 = 4		
10. Exponential Equations	10, 26	2 + 2 = 4		
11. Graphing	5, 8, 13, 17, 29	2 + 2 + 2 + 2 + 2 = 10		
12. Statistics	4	2		
13. Number Properties	1	2		

HOW TO CONVERT YOUR RAW SCORE TO YOUR ALGEBRA I REGENTS EXAMINATION SCORE

The accompanying conversion chart must be used to determine your final score on the August 2014 Regents Examination in Algebra I. To find your final exam score, locate in the column labeled "Raw Score" the total number of points you scored out of a possible 86 points. Since partial credit is allowed in Parts II, III, and IV of the test, you may need to approximate the credit you would receive for a solution that is not completely correct. Then locate in the adjacent column to the right the scale score that corresponds to your raw score. The scale score is your final Algebra I Regents Examination score.

Regents Examination in Algebra I—August 2014
Chart for Converting Total Test Raw Scores to Final Examination Scores (Scaled Scores)

Raw Score	Scale Score	Performance Level	Raw Score	Scale Score	Performance Level	Raw Score	Scale Score	Performance Level
86	100	5	57	74	4	28	63	2
85	99	5	56	73	3	27	62	2
84	98	5	55	73	3	26	61	2
83	96	5	54	73	3	25	60	2
82	94	5	53	73	3	24	59	2
81	93	5	52	72	3	23	58	2
80	91	5	51	72	3	22	57	2
79	90	5	50	72	3	21	56	2
78	88	5	49	72	3	20	55	2
77	87	5	48	71	3	19	53	1
76	86	5	47	71	3	18	51	1
75	85	5	46	71	3	17	49	1
74	83	4	45	71	3	16	48	1
73	82	4	44	70	3	15	46	1
72	82	4	43	70	3	14	44	1
71	81	4	42	70	3	13	41	1
70	80	4	41	69	3	12	39	1
69	79	4	40	69	3	11	37	1
68	79	4	39	69	3	10	34	1
67	78	4	38	68	3	9	31	1
66	77	4	37	68	3	8	28	1
65	77	4	36	68	3	7	25	1
64	76	4	35	67	3	6	22	1
63	76	4	34	67	3	5	19	1
62	75	4	33	66	3	4	15	1
61	75	4	32	66	3	3	12	1
60	75	4	31	65	3	2	8	1
59	74	4	30	64	2	1	4	1
58	74	4	29	64	2	0	0	1

Examination June 2015

Algebra I

HIGH SCHOOL MATH REFERENCE SHEET

Conversions

1 inch = 2.54 centimeters	1 cup = 8 fluid ounces
1 meter = 39.37 inches	1 pint = 2 cups
1 mile = 5280 feet	1 quart = 2 pints
1 mile = 1760 yards	1 gallon = 4 quarts
1 mile = 1.609 kilometers	1 gallon = 3.785 liters
	1 liter = 0.264 gallon
1 kilometer = 0.62 mile	1 liter = 1000 cubic centimeters
1 pound = 16 ounces	
1 pound = 0.454 kilogram	
1 kilogram = 2.2 pounds	
1 ton = 2000 pounds	

Formulas

Triangle	$A = \frac{1}{2}bh$
Parallelogram	$A = bh$
Circle	$A = \pi r^2$
Circle	$C = \pi d$ or $C = 2\pi r$

Formulas (continued)

General Prisms	$V = Bh$
Cylinder	$V = \pi r^2 h$
Sphere	$V = \frac{4}{3}\pi r^3$
Cone	$V = \frac{1}{3}\pi r^2 h$
Pyramid	$V = \frac{1}{3}Bh$
Pythagorean Theorem	$a^2 + b^2 = c^2$
Quadratic Formula	$x = \frac{-b \pm \sqrt{b^2 - 4ac}}{2a}$
Arithmetic Sequence	$a_n = a_1 + (n - 1)d$
Geometric Sequence	$a_n = a_1 r^{n-1}$
Geometric Series	$S_n = \frac{a_1 - a_1 r^n}{1 - r}$ where $r \neq 1$
Radians	1 radian = $\frac{180}{\pi}$ degrees
Degrees	1 degree = $\frac{\pi}{180}$ radians
Exponential Growth/Decay	$A = A_0 e^{k(t - t_0)} + B_0$

PART I

Answer all 24 questions in this part. Each correct answer will receive 2 credits. No partial credit will be allowed. For each statement or question, write in the space provided the numeral preceding the word or expression that best completes the statement or answers the question. [48 credits]

1 The cost of airing a commercial on television is modeled by the function $C(n) = 110n + 900$, where n is the number of times the commercial is aired. Based on this model, which statement is true?

(1) The commercial costs \$0 to produce and \$110 per airing up to \$900.
(2) The commercial costs \$110 to produce and \$900 each time it is aired.
(3) The commercial costs \$900 to produce and \$110 each time it is aired.
(4) The commercial costs \$1010 to produce and can air an unlimited of times.

1 ____

2 The graph below represents a jogger's speed during her 20-minute jog around her neighborhood.

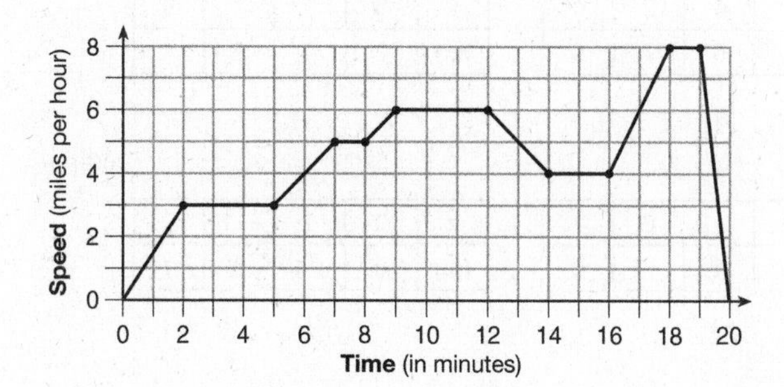

Which statement best describes what the jogger was doing during the 9–12 minute interval of her jog?

(1) She was standing still.
(2) She was increasing her speed.
(3) She was decreasing her speed.
(4) She was jogging at a constant rate.

2 _____

3 If the area of a rectangle is expressed as $x^4 - 9y^2$, then the product of the length and the width of the rectangle could be expressed as

(1) $(x - 3y)(x + 3y)$
(2) $(x^2 - 3y)(x^2 + 3y)$
(3) $(x^2 - 3y)(x^2 - 3y)$
(4) $(x^4 + y)(x - 9y)$

3 _____

4 Which table represents a function?

x	2	4	2	4
f(x)	3	5	7	9

(1)

x	3	5	7	9
f(x)	2	4	2	4

(3)

x	0	−1	0	1
f(x)	0	1	−1	0

(2)

x	0	1	−1	0
f(x)	0	−1	0	1

(4)

4 ____

5 Which inequality is represented in the graph below?

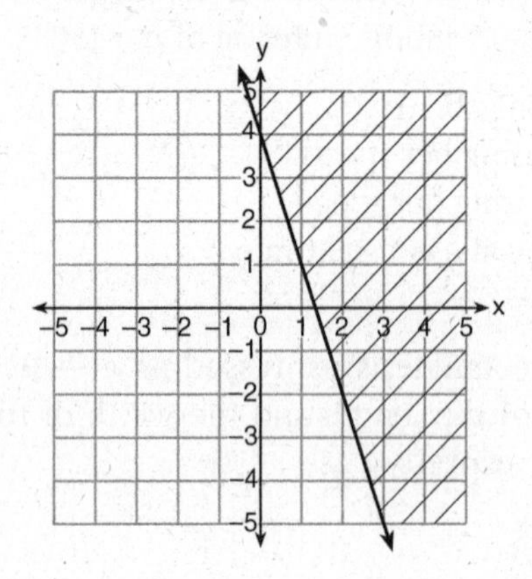

(1) $y \geq -3x + 4$

(2) $y \leq -3x + 4$

(3) $y \geq -4x - 3$

(4) $y \leq -4x - 3$

5 ____

6 Mo's farm stand sold a total of 165 pounds of apples and peaches. She sold apples for \$1.75 per pound and peaches for \$2.50 per pound. If she made \$337.50, how many pounds of peaches did she sell?

(1) 11

(2) 18

(3) 65

(4) 100

6 ____

7 Morgan can start wrestling at age 5 in Division 1. He remains in that division until his next odd birthday when he is required to move up to the next division level. Which graph correctly represents this information?

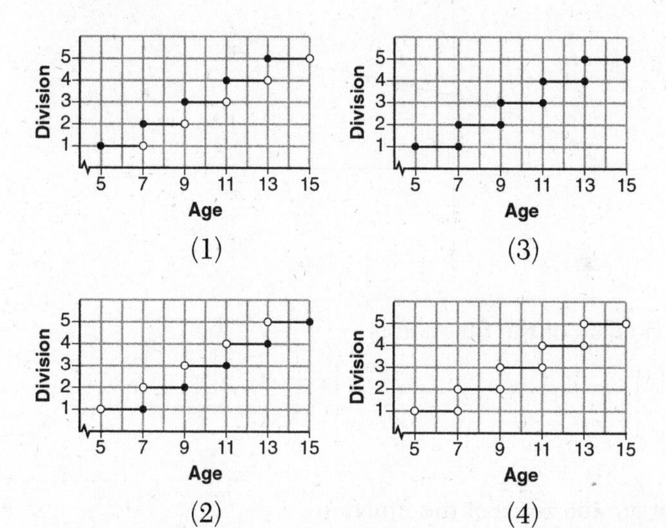

7 _____

8 Which statement is *not* always true?

(1) The sum of two rational numbers is rational.
(2) The product of two irrational numbers is rational.
(3) The sum of a rational number and an irrational number is irrational.
(4) The product of a nonzero rational number and an irrational number is irrational.

8 _____

9 The graph of the function $f(x) = \sqrt{x+4}$ is shown below.

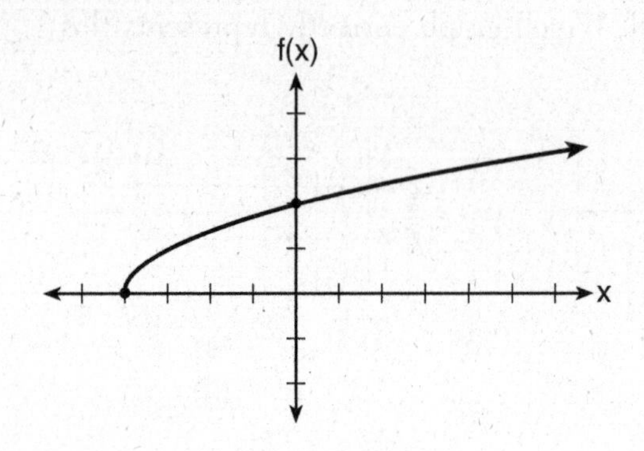

The domain of the function is

(1) $\{x \mid x > 0\}$
(2) $\{x \mid x \geq 0\}$
(3) $\{x \mid x > -4\}$
(4) $\{x \mid x \geq -4\}$

9 ____

10 What are the zeros of the function $f(x) = x^2 - 13x - 30$?

(1) −10 and 3
(2) 10 and −3
(3) −15 and 2
(4) 15 and −2

10 ____

11 Joey enlarged a 3-inch by 5-inch photograph on a copy machine. He enlarged it four times. The table below shows the area of the photograph after each enlargement.

Enlargement	0	1	2	3	4
Area (square inches)	15	18.8	23.4	29.3	36.6

What is the average rate of change of the area from the original photograph to the fourth enlargement, to the *nearest tenth*?

(1) 4.3 (3) 5.4

(2) 4.5 (4) 6.0 11 ____

12 Which equation(s) represent the graph below?

I. $y = (x + 2)(x^2 - 4x - 12)$

II. $y = (x - 3)(x^2 + x - 2)$

III. $y = (x - 1)(x^2 - 5x - 6)$

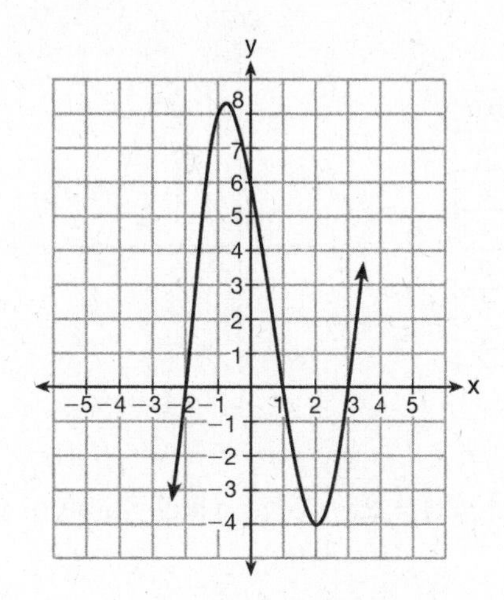

(1) I, only (3) I and II

(2) II, only (4) II and III 12 ____

13 A laboratory technician studied the population growth of a colony of bacteria. He recorded the number of bacteria every other day, as shown in the partial table below.

t (time, in days)	0	2	4
f(t) (bacteria)	25	15,625	9,765,625

Which function would accurately model the technician's data?

(1) $f(t) = 25^t$

(2) $f(t) = 25^{t+1}$

(3) $f(t) = 25t$

(4) $f(t) = 25(t + 1)$

13 ____

14 Which quadratic function has the largest maximum?

$h(x) = (3 - x)(2 + x)$

(1)

$k(x) = -5x^2 - 12x + 4$

(3)

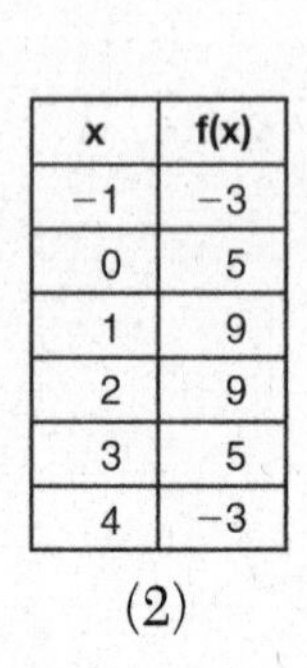

x	f(x)
−1	−3
0	5
1	9
2	9
3	5
4	−3

(2)

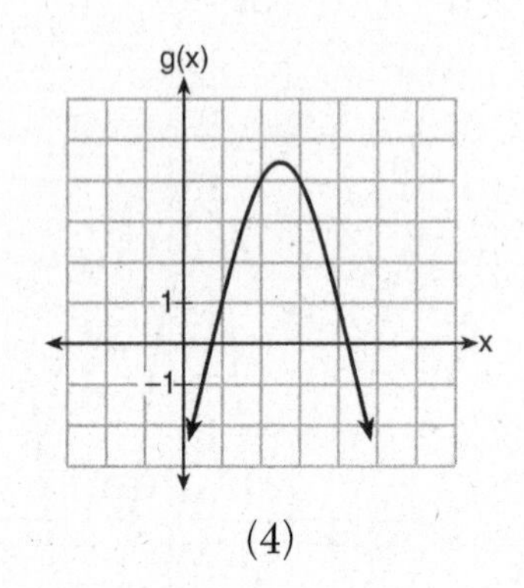

(4)

14 ____

15 If $f(x) = 3x$ and $g(x) = 2x + 5$, at which value of x is $f(x) < g(x)$?

(1) −1

(2) 2

(3) −3

(4) 4

15 ____

16 Beverly did a study this past spring using data she collected from a cafeteria. She recorded data weekly for ice cream sales and soda sales. Beverly found the line of best fit and the correlation coefficient, as shown in the diagram below.

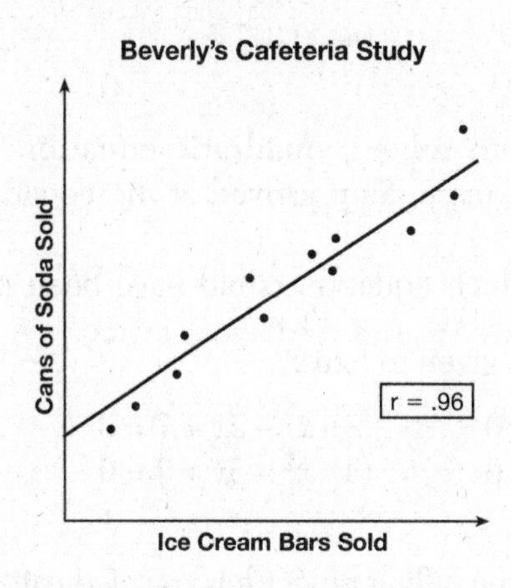

Given this information, which statement(s) can correctly be concluded?

I. Eating more ice cream causes a person to become thirsty.
II. Drinking more soda causes a person to become hungry.
III. There is a strong correlation between ice cream sales and soda sales.

(1) I, only
(2) III, only
(3) I and III
(4) II and III

16 ____

17 The function $V(t) = 1350(1.017)^t$ represents the value $V(t)$, in dollars, of a comic book t years after its purchase. The yearly rate of appreciation of the comic book is

(1) 17% (3) 1.017%
(2) 1.7% (4) 0.017% 17____

18 When directed to solve a quadratic equation by completing the square, Sam arrived at the equation $\left(x - \frac{5}{2}\right)^2 = \frac{13}{4}$. Which equation could have been the original equation given to Sam?

(1) $x^2 + 5x + 7 = 0$ (3) $x^2 - 5x + 7 = 0$
(2) $x^2 + 5x + 3 = 0$ (4) $x^2 - 5x + 3 = 0$ 18____

19 The distance a free falling object has traveled can be modeled by the equation $d = \frac{1}{2}at^2$, where a is acceleration due to gravity and t is the amount of time the object has fallen. What is t in terms of a and d?

(1) $t = \sqrt{\frac{da}{2}}$ (3) $t = \left(\frac{da}{d}\right)^2$

(2) $t = \sqrt{\frac{2d}{a}}$ (4) $t = \left(\frac{2d}{a}\right)^2$ 19____

20 The table below shows the annual salaries for the 24 members of a professional sports team in terms of millions of dollars.

0.5	0.5	0.6	0.7	0.75	0.8
1.0	1.0	1.1	1.25	1.3	1.4
1.4	1.8	2.5	3.7	3.8	4
4.2	4.6	5.1	6	6.3	7.2

The team signs an additional player to a contract worth 10 million dollars per year. Which statement about the median and mean is true?

(1) Both will increase.
(2) Only the median will increase.
(3) Only the mean will increase.
(4) Neither will change.

20 ____

21 A student is asked to solve the equation

$$4(3x - 1)^2 - 17 = 83.$$

The student's solution to the problem starts as

$$4(3x - 1)^2 = 100$$
$$(3x - 1)^2 = 25$$

A correct next step in the solution of the problem is

(1) $3x - 1 = \pm 5$
(2) $3x - 1 = \pm 25$
(3) $9x^2 - 1 = 25$
(4) $9x^2 - 6x + 1 = 5$

21 ____

22 A pattern of blocks is shown below.

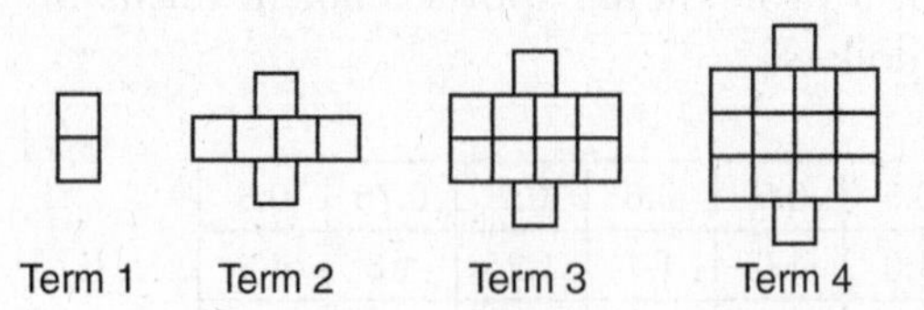

If the pattern of blocks continues, which formula(s) could be used to determine the number of blocks in the nth term?

I	II	III
$a_n = n + 4$	$a_1 = 2$ $a_n = a_{n-1} + 4$	$a_n = 4n - 2$

(1) I and II
(2) I and III
(3) II and III
(4) III, only

22 ____

23 What are the solutions to the equation $x^2 - 8x = 24$?

(1) $x = 4 \pm 2\sqrt{10}$
(2) $x = -4 \pm 2\sqrt{10}$
(3) $x = 4 \pm 2\sqrt{2}$
(4) $x = -4 \pm 2\sqrt{2}$

23 ____

24 Natasha is planning a school celebration and wants to have live music and food for everyone who attends. She has found a band that will charge her $750 and a caterer who will provide snacks and drinks for $2.25 per person. If her goal is to keep the average cost per person between $2.75 and $3.25, how many people, p, must attend?

(1) $225 < p < 325$
(2) $325 < p < 750$
(3) $500 < p < 1000$
(4) $750 < p < 1500$

24 ____

PART II

Answer all 8 questions in this part. Each correct answer will receive 2 credits. Clearly indicate the necessary steps, including appropriate formula substitutions, diagrams, graphs, charts, etc. For all questions in this part, a correct numerical answer with no work shown will receive only 1 credit. [16 credits]

25 Graph the function $y = |x - 3|$ on the set of axes below.

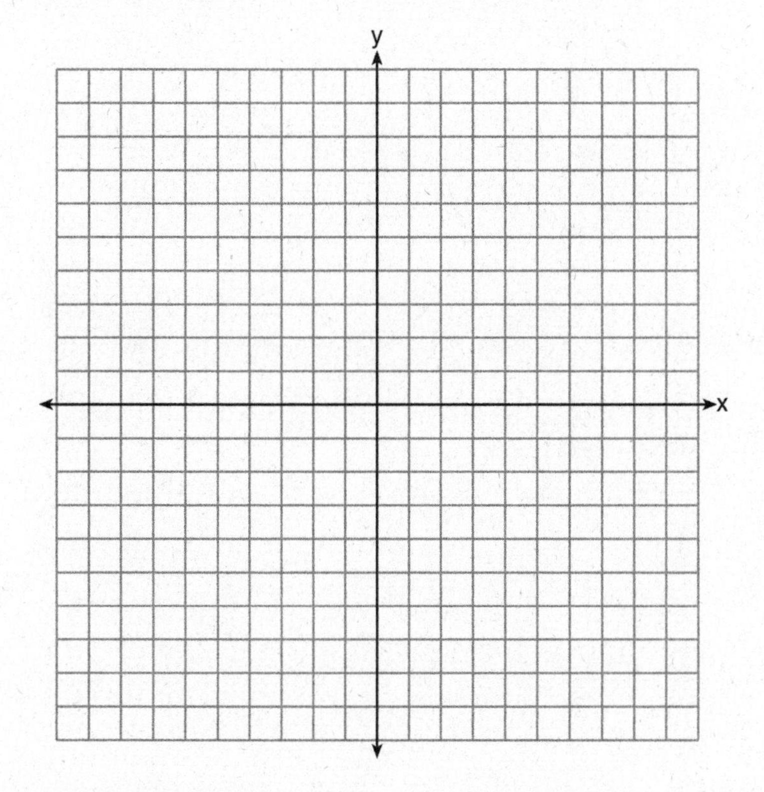

Explain how the graph of $y = |x - 3|$ has changed from the related graph $y = |x|$.

26 Alex is selling tickets to a school play. An adult ticket costs \$6.50 and a student ticket costs \$4.00. Alex sells x adult tickets and 12 student tickets. Write a function, $f(x)$, to represent how much money Alex collected from selling tickets.

27 John and Sarah are each saving money for a car. The total amount of money John will save is given by the function $f(x) = 60 + 5x$. The total amount of money Sarah will save is given by the function $g(x) = x^2 + 46$. After how many weeks, x, will they have the same amount of money saved? Explain how you arrived at your answer.

28 If the difference $(3x^2 - 2x + 5) - (x^2 + 3x - 2)$ is multiplied by $\frac{1}{2}x^2$, what is the result, written in standard form?

29 Dylan invested \$600 in a savings account at a 1.6% annual interest rate. He made no deposits or withdrawals on the account for 2 years. The interest was compounded annually. Find, to the *nearest cent*, the balance in the account after 2 years.

30 Determine the smallest integer that makes $-3x + 7 - 5x < 15$ true.

31 The residual plots from two different sets of bivariate data are graphed below.

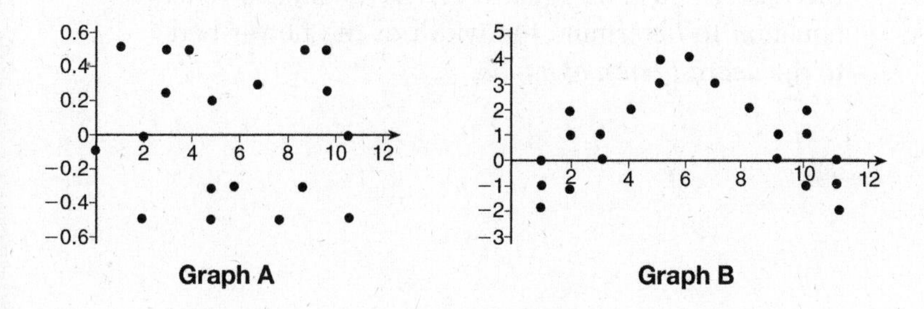

Graph A

Graph B

Explain, using evidence from graph A and graph B, which graph indicates that the model for the data is a good fit.

32 A landscaper is creating a rectangular flower bed such that the width is half of the length. The area of the flower bed is 34 square feet. Write and solve an equation to determine the width of the flower bed, to the *nearest tenth of a foot*.

PART III

Answer all 4 questions in this part. Each correct answer will receive 4 credits. Clearly indicate the necessary steps, including appropriate formula substitutions, diagrams, graphs, charts, etc. For all questions in this part, a correct numerical answer with no work shown will receive only 1 credit. [16 credits]

33 Albert says that the two systems of equations shown below have the same solutions.

First System	Second System
$8x + 9y = 48$ $12x + 5y = 21$	$8x + 9y = 48$ $-8.5y = -51$

Determine and state whether you agree with Albert. Justify your answer.

34 The equation to determine the weekly earnings of an employee at The Hamburger Shack is given by $w(x)$, where x is the number of hours worked.

$$w(x) = \begin{cases} 10x, & 0 \le x \le 40 \\ 15(x-40)+400, & x > 40 \end{cases}$$

Determine the difference in salary, *in dollars*, for an employee who works 52 hours versus one who works 38 hours.

Determine the number of hours an employee must work in order to earn \$445. Explain how you arrived at this answer.

35 An on-line electronics store must sell at least \$2500 worth of printers and computers per day. Each printer costs \$50 and each computer costs \$500. The store can ship a maximum of 15 items per day.

On the set of axes below, graph a system of inequalities that models these constraints.

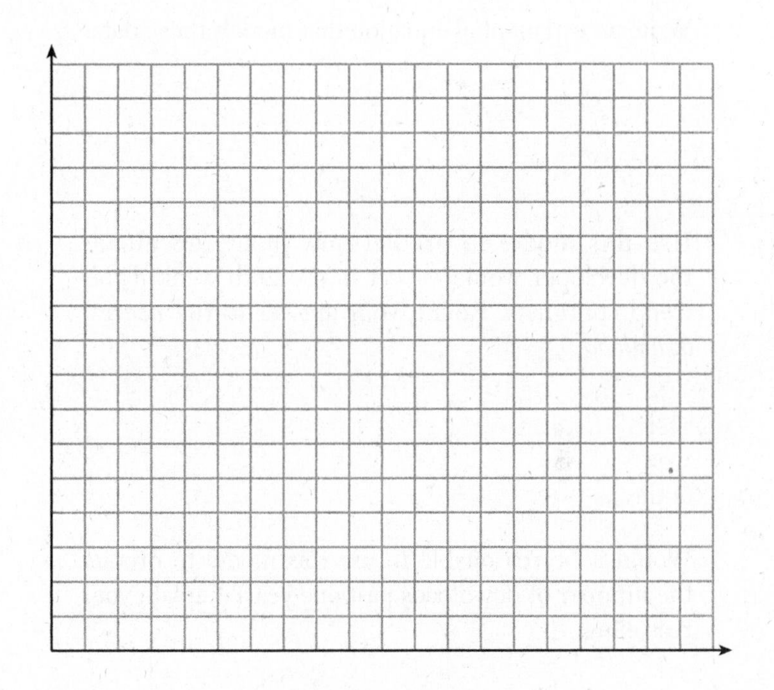

Determine a combination of printers and computers that would allow the electronics store to meet all of the constraints. Explain how you obtained your answer.

36 An application developer released a new app to be downloaded. The table below gives the number of downloads for the first four weeks after the launch of the app.

Number of Weeks	1	2	3	4
Number of Downloads	120	180	270	405

Write an exponential equation that models these data.

Use this model to predict how many downloads the developer would expect in the 26th week if this trend continues. Round your answer to the *nearest download.*

Would it be reasonable to use this model to predict the number of downloads past one year? Explain your reasoning.

PART IV

Answer the question in this part. A correct answer will receive 6 credits. Clearly indicate the necessary steps, including appropriate formula substitutions, diagrams, graphs, charts, etc. A correct numerical answer with no work shown will receive only 1 credit. [6 credits]

37 A football player attempts to kick a football over a goal post. The path of the football can be modeled by the function $h(x) = -\frac{1}{225}x^2 + \frac{2}{3}x$, where x is the horizontal distance from the kick, and $h(x)$ is the height of the football above the ground, when both are measured in feet.

On the set of axes below, graph the function $y = h(x)$ over the interval $0 \leq x \leq 150$.

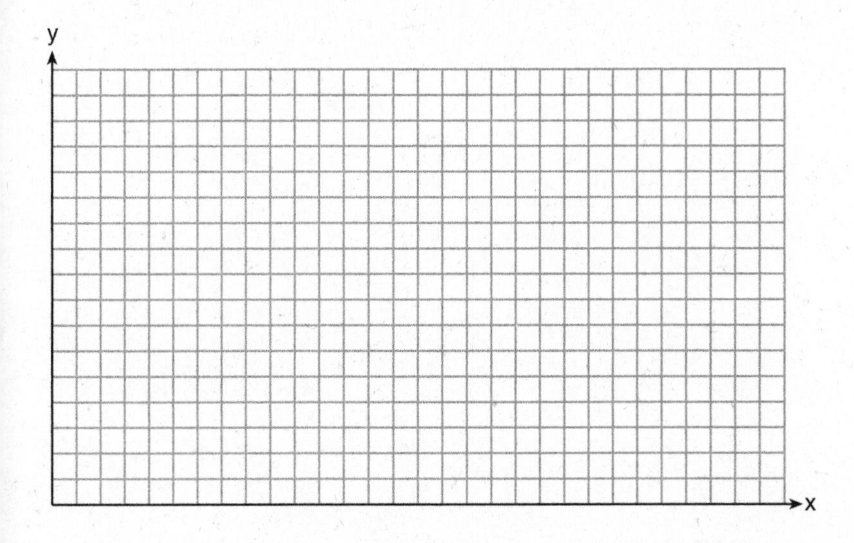

Question 37 is continued on the next page.

Question 37 continued

Determine the vertex of $y = h(x)$. Interpret the meaning of this vertex in the context of the problem.

The goal post is 10 feet high and 45 yards away from the kick. Will the ball be high enough to pass over the goal post? Justify your answer.

Answers June 2015

Algebra I

Answer Key

PART I

1. (3)	**5.** (1)	**9.** (4)	**13.** (2)	**17.** (2)	**21.** (1)
2. (4)	**6.** (3)	**10.** (4)	**14.** (3)	**18.** (4)	**22.** (3)
3. (2)	**7.** (1)	**11.** (3)	**15.** (1)	**19.** (2)	**23.** (1)
4. (3)	**8.** (2)	**12.** (2)	**16.** (2)	**20.** (3)	**24.** (4)

PART II

25. It is the same graph as $y = |x|$ but shifted to the right 3 units.

26. $f(x) = 6.50x + 4(12)$

27. 7

28. $x^4 - \frac{5}{2}x^3 + \frac{7}{2}x^2$

29. 619.35

30. 0

31. Graph *A*

32. 4.1

PART III

33. Agree

34. 200, 43

35. One possible answer is 5 printers and 8 computers.

36. $y = 80 \cdot 1.5^x$ or $y = 120 \cdot 1.5^{x-1}$ and 3,030,140

PART IV

37. The vertex is at (75,25). No, it will not pass over the goal post.

In Parts II–IV, you are required to show how you arrived at your answers. For sample methods of solutions, see the *Answers Explained* section.

Answers Explained

PART I

1. If the cost is modeled by the function $C(n) = 110n + 900$, then $C(0) = 110 \cdot 0 + 900 = 900$ and $C(1) = 110 \cdot 1 + 900 = 1010$. This means that if the commercial is aired 0 times, the cost is \$900. If the commercial is aired 1 time, the cost is \$1010. Since the difference between the two costs is \$110, that is the cost for each time the commercial airs. Since the commercial costs \$900 even if it is not aired at all, \$900 must be the cost of producing the commercial.

   In general, when a real-world scenario is modeled by a function like $f(x) = mx + b$ or even $f(x) = b + mx$, the m will represent the amount of increase for each time that x increases by 1 and the b will represent the starting value, which is the value when x is 0.

   The correct choice is **(3)**.

2. Between minutes 9–12, she was jogging at a speed of 6 miles per hour. Since she was jogging at that same speed for the entire time, her rate is said to be *constant*, or not changing.

   Between minutes 0–2, the jogger's speed was increasing. Between minutes 12–14, her speed was decreasing. She was standing still only when her speed was 0, which was at the start of minute 0 and at the start of minute 20.

   If this same graph had the y-axis representing distance instead of speed, then minutes 9–12 would indicate that the jogger was standing still since her distance would not change. However, as the graph is shown, the y-axis represents speed, not distance.

   The correct choice is **(4)**.

3. This question is asking which of the choices is a possible factorization of $x^4 - 9y^4$. Since both x^4 and $9y^4$ are perfect squares, $x^4 = (x^2)^2$ and $9y^4 = (3y^2)^2$, the expression $x^4 - 9y^4$ can be written as $(x^2)^2 - (3y^2)^2$. The factoring pattern called the difference of perfect squares says that an expression in the form $a^2 - b^2$ is equivalent to $(a - b)(a + b)$. So $(x^2)^2 - (3y^2)^2$ is equivalent to $(x^2 - 3y^2)(x^2 + 3y^2)$.

Since this is a multiple-choice question, an alternative way to get the solution is to multiply each of the four choices to see which one is equivalent to $x^4 - 9y^4$.

Testing choice (1):

$(x - 3y)(x + 3y) = x^2 + 3xy - 3xy - 9y^2 = x^2 - 9y^2$ No

Testing choice (2):

$(x^2 - 3y^2)(x^2 + 3y^2) = x^4 + 3x^2y^2 - 3x^2y^2 - 9y^4 = x^4 - 9y^4$ Yes

Testing choice (3):

$(x^2 - 3y)(x^2 - 3y) = x^4 - 3x^2y - 3x^2y - 9y^2 = x^4 - 6x^2y^2 - 9y^2$ No

Testing choice (4):

$(x^4 + y)(x - 9y) = x^5 - 9x^4y + xy - 9y^2$ No

The correct choice is **(2)**.

4. In a function, each x-value can be mapped to only one $f(x)$-value. In choice (1), the number 2 is mapped to two different values, 3 and 7. In choice (2), the number 0 is mapped to two different values, 0 and -1. In choice (4), the number 0 is mapped to two different values, 0 and 1. Only in choice (3) does each x get mapped to one $f(x)$. It is fine that in choice (3) the numbers 3 and 7 are both mapped to the same value, 2, and that 5 and 9 are both mapped to the same value, 4. As long as each x-value maps to only one $f(x)$-value, the table represents a function. Repeated values in the $f(x)$ row are permitted.

   The correct choice is **(3)**.

5. Since the y-intercept of the line is (0,4), the equation of that line in $y = mx + b$ form must have a b-value of 4. This allows you to eliminate choices (3) and (4). Since the side of the line that does not contain (0,0) is shaded in, the correct choice is the one that is false when 0 is substituted for x and 0 is substituted for y.

   Since (0,0) is not shaded on the graph, it is not part of the solution set of the inequality. So the answer is the choice for which (0,0) does not make it true. Substittue (0,0) into the two remaining choices.

Testing choice (1): $0 \geq -3(0) + 4 = 4$ False

Testing choice (2): $0 \leq -3(0) + 4 = 4$ True

Using a graphing calculator can help.

On the TI-84:

You can graph the equations $y = -3x + 4$ and $y = -4x - 3$ to see which is the equation for the boundary line between the nonshaded and shaded regions.

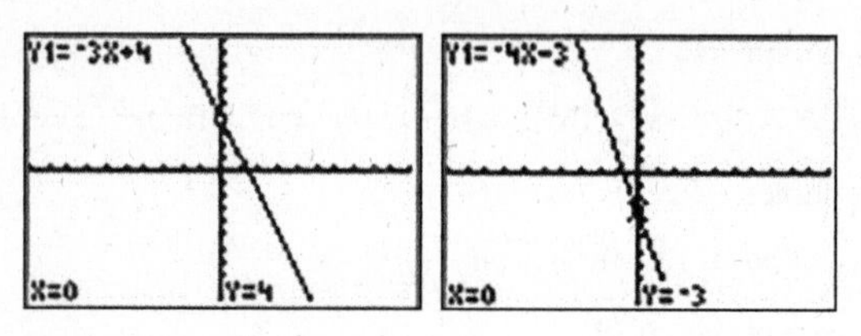

On the TI-Nspire:

You can graph the four inequalities to see which looks like the graph.

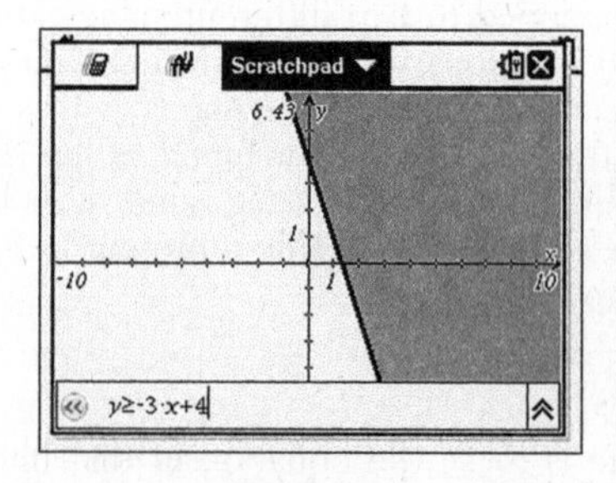

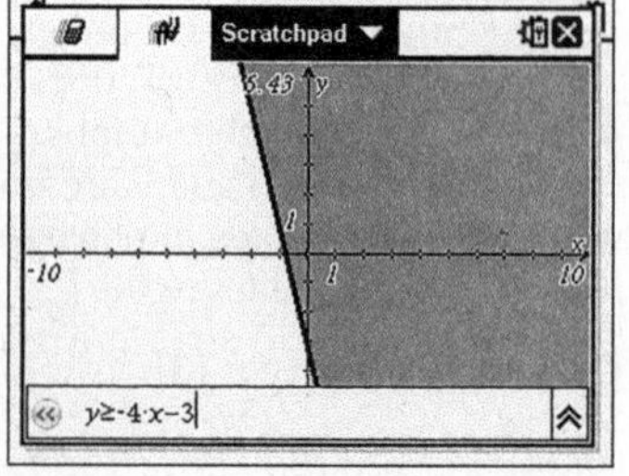

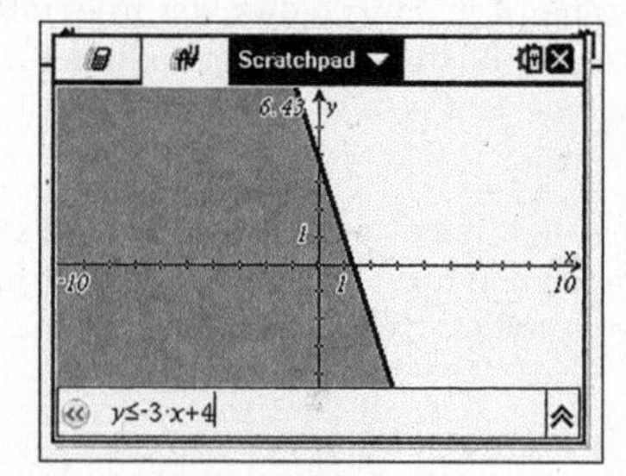

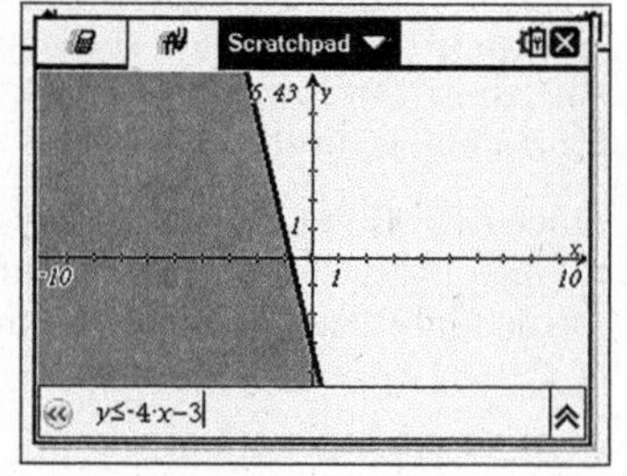

The correct choice is **(1)**.

6. If a is the number of pounds of apples and if p is the number of pounds of peaches, this scenario can be modeled with the following system of equations.

$$a + p = 165$$
$$1.75a + 2.50p = 337.50$$

This system of equations can be solved several different ways. The quickest way is to use the substitution method.

First solve for a in terms of p by subtracting p from both sides of the first equation.

$$\begin{array}{r} a + p = 165 \\ -p = \quad -p \\ \hline a = 165 - p \end{array}$$

Next substitute $165 - p$ for the a in the second equation and solve for p.

$$1.75(165-p)+2.50p = 337.50$$
$$288.75-1.75p+2.50p = 337.50$$
$$288.75+0.75p = 337.50$$
$$288.75+0.75p-288.75 = 337.50-288.75$$
$$\frac{0.75p}{0.75} = \frac{48.75}{0.75}$$
$$p = 65$$

Since this is a multiple-choice question, an alternative method is to test each of the four answer choices.

Testing choice (1): To see if there were 11 pounds of peaches, subtract 11 from 165 to get 154 pounds of apples. Now calculate the combined cost of 11 pounds of peaches and 154 pounds of apples.

$$11 \cdot 2.50 + 154 \cdot 1.75 = 297.00 \neq 337.50 \quad \text{No}$$

Testing choice (2): To see if there were 18 pounds of peaches, subtract 18 from 165 to get 147 pounds of apples. Now calculate the combined cost of 18 pounds of peaches and 147 pounds of apples.

$$18 \cdot 2.50 + 147 \cdot 1.75 = 302.25 \neq 337.50 \quad \text{No}$$

Testing choice (3): To see if there were 65 pounds of peaches, subtract 65 from 165 to get 100 pounds of apples. Now calculate the combined cost of 65 pounds of peaches and 100 pounds of apples.

$$65 \cdot 2.50 + 100 \cdot 1.75 = 337.50 \quad \text{Yes}$$

Testing choice (4): To see if there were 100 pounds of peaches, subtract 100 from 165 to get 65 pounds of apples. Now calculate the combined cost of 100 pounds of peaches and 65 pounds of apples.

$$100 \cdot 2.50 + 65 \cdot 1.75 = 363.75 \neq 337.50 \quad \text{No}$$

The correct choice is **(3)**.

7. The only difference between the graphs in the four answer choices is the endpoints of the line segments. Some have an open circle on one or both of the endpoints. Since Morgan can start wrestling at age 5 in Division 1, there must be a filled circle at (5,1). Only choices (1) and (3) have this. At age 7, which is Morgan's next odd birthday, he moves to Division 2. So there must be a filled circle at (7,2). Both choices (1) and (3) have this, but choice (3) has an additional filled circle at (7,1). Since Morgan cannot be in both Division 1 and Division 2, there must be an open circle at (7,1), leaving just choice (1).

   The correct choice is **(1)**.

8. The product of two irrational numbers can be either irrational or rational. $\sqrt{2} \cdot \sqrt{5} = \sqrt{10}$ is irrational. However, $\sqrt{5} \cdot \sqrt{5} = 5$, which is rational. Choices (1), (3), and (4) are always true.

   The correct choice is **(2)**.

9. The domain of a function can be determined from the graph of that function by finding all the possible x-coordinates of all the points on the graph. The point farthest to the left has the coordinates (–4,0). No points are to the left of that point. Since there is an arrow at the right end, this curve continues to the right infinitely. This graph has points with every x-coordinate greater than or equal to –4. Since the domain is the set of all the possible x-coordinates on the graph, the domain of this function can be described as $\{x|x \geq -4\}$. You must have the $\geq$ sign rather than just the $>$ sign because there is a filled circle at (–4,0).

   The correct choice is **(4)**.

10. The zeros of a function are the values that, when substituted in for x in the function, result in a value of zero. One way to answer this is to set the equation equal to zero and solve by factoring.

$$0 = x^2 - 13x - 30$$
$$0 = (x - 15)(x + 2)$$
$$x - 15 = 0 \quad \text{or } x + 2 = 0$$
$$x = 15 \text{ or } x = -2$$

Since this is a multiple-choice question, it can also be solved by substituting the four answer choices into the function to determine which evaluates to zero. If either of the numbers in the choices does not make the function evaluate to zero, that choice must be eliminated. If just one of the numbers makes the function evaluate to zero, the other number still needs to be checked since they both must work for the choice to be the answer.

Testing choice (1): $(-10)^2 - 13(-10) - 30 = 100 + 130 - 30 = 200 \neq 0$ No

Testing choice (2): $10^2 - 13(10) - 30 = 100 - 130 - 30 = -60 \neq 0$ No

Testing choice (3): $2^2 - 13(2) - 30 = 4 - 26 - 30 = -52 \neq 0$ No

Testing choice (4): $15^2 - 13(15) - 30 = 225 - 195 - 30 = 0$
$(-2)^2 - 13(-2) - 30 = 4 + 26 - 30 = 0$ Yes

The correct choice is **(4)**.

11. To calculate average rate of change of the area, divide the change in the area by the change in the enlargement number:

$$\text{Average rate of change} = \frac{36.6-15}{4-0} = \frac{21.6}{4} = 5.4$$

The correct choice is **(3)**.

12. If a graph has x-intercepts at -2, 1, and 3 then its equation is $y = a(x + 2)(x - 1)(x - 3)$ since substituting -2, 1, or 3 in for the x-values will make y equal to zero. Since the y-intercept is 6, the value of a must be 1. If you substitute (0,6) into the equation, it becomes

$$\begin{aligned} 6 &= a(0 + 2)(0 - 1)(0 - 3) \\ &= a(2)(-1)(-3) \\ &= 6a \\ 1 &= a \end{aligned}$$

So the equation $y = (x + 2)(x - 1)(x - 3)$ would be one answer, though this does not exactly match either of the options. If possible, factor each of the options to see if any are equivalent to $y = (x + 2)(x - 1)(x - 3)$.

Option I: $y = (x + 2)(x^2 - 4x - 12) = (x + 2)(x + 2)(x - 6)$ No

Option II: $y = (x - 3)(x^2 + x - 2) = (x - 3)(x + 2)(x - 1)$ Yes

Option III: $y = (x - 1)(x^2 - 5x - 6) = (x - 1)(x + 1)(x - 6)$ No

Another method is to graph the three options on the graphing calculator to see which have graphs equivalent to the given graph.

On the Ti-84:

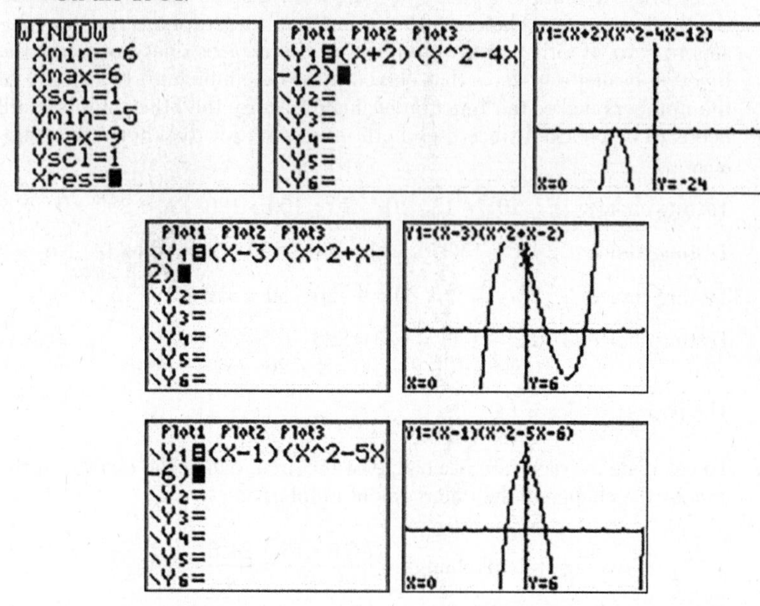

On the TI-Nspire:

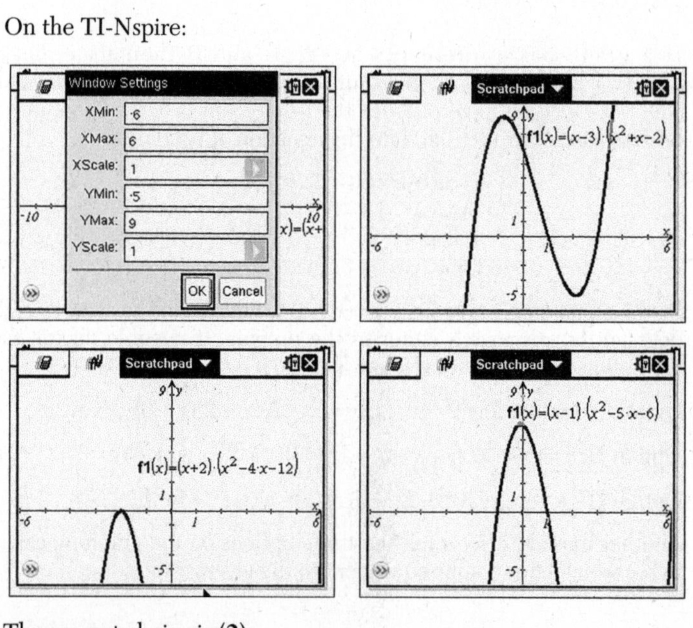

The correct choice is **(2)**.

13. Substitute 0 for t into each of the equations to see if any of them do not evaluate to 25.

    Testing choice (1): $f(0) = 25^0 = 1 \neq 25$ No

    Testing choice (2): $f(0) = 25^{0+1} = 25^1 = 25$ Yes

    Testing choice (3): $f(0) = 25(0) = 0 \neq 25$ No

    Testing choice (4): $f(0) = 25(0 + 1) = 25(1) = 25$ Yes

    Choices (1) and (3) can be eliminated.

    For the two remaining choices, substitute 2 for t to see which evaluates to 15,626.

    Testing choice (2): $f(2) = 25^{2+1} = 25^3 = 15{,}626$ Yes

    Testing choice (4): $f(2) = 25(2 + 1) = 25(3) = 75 \neq 15{,}626$ No

    Another thing to notice is that the bacteria are growing at an increasing rate. This suggests exponential growth, which requires a variable in the exponent as in choice (2). In contrast, choice (4) represents linear growth since the variable is neither an exponent nor raised to a power greater than 1.

    The correct choice is **(2)**.

14. The maximum of a quadratic function is the greatest value of a function. When you graph a quadratic function with a negative coefficient in front of the x^2, the maximum will be the y-coordinate of the vertex. In this example, choice (4) is already a graph. Its vertex has a y-coordinate of about 4.5. Here are the graphs of the other three choices:

    Choice (1):

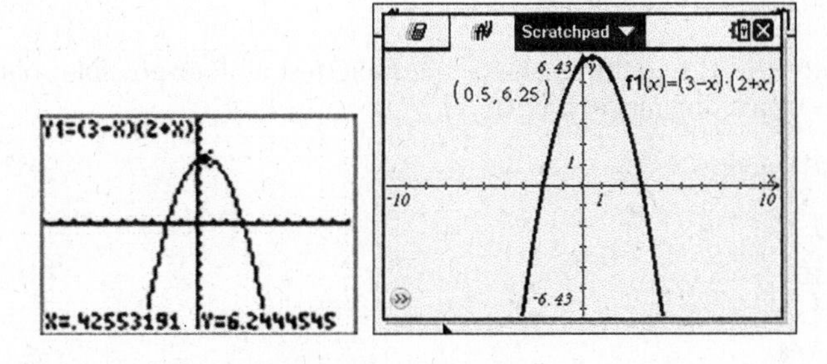

    The vertex has a y-coordinate of 6.25.

Choice (2):

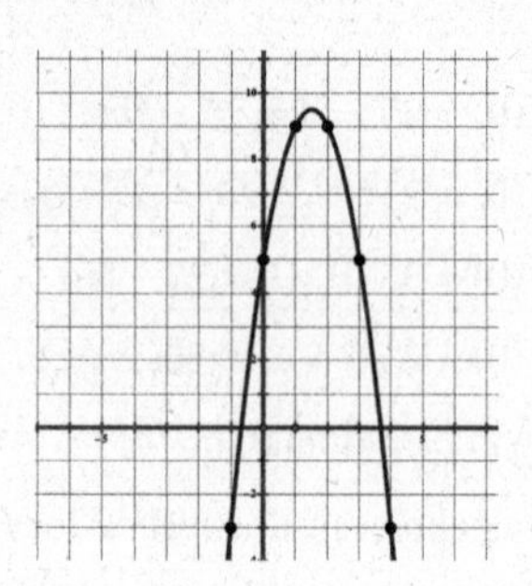

Though the chart does not have the vertex as one of its points, it appears that the y-coordinate of the vertex is a little more than 9.

Choice (3):

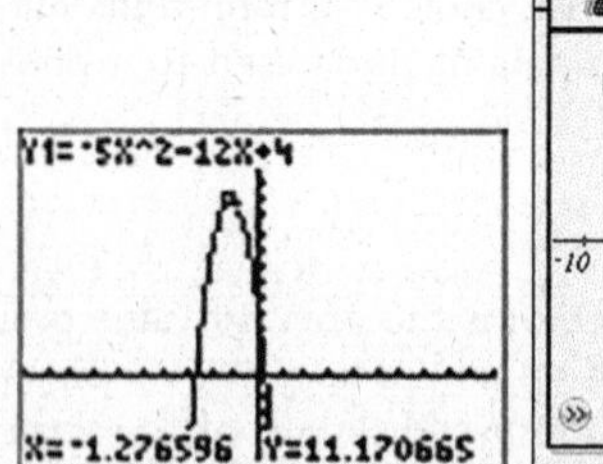

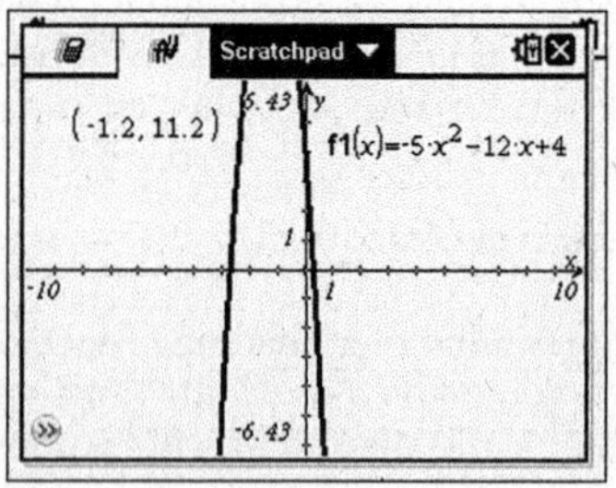

The vertex has a y-coordinate of 11.2.

The correct choice is **(3)**.

15. Since this is a multiple-choice question, test all four possible solutions to see which one makes $f(x) < g(x)$.

Testing choice (1):

$$f(-1) = 3^{-1} \approx 0.333$$
$$g(-1) = 2(-1) + 5 = 3$$
$$f(-1) < g(-1)$$

Yes

Testing choice (2):

$f(2) = 3^2 = 9$
$g(2) = 2(2) + 5 = 9$
$f(2) = g(2)$

No

Testing choice (3):

$f(-3) = 3^{-3} \approx 0.037$
$g(-3) = 2(-3) + 5 = -1$
$f(-3) > g(-3)$

No

Testing choice (4):

$f(4) = 3^4 = 81$
$g(4) = 2(4) + 5 = 13$
$f(4) > g(4)$

No

You can also graph the two functions on a graphing calculator to see where the graph of $f(x)$ is below the graph of $g(x)$.

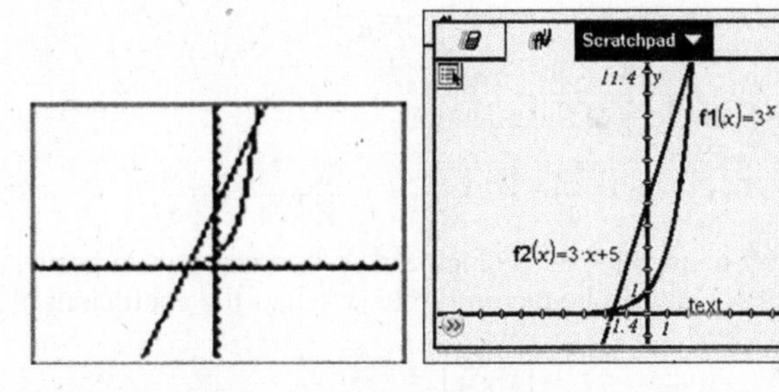

Based on the graph, for x-values between around –2.5 and 2, the graph of $f(x)$ is below the graph of $g(x)$. Of the four choices, only –1 is between –2.5 and 2.

The correct choice is **(1)**.

16. When the r-value for a correlation coefficient is close to 1, there is a strong correlation between the two variables. In this case since 0.96 is very close to 1, option III is true. Options I and II could be true. However, a strong correlation does not mean that an increase in one of the variables must have caused the increase (or decrease) in the other variable.

    The correct choice is **(2)**.

17. When an exponential function is in the form $f(x) = a(1 + r)^x$, the a is the starting value and the r is the growth rate. In this example, $1.017 = 1 + 0.017$. So the r-value is 0.017, which represents 1.7%.

 The correct choice is **(2)**.

18. The completing the square process involves using algebra to convert the left side of a quadratic equation into a perfect square trinomial. By solving each of the four choices with the completing the square process, you can see which matches the expression in the question.

 Testing choice (1):

$$x^2 + 5x + 7 = 0$$

 The first step is to eliminate the constant from the left hand side of the equation. Do this by subtracting 7 from both sides of the equation.

$$x^2 + 5x = -7$$

 A perfect square trinomial is one where the constant is equal to the square of half the coefficient of the x. In this example, to find the value of the constant that would need to be added to the left hand side to make it a perfect square trinomial, calculate the square of half the coefficient of the x which is 5.

$$\left(\frac{5}{2}\right)^2 = \frac{25}{4}$$

 Add this to both sides of the equation

$$x^2 + 5x + \frac{25}{4} = -7 + \frac{25}{4} = -\frac{28}{4} + \frac{25}{4} = -\frac{3}{4}$$

 Now the left hand side can be factored. When you have a perfect square trinomial, it can always be factored into $(x + \text{half the coefficient of the } x)^2$

$$\left(x + \frac{5}{2}\right)^2 = -\frac{3}{4}$$

 This does not match the equation from the question so choice (1) is not correct.

 Though it is not necessary for this question, to solve for x you would then take the square root of both sides and then eliminate the constant from the right hand side.

$$x + \frac{5}{2} = \pm\sqrt{-\frac{3}{4}}$$

$$x = -\frac{5}{2} \pm \sqrt{-\frac{3}{4}}$$

Testing choice (2):

$$x^2+5x+3=0$$
$$-3=-3$$
$$x^2+5x=-3$$
$$x^2+5x+\frac{25}{4}=-3+\frac{25}{4}$$
$$\left(x+\frac{5}{2}\right)^2=\frac{13}{4}$$

Testing choice (3):

$$x^2-5x+7=0$$
$$-7=-7$$
$$x^2-5x=-7$$
$$x^2-5x+\frac{25}{4}=-7+\frac{25}{4}$$
$$\left(x-\frac{5}{2}\right)^2=-\frac{3}{4}$$

Testing choice (4):

$$x^2-5x+3=0$$
$$-3=-3$$
$$x^2-5x=-3$$
$$x^2-5x+\frac{25}{4}=-3+\frac{25}{4}$$
$$\left(x-\frac{5}{2}\right)^2=\frac{13}{4}$$

The correct choice is **(4)**.

An alternative way to solve this question is to solve the equation $\left(x-\frac{5}{2}\right)^2=\frac{13}{4}$ and compare the solutions to the solutions for the equations in the answer choices.

$\left(x-\frac{5}{2}\right)^2=\frac{13}{4}$ can be solved by taking the square root of both sides of the equation and then adding $\frac{5}{2}$ to both sides of the equation.

$$\left(x-\frac{5}{2}\right)^2=\frac{13}{4}$$
$$\sqrt{\left(x-\frac{5}{2}\right)^2}=\pm\sqrt{\frac{13}{4}}$$
$$x-\frac{5}{2}=\pm\sqrt{\frac{13}{4}}$$
$$x=\frac{5}{2}\pm\frac{\sqrt{13}}{2}=\frac{5\pm\sqrt{13}}{2}$$
$$x\approx 4.3 \text{ or } x\approx .70$$

19. Isolating the variable t in the expression $d=\frac{1}{2}at^2$ can be done in three steps.

Step 1: Multiply both sides by 2 to eliminate the $\frac{1}{2}$ from the right side.

$$2\cdot d=2\cdot\frac{1}{2}at^2$$
$$2d=at^2$$

Step 2: Divide both sides by a to eliminate the a from the right side.

$$\frac{2d}{a}=\frac{at^2}{a}$$
$$\frac{2d}{a}=t^2$$

Step 3: Take the square root of both sides to eliminate the exponent from the right side.

$$\sqrt{\frac{2d}{a}}=\sqrt{t^2}$$
$$\sqrt{\frac{2d}{a}}=t$$

The correct choice is **(2)**.

20. To get the mean for the original 24 players, add up the 24 numbers and divide by 24. The sum of the numbers is 61.5, so the mean is $61.5 \div 24 = 2.5625$. When 10 is added to the total, the new total is 71.5. To find the mean for the 25 numbers, divide the new total by 25. The mean is $71.5 \div 25 = 2.86$. So the mean increases. In general, if you add something greater than the mean to a set of numbers, the mean will increase.

    The median is the middle number when the numbers are arranged from smallest to largest. Since 24 is an even number, there is no single middle number. So you have to take the average of the two middle numbers, which are the 12th and 13th numbers. They are both 1.4. So the median is $\frac{1.4+1.4}{2} = 1.4$. When the number 10 is added to the set, the set will now have 25 numbers. So the median is the 13th smallest. The 13th smallest number is 1.4. So the median of the new set is the same as the median of the old set.

    The correct choice is **(3)**.

21. One way to isolate the x in the equation $(3x - 1)^2 = 25$ is to take the square root of both sides of the equation to eliminate the exponent. On the right side of the equation, include a ± sign because $(+5)^2 = 25$ and also $(-5)^2 = 25$.

$$\sqrt{(3x-1)^2} = \pm\sqrt{25}$$
$$3x - 1 = \pm 5$$

    The correct choice is **(1)**.

22. The number of squares follows the pattern 2, 6, 10, 14. Test each of the three rules to see which generates the numbers 2, 6, 10, 14 for n-values 1, 2, 3, 4.

    Testing option I:

    For $n = 1$, $a_1 = 1 + 4 = 5$. Since the first value should be 2, this is not a formula for describing the sequence.

    Testing option II:

    For $n = 1$, $a_1 = 2$ according to the first line of the formula.

    For $n = 2$, $a_2 = a_{2-1} + 4 = a_1 + 4 = 2 + 4 = 6$.

    For $n = 3$, $a_3 = a_{3-1} + 4 = a_2 + 4 = 6 + 4 = 10$.

    For $n = 4$, $a_4 = a_{4-1} + 4 = a_3 + 4 = 10 + 4 = 14$.

Testing option III:

For $n = 1$, $a_2 = 4 \cdot 1 - 2 = 2$.

For $n = 2$, $a_2 = 4 \cdot 2 - 2 = 6$.

For $n = 3$, $a_3 = 4 \cdot 3 - 2 = 10$.

For $n = 4$, $a_4 = 4 \cdot 4 - 2 = 14$.

Both options II and III generate the sequence 2, 6, 10, 14.

The correct choice is **(3)**.

23. One way to solve this equation is to use completing the square. Since $\left(\frac{-8}{2}\right)^2 = 16$, add 16 to both sides of the equation.

$$x^2 - 8x + 16 = 24 + 16$$
$$(x-4)^2 = 40$$
$$\sqrt{(x-4)^2} = \pm\sqrt{40}$$
$$x - 4 = \pm\sqrt{40}$$
$$x = 4 \pm \sqrt{40}$$
$$x = 4 \pm \sqrt{4}\sqrt{10}$$
$$x = 4 \pm 2\sqrt{10}$$

Another way to find the solution is to get the equation into the form $x^2 - 8x - 24 = 0$ and solve using the quadratic formula $x = \frac{-b \pm \sqrt{b^2 - 4ac}}{2a}$. For this equation, $a = 1$, $b = -8$, $c = -24$.

$$x = \frac{-(-8) \pm \sqrt{(-8)^2 - 4(1)(-24)}}{2(1)} = \frac{8 \pm \sqrt{64+96}}{2} = \frac{8 \pm \sqrt{160}}{2}$$
$$x = \frac{8 \pm \sqrt{16}\sqrt{10}}{2} = \frac{8 \pm 4\sqrt{10}}{2} = 4 \pm 2\sqrt{10}$$

The correct choice is **(1)**.

24. The total cost of the party if p people attend is $2.25p + 750$. The price per person is the total cost divided by the number of people. In this case, the price per person is $\frac{2.25p+750}{p}$.

The number of people needed to make the price per person \$2.75 can be determined with the equation $\frac{2.25p+750}{p} = 2.75$.

Multiply both sides by p and solve.

$$\begin{aligned} p \cdot \frac{2.25p+750}{p} &= p \cdot 2.75 \\ 2.25p+750 &= 2.75p \\ -2.25p \qquad &= -2.25p \\ \frac{750}{0.5} &= \frac{0.5p}{0.5} \\ 1500 &= p \end{aligned}$$

The number of people needed to make the price per person \$3.25 can be determined with the equation $\frac{2.25p+750}{p} = 3.25$.

Multiply both sides by p and solve.

$$\begin{aligned} p \cdot \frac{2.25p+750}{p} &= p \cdot 3.25 \\ 2.25p+750 &= 3.25p \\ -2.25p \qquad &= -2.25p \\ 750 &= 1p \\ 750 &= p \end{aligned}$$

So the number of people needed to keep the cost between \$2.25 and \$3.25 must be between 750 and 1500.

The correct choice is **(4)**.

PART II

25. One way to create a graph is to make a chart starting with x-values between –5 and 5.

x	$\lvert x - 3\rvert$	y
–5	$\lvert -5 - 3\rvert = \lvert -8\rvert$	8
–4	$\lvert -4 - 3\rvert = \lvert -7\rvert$	7
–3	$\lvert -3 - 3\rvert = \lvert -6\rvert$	6
–2	$\lvert -2 - 3\rvert = \lvert -5\rvert$	5
–1	$\lvert -1 - 3\rvert = \lvert -4\rvert$	4
0	$\lvert 0 - 3\rvert = \lvert -3\rvert$	3
1	$\lvert 1 - 3\rvert = \lvert -2\rvert$	2
2	$\lvert 2 - 3\rvert = \lvert -1\rvert$	1
3	$\lvert 3 - 3\rvert = \lvert 0\rvert$	0
4	$\lvert 4 - 3\rvert = \lvert 1\rvert$	1
5	$\lvert 5 - 3\rvert = \lvert 2\rvert$	2

When these eleven values are plotted on the axes, the graph looks like this.

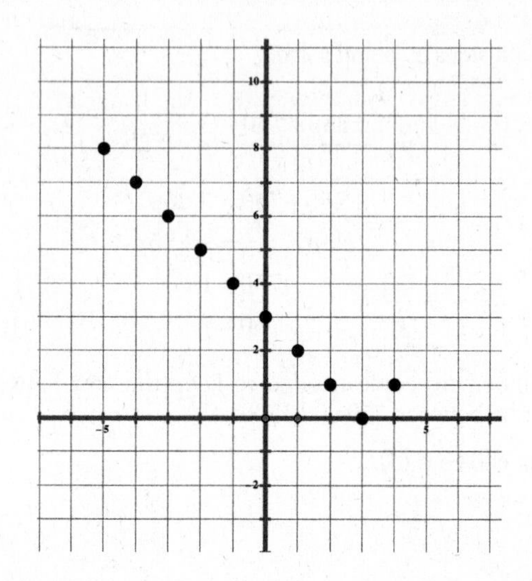

When the points are connected, the graph looks like this.

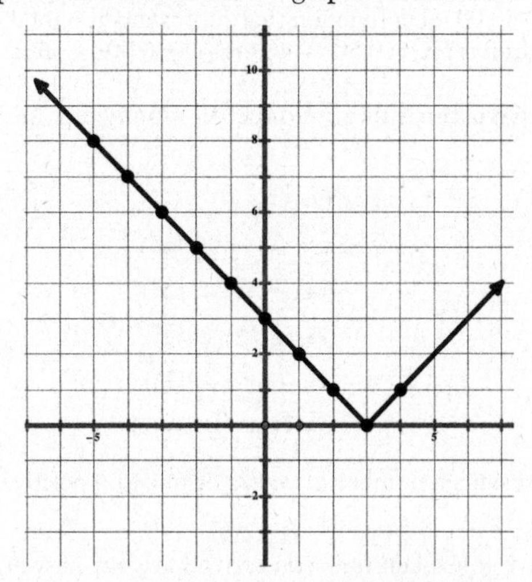

The graphs can also be created on the graphing calculator.

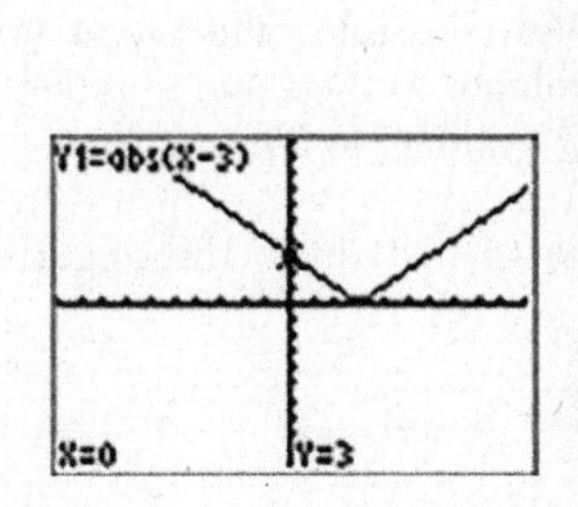

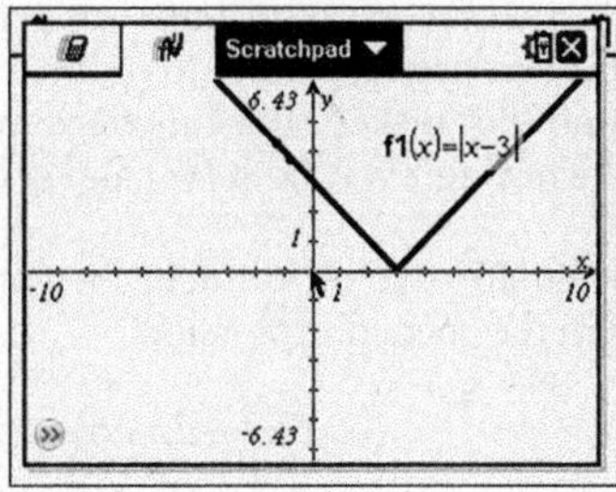

The graph of $y = |x - 3|$ has a graph just like $y = |x|$ but is translated to the right 3 units.

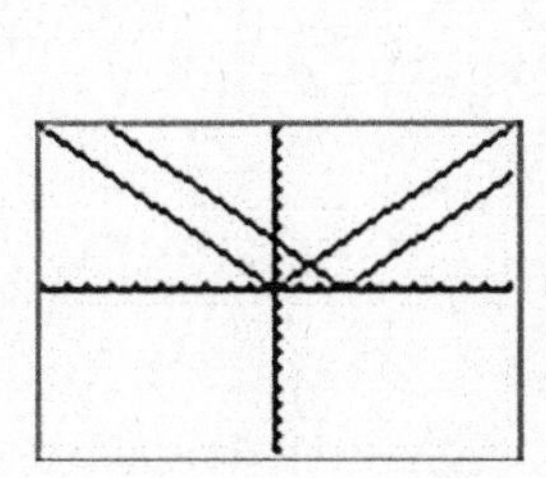

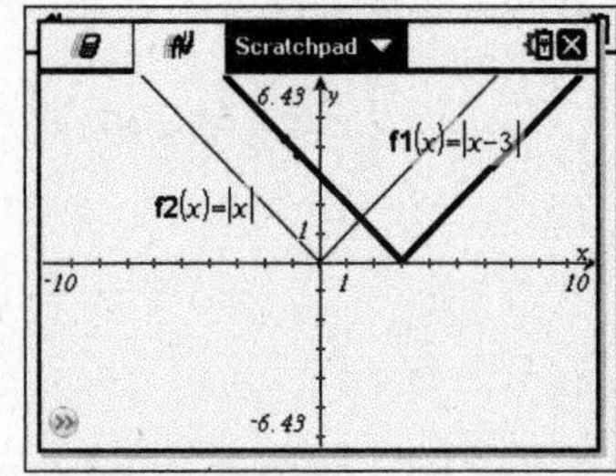

26. The cost of x adult tickets is $6.50x$. The cost of 12 student tickets is $(4.00)(12) = 48.00$. So a function that represents the total cost for x adults and 12 children is $f(x) = 6.50x + 48$ or $f(x) = 6.50x + 4(12)$.

27. They will have the same amount of money for an x-value where $f(x) = g(x)$.

$$
\begin{aligned}
60 + 5x &= x^2 + 46 \\
-60 - 5x &= \quad -60 - 5x \\
0 &= x^2 - 5x - 14 \\
0 &= (x-7)(x+2) \\
x - 7 = 0 &\text{ or } x + 2 = 0 \\
x = 7 &\text{ or } \quad x = -2
\end{aligned}
$$

Since x represents a number of weeks, it must be positive. So the answer is $x = 7$.

You also have to explain how you arrived at your answer. If you did the equation as described above, you would need to write something like "I set the two expressions equal to each other and solved for x."

Another way you could have solved this question, since your are not required to use algebra is to guess a few possible values for x until you find one that makes both functions evaluate to the same number. In this case by testing $x = 7$, both $f(7)$ and $g(7)$ evaluate to 95.

28. To calculate the difference, be sure to distribute the negative sign through the second expression.

$$
\begin{aligned}
&(3x^2 - 2x + 5) - (x^2 + 3x - 2) \\
&3x^2 - 2x + 5 - x^2 - 3x + 2 \\
&2x^2 - 5x + 7
\end{aligned}
$$

Then multiply this difference by $\frac{1}{2}x^2$ by using the distributive property.

$$\frac{1}{2}x^2(2x^2 - 5x + 7) = x^4 - \frac{5}{2}x^3 + \frac{7}{2}x^2$$

29. At the end of the first year, Dylan earned \$600(0.016) interest. So he had \$600 + \$9.60 = \$609.60 after 1 year. At the end of the second year, Dylan earned \$609.60(0.016) = \$9.7536. So he had \$609.60 + \$9.7536 = \$619.3536, which rounds to \$619.35.

    A faster way to get to this answer is to calculate $\$600(1.016)^2 \approx \$619.35$.

30. First isolate the x by combining like terms.

$$-3x + 7 - 5x < 15$$
$$-8x + 7 < 15$$

    Then subtract 7 from both sides of the inequality.

$$-8x + 7 < 15$$
$$-8x + 7 - 7 < 15 - 7$$
$$-8x < 8$$

    In an inequality when you divide both sides by a negative, the direction of the inequality sign must be changed.

$$\frac{-8x}{-8} < \frac{8}{-8}$$
$$x > -1$$

    This inequality is true for all numbers greater than –1. Of all the integers greater than –1, 0 is the smallest. So the answer to this question is 0.

    If the > sign were instead a ≥ sign, then the answer would be –1.

31. If a scatter plot is a good fit for a set of bivariate data, the residual plot will look like a random collection of points with no obvious pattern. Graph *A* does look like a random collection of points. In constrast, graph *B* seems to have some kind of upside-down "V" shape. So graph *A* is the one for which the model for the data is a good fit.

32. Since the question is asking you to find the width, call the width x. If the width is represented by x then the length can be represented by $2x$. The formula for area of a rectangle is $A = lw$ and the area is given as 34.

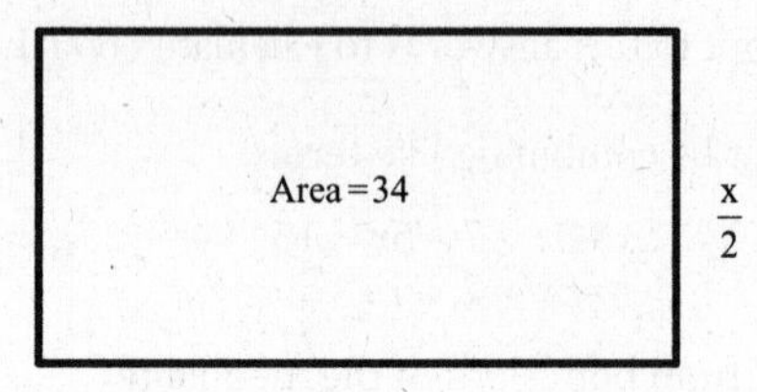

The length can be calculated with the equation

$$34 = 2x \cdot x$$

$$34 = 2x^2$$

$$\frac{34}{2} = \frac{2x^2}{2}$$

$$17 = x^2$$

$$\pm\sqrt{17} = \sqrt{x^2}$$

$$\pm 4.1 \approx x$$

The width is approximately 4.1 feet.

PART III

33. The first system can be solved by the elimination method.

$$8x + 9y = 48$$
$$12x + 5y = 21$$

To eliminate the x-variable, multiply both sides of the first equation by –12, which is the opposite of the coefficient of the x in the second equation. Multiply both sides of the second equation by 8, which is the coefficient of the x in the first equation.

$$-12(8x + 9y) = -12(48)$$
$$8(12x + 5y = 8(21)$$

$$-96x - 108y = -576$$
$$96x + 40y = 168$$

Add the two equations together to get an equation with just one variable. Solve for that variable.

$$\frac{-68y}{-68} = \frac{-408}{-68}$$
$$y = 6$$

Substitute $y = 6$ into either of the original equations to solve for x.

$$8x + 9(6) = 48$$
$$8x + 54 = 48$$
$$-54 = -54$$
$$\frac{8x}{8} = \frac{-6}{8}$$
$$x = -\frac{3}{4}$$

For the second system of equations, notice that the first equation is the same as the first equation in the first system. The second equation can be solved by dividing both sides of the equation by –8.5.

$$\frac{-8.5y}{-8.5} = \frac{-51}{-8.5}$$
$$y = 6$$

It is not necessary to substitute this y-value back into the first equation to solve for x. We already know that the answer will be $x = -\frac{3}{4}$ because it is the same equation as in the first system.

So Albert is correct. These two systems of equations have the same solution.

34. Function $w(x)$ is known as a piecewise function. Which rule is used to evaluate the function depends on which number is input.

To calculate $w(52)$, use the second rule. To calculate $w(38)$, use the first rule.

$$\begin{aligned} w(52) &= 15(52-40)+400 \\ &= 15(12)+400 \\ &= 580 \\ w(38) &= 10(38) \\ &= 380 \end{aligned}$$

The difference is $580 - 380 = 200$.

To make \$445, the employee would have to work for more than 40 hours since working exactly 40 hours would earn \$400 using the first rule in the piecewise function.

To find the number of hours needed to make \$445, set up the equation using the second rule.

$$\begin{aligned} 445 &= 15(x-40)+400 \\ 445 &= 15x-600+400 \\ 445 &= 15x-200 \\ +200 &= \quad +200 \\ \frac{645}{15} &= \frac{15x}{15} \\ 43 &= x \end{aligned}$$

The employee needs to work 43 hours to earn \$445.

35. Call the number of printers sold x and the number of computers sold y since that is the way the axes are labeled.

Since the store can sell at most 15 items, one of the inequalities is

$$x + y \leq 15$$

The cost of x printers is $50x$. The cost of y computers is $500y$. Together, the cost of x printers and y computers is $50x + 500y$. Since the store needs to sell at least \$2500 worth of items, the other inequality is

$$50x + 500y \geq 2500$$

To graph the first inequality, graph the line $x + y = 15$. First change to slope-intercept form $y = -x + 15$. Make the line solid (not dotted) since the sign is a $\leq$ (rather than a $<$). To decide which side of the line to shade, test to see if the point (0,0) makes $x + y \leq 15$ true. It is true that $0 + 0 \leq 15$, so shade the side of the line that contains (0,0).

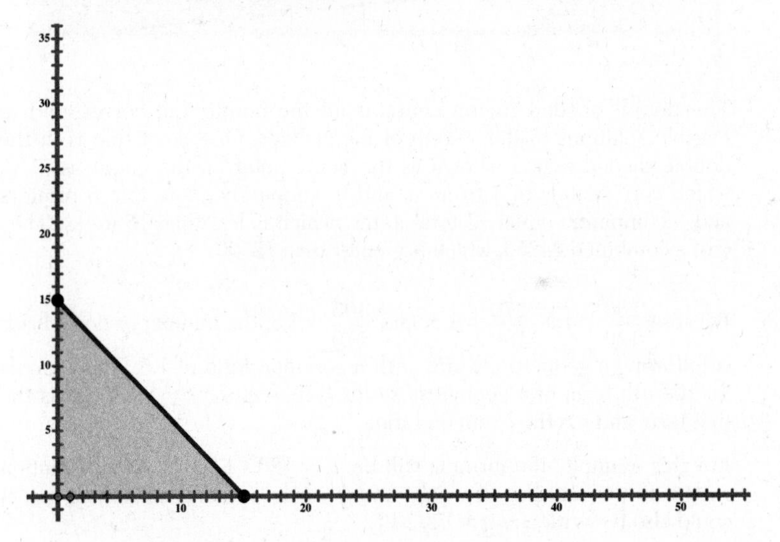

To graph the second inequality, graph the line $50x + 500y = 2500$. First change to slope-intercept form $y = -\frac{1}{10}x + 5$. Make the line solid. Then test to see if the point (0, 0) makes $50x + 500y \geq 2500$ true. Since it is not true that $0 + 0 \geq 2500$, shade the side of the line that does not contain (0,0).

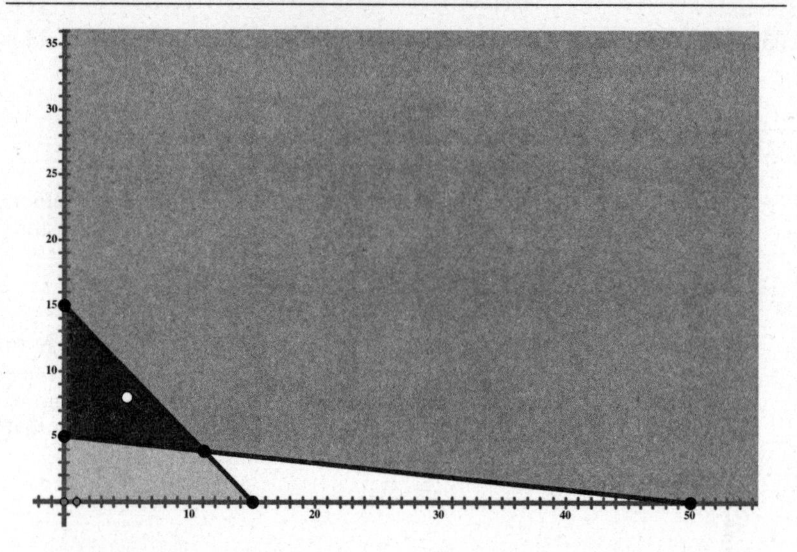

The double-shaded region contains all the points that correspond to feasible solutions to the system of inequalities. One point that is in the double-shaded region (shown as the white point on the graph) is (5,8), which corresponds to 5 printers and 8 computers. Note that 5 printers and 8 computers equal 13 total items, which is less than 15 items. They cost a combined \$4250, which is greater than \$2500.

36. Because $\frac{180}{120} = 1.5$, $\frac{270}{180} = 1.5$, and $\frac{405}{270} = 1.5$, the number of down loads is following a geometric series with a common ratio of 1.5. The formula for the nth term of a geometric series is $a_n = a_1 \cdot r^{n-1}$ where a_1 is the first term and r is the common ratio.

For this example, the formula will be $a_n = 120 \cdot 1.5^{n-1}$. As an equation with x as the number of weeks and y as the number of downloads, it could also be written as $y = 120 \cdot 1.5^{x-1}$.

Using this formula with $x = 26$ becomes

$$y = 120 \cdot 1.5^{26-1} = 120 \cdot 1.5^{25} \approx 3{,}030{,}140.$$

You could also use the exponential regression feature of your calculator. You would get a different but equivalent equation, $y = 80 \cdot 1.5^x$. You would still get the correct answer for $x = 26$; $y = 80 \cdot 1.5^{26} \approx 3{,}030{,}140$.

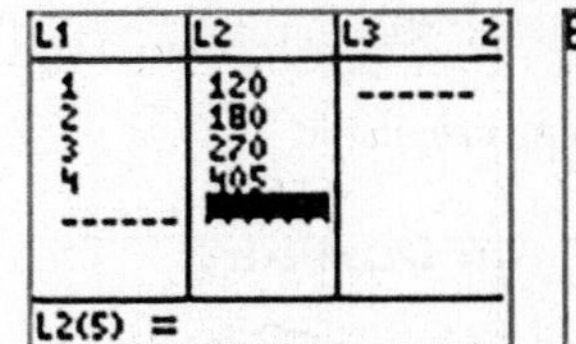

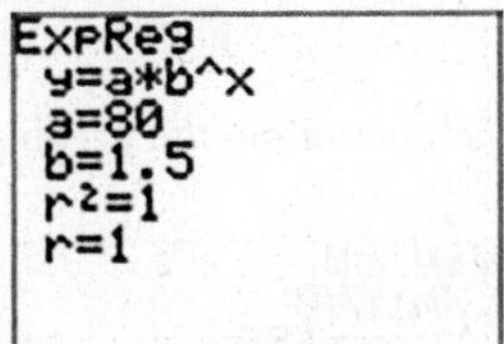

1.1 *Unsaved

	B y	C	D	E
◆				=ExpReg(
1	120		Title	Exponen...
2	180		RegEqn	a*b^x
3	270		a	80.
4	405		b	1.5
5			r²	1.

E1 ="Exponential Regression"

It is not reasonable to use this model for large numbers since viral apps tend to lose popularity after an initial wave of interest.

PART IV

37. Graph this function on the graphing calculator.

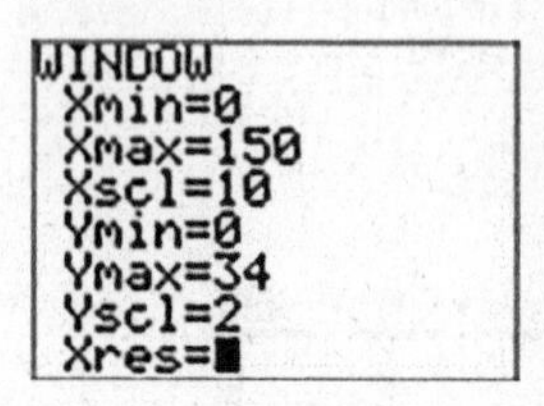

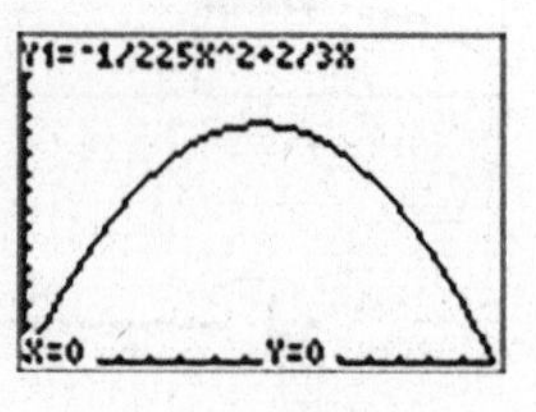

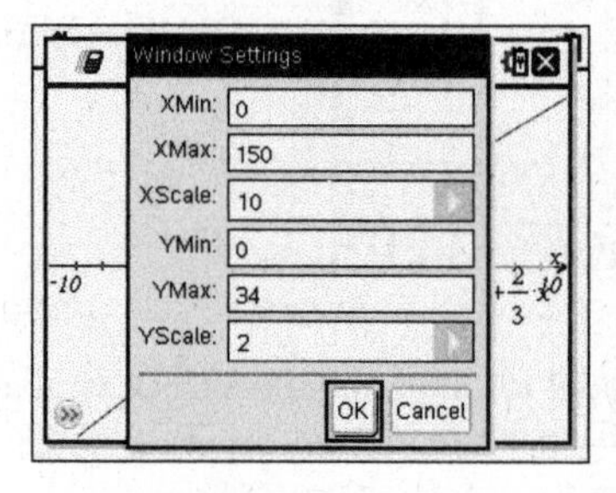

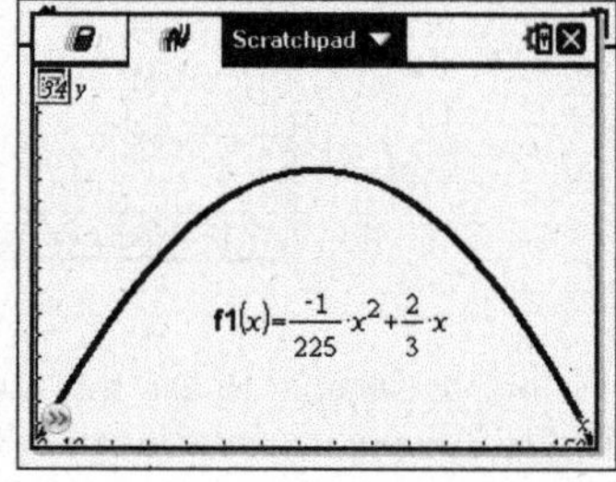

To draw this graph on the axes provided, use the table feature on the graphing calculator to find the location of all the points with x-coordinates 10, 20, 30, up to 150.

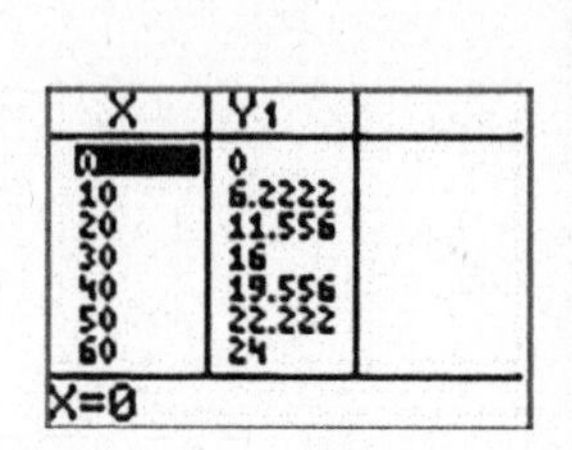

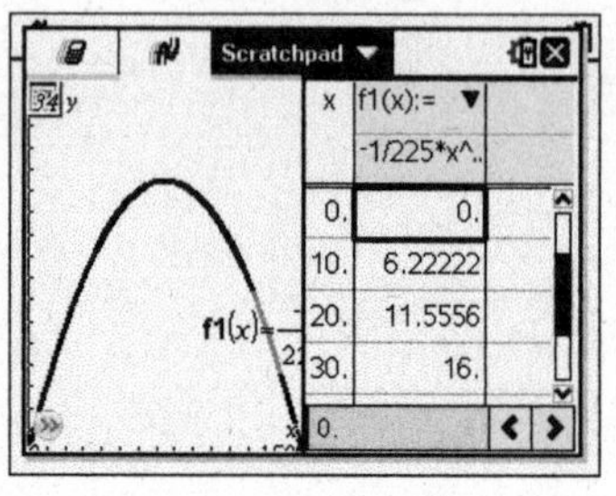

The vertex can be found by using the "maximum" feature of the calculator.

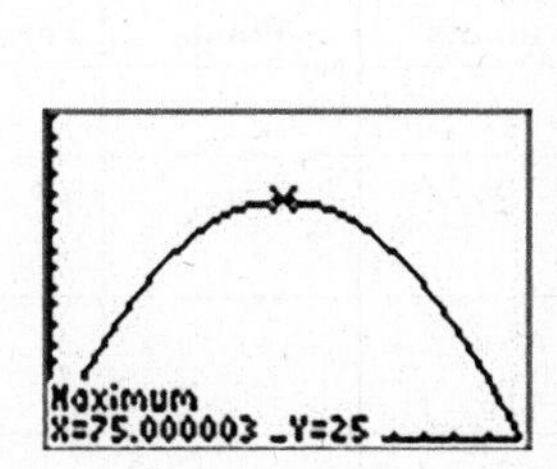

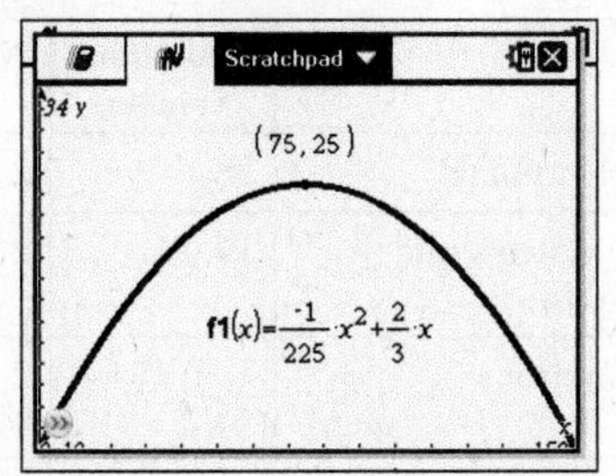

The x-coordinate of the vertex can also be found by using the formula $x = -\frac{b}{2a}$, where a is the coefficient of the x^2-term and b is the coefficient of the x-term. For this example, $x = \frac{\frac{2}{3}}{2\left(-\frac{1}{225}\right)} = 75$. Substitute the 75 for x into the $h(x)$ equation to find $h(75) = 25$, which is the maximum height. This means that the football is 25 feet high when it has traveled 75 feet horizontally.

To see if the ball will be high enough to pass over the goal post, remember that the function measures distance in feet. So 45 yards away is 135 feet. Calculate $h(135)$ and compare it to 10 feet.

$$\begin{aligned} h(135) &= -\frac{1}{225} \cdot 135^2 + \frac{2}{3} \cdot 135 \\ &= 9 \end{aligned}$$

The football will be 9 feet high after it has traveled 135 feet horizontally. So it will not be higher than the goal post, which is 10 feet high.

Topic	Question Numbers	Number of Points	Your Points	Your Percentage
1. Polynomials	28	2		
2. Properties of Algebra	19, 21	2 + 2 =4		
3. Functions	4, 9, 15, 34	2 +2 + 2 + 4 =10		
4. Creating and Interpreting Equations	1, 17, 26, 29	2 + 2 + 2 + 2 =8		
5. Inequalities	5, 24, 30, 35	2 + 2 + 2 + 4 =10		
6. Sequences and Series	22	2		
7. Systems of Equations	6, 33	2 + 4 =6		
8. Quadratic Equations and Factoring	3, 10, 14, 18, 23, 27, 32, 37	2 + 2 + 2 + 2 + 2 + 2 + 2 + 6 =20		
9. Regression	16, 31	2 + 2 =4		
10. Exponential Equations	13, 36	2 + 4 =6		
11. Graphing	2, 7, 11, 12, 25	2 + 2 + 2 + 2 + 2 = 10		
12. Statistics	20	2		
13. Number Properties	8	2		

HOW TO CONVERT YOUR RAW SCORE TO YOUR ALGEBRA I REGENTS EXAMINATION SCORE

The accompanying conversion chart must be used to determine your final score on the June 2015 Regents Examination in Algebra I. To find your final exam score, locate in the column labeled "Raw Score" the total number of points you scored out of a possible 86 points. Since partial credit is allowed in Parts II, III, and IV of the test, you may need to approximate the credit you would receive for a solution that is not completely correct. Then locate in the adjacent column to the right the scale score that corresponds to your raw score. The scale score is your final Algebra I Regents Examination score.

Regents Examination in Algebra I—June 2015 Chart for Converting Total Test Raw Scores to Final Examination Scores (Scaled Scores)

Raw Score	Scale Score	Performance Level	Raw Score	Scale Score	Performance Level	Raw Score	Scale Score	Performance Level
86	100	5	57	74	4	28	63	2
85	99	5	56	73	3	27	62	2
84	98	5	55	73	3	26	61	2
83	96	5	54	73	3	25	60	2
82	94	5	53	73	3	24	59	2
81	93	5	52	72	3	23	58	2
80	91	5	51	72	3	22	57	2
79	90	5	50	72	3	21	56	2
78	88	5	49	72	3	20	55	2
77	87	5	48	71	3	19	53	1
76	86	5	47	71	3	18	51	1
75	85	5	46	71	3	17	49	1
74	83	4	45	71	3	16	48	1
73	82	4	44	70	3	15	46	1
72	82	4	43	70	3	14	44	1
71	81	4	42	70	3	13	41	1
70	80	4	41	69	3	12	39	1
69	79	4	40	69	3	11	37	1
68	79	4	39	69	3	10	34	1
67	78	4	38	68	3	9	31	1
66	77	4	37	68	3	8	28	1
65	77	4	36	68	3	7	25	1
64	76	4	35	67	3	6	22	1
63	76	4	34	67	3	5	19	1
62	75	4	33	66	3	4	15	1
61	75	4	32	66	3	3	12	1
60	75	4	31	65	3	2	8	1
59	74	4	30	64	2	1	4	1
58	74	4	29	64	2	0	0	1

Examination August 2015

Algebra I

HIGH SCHOOL MATH REFERENCE SHEET

Conversions

1 inch = 2.54 centimeters
1 meter = 39.37 inches
1 mile = 5280 feet
1 mile = 1760 yards
1 mile = 1.609 kilometers

1 kilometer = 0.62 mile
1 pound = 16 ounces
1 pound = 0.454 kilogram
1 kilogram = 2.2 pounds
1 ton = 2000 pounds

1 cup = 8 fluid ounces
1 pint = 2 cups
1 quart = 2 pints
1 gallon = 4 quarts
1 gallon = 3.785 liters
1 liter = 0.264 gallon
1 liter = 1000 cubic centimeters

Formulas

Triangle	$A = \frac{1}{2}bh$
Parallelogram	$A = bh$
Circle	$A = \pi r^2$
Circle	$C = \pi d$ or $C = 2\pi r$

Formulas (continued)

General Prisms	$V = Bh$
Cylinder	$V = \pi r^2 h$
Sphere	$V = \frac{4}{3}\pi r^3$
Cone	$V = \frac{1}{3}\pi r^2 h$
Pyramid	$V = \frac{1}{3}Bh$
Pythagorean Theorem	$a^2 + b^2 = c^2$
Quadratic Formula	$x = \frac{-b \pm \sqrt{b^2 - 4ac}}{2a}$
Arithmetic Sequence	$a_n = a_1 + (n - 1)d$
Geometric Sequence	$a_n = a_1 r^{n-1}$
Geometric Series	$S_n = \frac{a_1 - a_1 r^n}{1 - r}$ where $r \neq 1$
Radians	1 radian $= \frac{180}{\pi}$ degrees
Degrees	1 degree $= \frac{\pi}{180}$ radians
Exponential Growth/Decay	$A = A_0 e^{k(t - t_0)} + B_0$

PART I

Answer all 24 questions in this part. Each correct answer will receive 2 credits. No partial credit will be allowed. For each statement or question, write in the space provided the numeral preceding the word or expression that best completes the statement or answers the question. [48 credits]

1 Given the graph of the line represented by the equation $f(x) = -2x + b$, if b is increased by 4 units, the graph of the new line would be shifted 4 units

(1) right (3) left
(2) up (4) down 1 ____

2 Rowan has $50 in a savings jar and is putting in $5 every week. Jonah has $10 in his own jar and is putting in $15 every week. Each of them plots his progress on a graph with time on the horizontal axis and amount in the jar on the vertical axis. Which statement about their graphs is true?

(1) Rowan's graph has a steeper slope than Jonah's.
(2) Rowan's graph always lies above Jonah's.
(3) Jonah's graph has a steeper slope than Rowan's.
(4) Jonah's graph always lies above Rowan's. 2 ____

3 To watch a varsity basketball game, spectators must buy a ticket at the door. The cost of an adult ticket is $3.00 and the cost of a student ticket is $1.50. If the number of adult tickets sold is represented by a and student tickets sold by s, which expression represents the amount of money collected at the door from the ticket sales?

(1) $4.50as$ (3) $(3.00a)(1.50s)$
(2) $4.50(a + s)$ (4) $3.00a + 1.50s$ 3 ____

4 The graph of $f(x)$ is shown below.

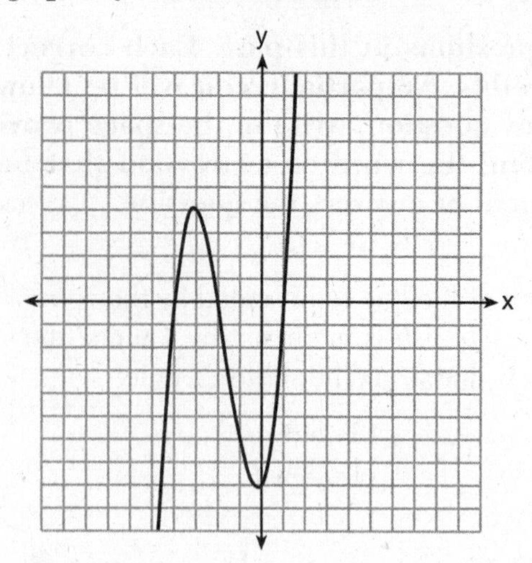

Which function could represent the graph of $f(x)$?

(1) $f(x) = (x + 2)(x^2 + 3x - 4)$
(2) $f(x) = (x - 2)(x^2 + 3x - 4)$
(3) $f(x) = (x + 2)(x^2 + 3x + 4)$
(4) $f(x) = (x - 2)(x^2 + 3x + 4)$ 4 ____

5 The cost of a pack of chewing gum in a vending machine is \$0.75. The cost of a bottle of juice in the same machine is \$1.25. Julia has \$22.00 to spend on chewing gum and bottles of juice for her team and she must buy seven packs of chewing gum. If b represents the number of bottles of juice, which inequality represents the maximum number of bottles she can buy?

(1) $0.75b + 1.25(7) \geq 22$
(2) $0.75b + 1.25(7) \leq 22$
(3) $0.75(7) + 1.25b \geq 22$
(4) $0.75(7) + 1.25b \leq 22$ 5 ____

6 Which graph represents the solution of $y \leq x + 3$ and $y \geq -2x - 2$?

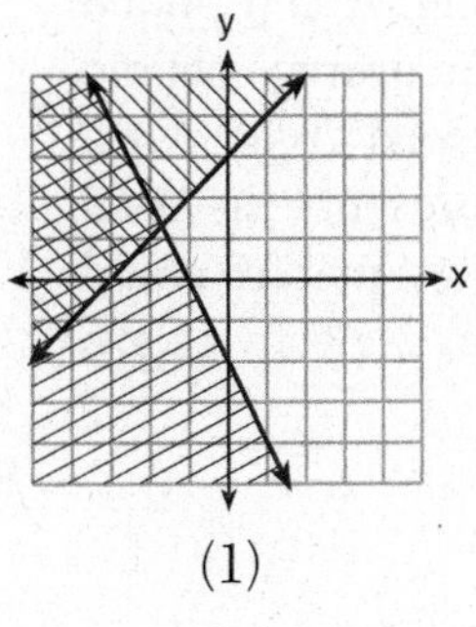

(1)

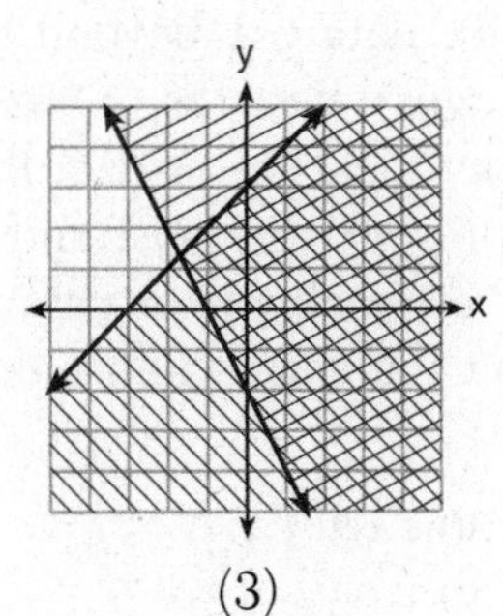

(3)

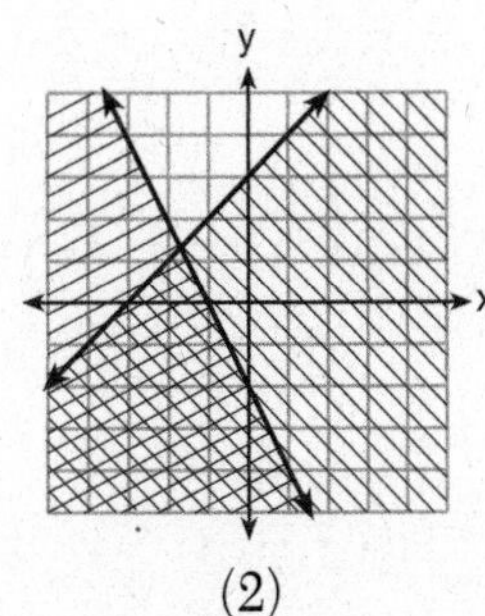

(2)

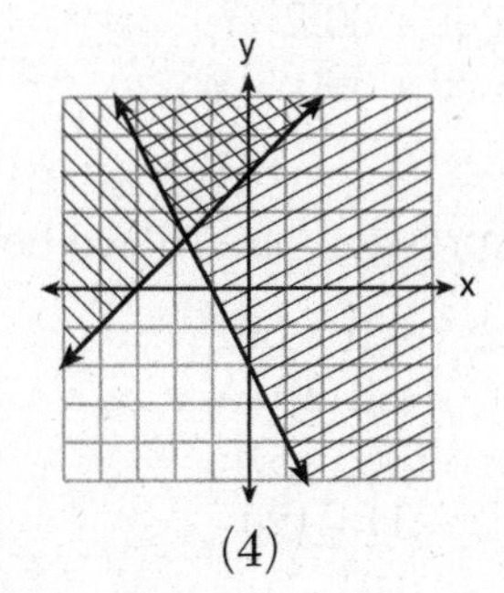

(4)

6 ____

7 The country of Benin in West Africa has a population of 9.05 million people. The population is growing at a rate of 3.1% each year. Which function can be used to find the population 7 years from now?

(1) $f(t) = (9.05 \times 10^6)(1 - 0.31)^7$
(2) $f(t) = (9.05 \times 10^6)(1 + 0.31)^7$
(3) $f(t) = (9.05 \times 10^6)(1 + 0.031)^7$
(4) $f(t) = (9.05 \times 10^6)(1 - 0.031)^7$

7 ____

8 A typical cell phone plan has a fixed base fee that includes a certain amount of data and an overage charge for data use beyond the plan. A cell phone plan charges a base fee of \$62 and an overage charge of \$30 per gigabyte of data that exceed 2 gigabytes. If C represents the cost and g represents the total number of gigabytes of data, which equation could represent this plan when more than 2 gigabytes are used?

(1) $C = 30 + 62(2 - g)$
(2) $C = 30 + 62(g - 2)$
(3) $C = 62 + 30(2 - g)$
(4) $C = 62 + 30(g - 2)$

8 _____

9 Four expressions are shown below.

I. $2(2x^2 - 2x - 60)$
II. $4(x^2 - x - 30)$
III. $4(x + 6)(x - 5)$
IV. $4x(x - 1) - 120$

The expression $4x^2 - 4x - 120$ is equivalent to

(1) I and II, only
(2) II and IV, only
(3) I, II, and IV
(4) II, III, and IV

9 _____

10 Last week, a candle store received \$355.60 for selling 20 candles. Small candles sell for \$10.98 and large candles sell for \$27.98. How many large candles did the store sell?

(1) 6
(2) 8
(3) 10
(4) 12

10 _____

11 Which representations are functions?

I

x	y
2	6
3	–12
4	7
5	5
2	–6

III

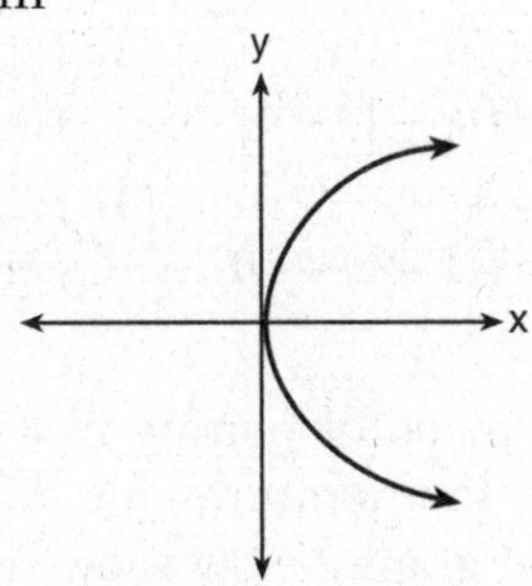

II {(1,1), (2,1), (3,2), (4,3), (5,5), (6,8), (7,13)} IV $y = 2x + 1$

(1) I and II (3) II, only
(2) II and IV (4) IV, only 11 ____

12 If $f(x) = \dfrac{\sqrt{2x+3}}{6x-5}$, then $f\left(\dfrac{1}{2}\right) =$

(1) 1 (3) –1
(2) –2 (4) $-\dfrac{13}{3}$ 12 ____

13 The zeros of the function $f(x) = 3x^2 - 3x - 6$ are

(1) –1 and –2 (3) 1 and 2
(2) 1 and –2 (4) –1 and 2 13 ____

14 Which recursively defined function has a first term equal to 10 and a common difference of 4?

(1) $f(1) = 10$
$f(x) = f(x - 1) + 4$

(2) $f(1) = 4$
$f(x) = f(x - 1) + 10$

(3) $f(1) = 10$
$f(x) = 4f(x - 1)$

(4) $f(1) = 4$
$f(x) = 10f(x - 1)$

14 ____

15 Firing a piece of pottery in a kiln takes place at different temperatures for different amounts of time. The graph below shows the temperatures in a kiln while firing a piece of pottery after the kiln is preheated to 200°F.

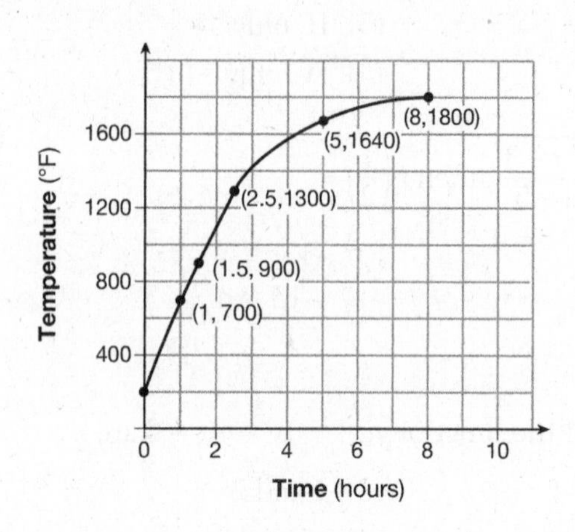

During which time interval did the temperature in the kiln show the greatest average rate of change?

(1) 0 to 1 hour
(2) 1 hour to 1.5 hours
(3) 2.5 hours to 5 hours
(4) 5 hours to 8 hours

15 ____

16 Which graph represents $f(x) = \begin{cases} |x| & x < 1 \\ \sqrt{x} & x \geq 1 \end{cases}$?

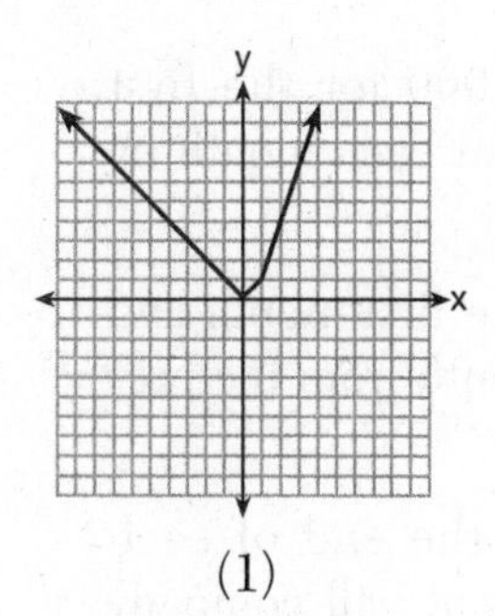

(1)

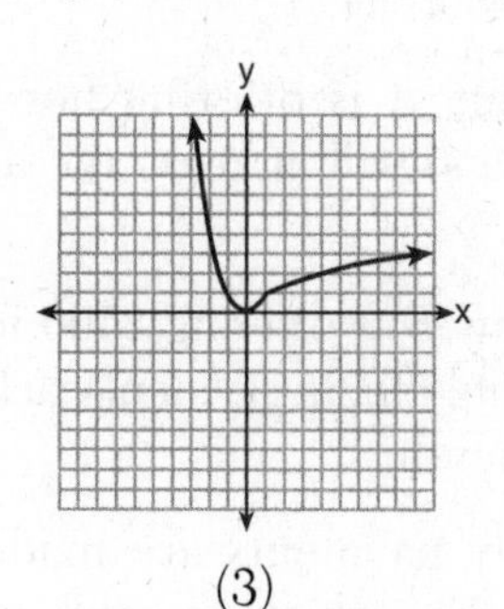

(3)

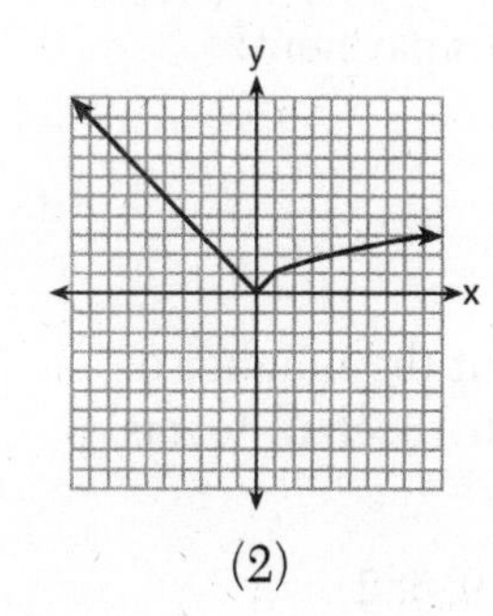

(2)

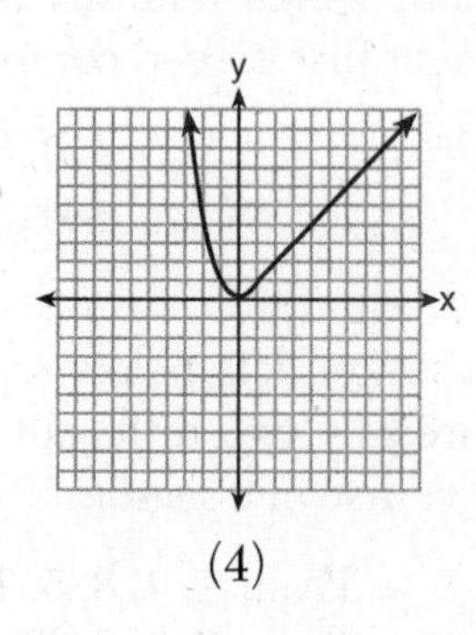

(4)

16 ____

17 If $f(x) = x^2 - 2x - 8$ and $g(x) = \frac{1}{4}x - 1$, for which values of x is $f(x) = g(x)$?

(1) −1.75 and −1.438 (3) −1.438 and 0
(2) −1.75 and 4 (4) 4 and 0

17 ____

18 Alicia has invented a new app for smart phones that two companies are interested in purchasing for a 2-year contract.

Company *A* is offering her \$10,000 for the first month and will increase the amount each month by \$5000.

Company *B* is offering \$500 for the first month and will double their payment each month from the previous month.

Monthly payments are made at the end of each month. For which monthly payment will company *B*'s payment first exceed company *A*'s payment?

(1) 6 (3) 8
(2) 7 (4) 9

18 ____

19 The two sets of data below represent the number of runs scored by two different youth baseball teams over the course of a season.

Team *A*: 4, 8, 5, 12, 3, 9, 5, 2
Team *B*: 5, 9, 11, 4, 6, 11, 2, 7

Which set of statements about the mean and standard deviation is true?

(1) mean $A <$ mean B
standard deviation $A >$ standard deviation B
(2) mean $A >$ mean B
standard deviation $A <$ standard deviation B
(3) mean $A <$ mean B
standard deviation $A <$ standard deviation B
(4) mean $A >$ mean B
standard deviation $A >$ standard deviation B

19 ____

20 If Lylah completes the square for $f(x) = x^2 - 12x + 7$ in order to find the minimum, she must write $f(x)$ in the general form $f(x) = (x - a)^2 + b$. What is the value of a for $f(x)$?

(1) 6
(2) –6
(3) 12
(4) –12

20 _____

21 Given the following quadratic functions:

$$g(x) = -x^2 - x + 6$$

and

x	–3	–2	–1	0	1	2	3	4	5
n(x)	–7	0	5	8	9	8	5	0	–7

Which statement about these functions is true?

(1) Over the interval $-1 \leq x \leq 1$, the average rate of change for $n(x)$ is less than that for $g(x)$.
(2) The y-intercept of $g(x)$ is greater than the y-intercept for $n(x)$.
(3) The function $g(x)$ has a greater maximum value than $n(x)$.
(4) The sum of the roots of $n(x) = 0$ is greater than the sum of the roots of $g(x) = 0$.

21 _____

22 For which value of P and W is $P + W$ a rational number?

(1) $P = \frac{1}{\sqrt{3}}$ and $W = \frac{1}{\sqrt{6}}$

(2) $P = \frac{1}{\sqrt{4}}$ and $W = \frac{1}{\sqrt{9}}$

(3) $P = \frac{1}{\sqrt{6}}$ and $W = \frac{1}{\sqrt{10}}$

(4) $P = \frac{1}{\sqrt{25}}$ and $W = \frac{1}{\sqrt{2}}$

22 ____

23 The solution of the equation $(x + 3)^2 = 7$ is

(1) $3 \pm \sqrt{7}$
(3) $-3 \pm \sqrt{7}$
(2) $7 \pm \sqrt{3}$
(4) $-7 \pm \sqrt{3}$

23 ____

24 Which trinomial is equivalent to $3(x - 2)^2 - 2(x - 1)$?

(1) $3x^2 - 2x - 10$
(3) $3x^2 - 14x + 10$
(2) $3x^2 - 2x - 14$
(4) $3x^2 - 14x + 14$

24 ____

PART II

Answer all 8 questions in this part. Each correct answer will receive 2 credits. Clearly indicate the necessary steps, including appropriate formula substitutions, diagrams, graphs, charts, etc. For all questions in this part, a correct numerical answer with no work shown will receive only 1 credit. [16 credits]

25 Each day Toni records the height of a plant for her science lab. Her data are shown in the table below.

Day (n)	1	2	3	4	5
Height (cm)	3.0	4.5	6.0	7.5	9.0

The plant continues to grow at a constant daily rate. Write an equation to represent $h(n)$, the height of the plant on the nth day.

26 On the set of axes below, graph the inequality $2x + y > 1$.

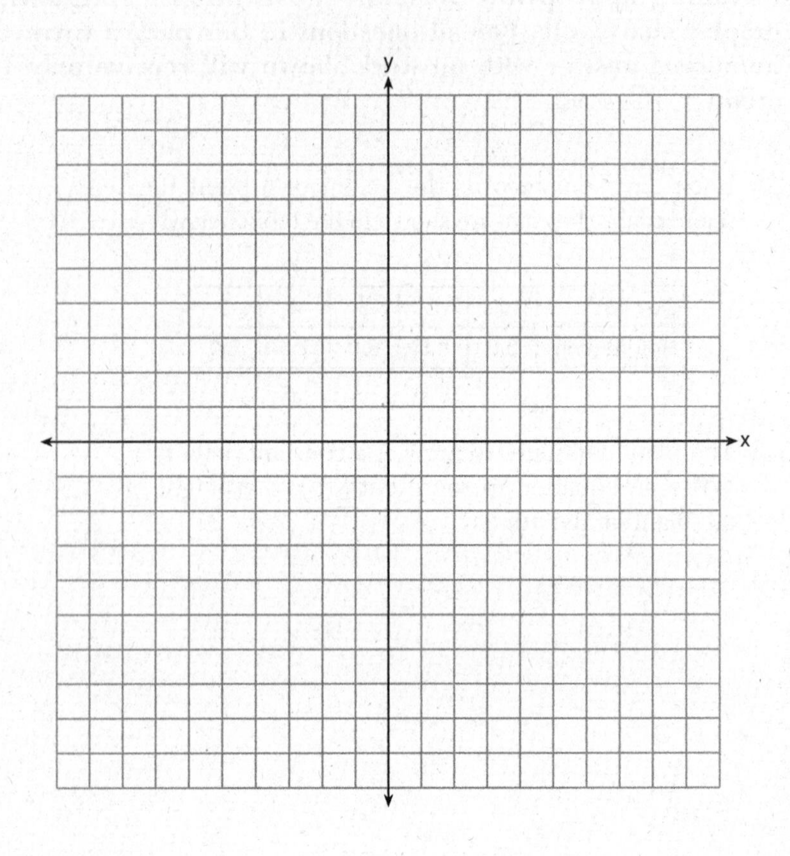

27 Rachel and Marc were given the information shown below about the bacteria growing in a Petri dish in their biology class.

Number of Hours, x	1	2	3	4	5	6	7	8	9	10
Number of Bacteria, $B(x)$	220	280	350	440	550	690	860	1070	1340	1680

Rachel wants to model this information with a linear function. Marc wants to use an exponential function. Which model is the better choice? Explain why you chose this model.

28 A driver leaves home for a business trip and drives at a constant speed of 60 miles per hour for 2 hours. Her car gets a flat tire, and she spends 30 minutes changing the tire. She resumes driving and drives at 30 miles per hour for the remaining one hour until she reaches her destination.

On the set of axes below, draw a graph that models the driver's distance from home.

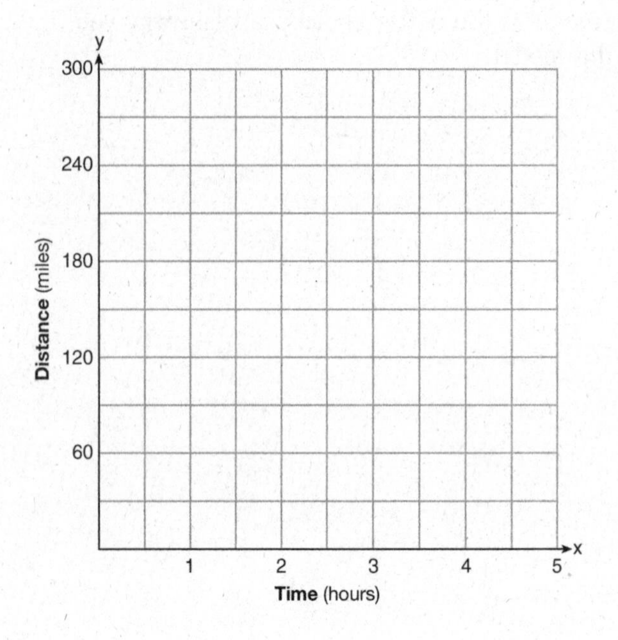

29 How many real solutions does the equation $x^2 - 2x + 5 = 0$ have? Justify your answer.

30 The number of carbon atoms in a fossil is given by the function $y = 5100(0.95)^x$, where x represents the number of years since being discovered.

What is the percent of change each year? Explain how you arrived at your answer.

31 A toy rocket is launched from the ground straight upward. The height of the rocket above the ground, in feet, is given by the equation $h(t) = -16t^2 + 64t$, where t is the time in seconds.

Determine the domain for this function in the given context. Explain your reasoning.

32 Jackson is starting an exercise program. The first day he will spend 30 minutes on a treadmill. He will increase his time on the treadmill by 2 minutes each day. Write an equation for $T(d)$, the time, in minutes, on the treadmill on day d.

Find $T(6)$, the minutes he will spend on the treadmill on day 6.

PART III

Answer all 4 questions in this part. Each correct answer will receive 4 credits. Clearly indicate the necessary steps, including appropriate formula substitutions, diagrams, graphs, charts, etc. For all questions in this part, a correct numerical answer with no work shown will receive only 1 credit. [16 credits]

33 Graph $f(x) = x^2$ and $g(x) = 2^x$ for $x \geq 0$ on the set of axes below.

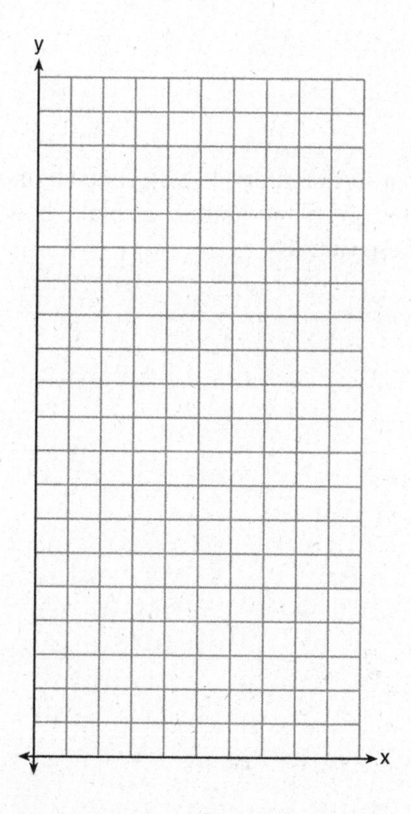

State which function, $f(x)$ or $g(x)$, has a greater value when $x = 20$. Justify your reasoning.

34 Solve for x algebraically: $7x - 3(4x - 8) \leq 6x + 12 - 9x$

If x is a number in the interval [4,8], state all integers that satisfy the given inequality. Explain how you determined these values.

35 The volume of a large can of tuna fish can be calculated using the formula $V = \pi r^2 h$. Write an equation to find the radius, r, in terms of V and h.

Determine the diameter, to the *nearest inch*, of a large can of tuna fish that has a volume of 66 cubic inches and a height of 3.3 inches.

36 The table below shows the attendance at a museum in select years from 2007 to 2013.

Attendance at Museum

Year	2007	2008	2009	2011	2013
Attendance (millions)	8.3	8.5	8.5	8.8	9.3

State the linear regression equation represented by the data table when $x = 0$ is used to represent the year 2007 and y is used to represent the attendance. Round all values to the *nearest hundredth*.

State the correlation coefficient to the *nearest hundredth* and determine whether the data suggest a strong or weak association.

PART IV

Answer the question in this part. A correct answer will receive 6 credits. Clearly indicate the necessary steps, including appropriate formula substitutions, diagrams, graphs, charts, etc. A correct numerical answer with no work shown will receive only 1 credit. [6 credits]

37 A rectangular picture measures 6 inches by 8 inches. Simon wants to build a wooden frame for the picture so that the framed picture takes up a maximum area of 100 square inches on his wall. The pieces of wood that he uses to build the frame all have the same width.

Write an equation or inequality that could be used to determine the maximum width of the pieces of wood for the frame Simon could create.

Explain how your equation or inequality models the situation.

Solve the equation or inequality to determine the maximum width of the pieces of wood used for the frame to the *nearest tenth of an inch*.

Answers August 2015

Algebra I

Answer Key

PART I

1. (2)	**5.** (4)	**9.** (3)	**13.** (4)	**17.** (2)	**21.** (4)
2. (3)	**6.** (3)	**10.** (2)	**14.** (1)	**18.** (3)	**22.** (2)
3. (4)	**7.** (3)	**11.** (2)	**15.** (1)	**19.** (1)	**23.** (3)
4. (1)	**8.** (4)	**12.** (3)	**16.** (2)	**20.** (1)	**24.** (4)

PART II

25. $h(n) = 3 + (n - 1)1.5$

26. Dotted line of $y = -2x + 1$ with shading above the line

27. Exponential model is more appropriate

28. A correct graph is drawn (see page 389)

29. No real solutions

30. 5%

31. Rocket lands after 4 seconds, domain is $0 \leq t \leq 4$

32. $T(d) = 30 + 2\,(d - 1)$, $T(6) = 40$

PART III

33. $g(20) > f(20)$

34. $x \geq 6$, 7, 8

35. $r = \sqrt{\frac{V}{\pi h}}$, 5 inches

36. $y = 0.16x + 8.27$, $r = 0.97$, strong correlation since r is close to 1

PART IV

37. $(2x + 6)(2x + 8) \leq 100$, $x = 1.5$

In Parts II–IV, you are required to show how you arrived at your answers. For sample methods of solutions, see the *Answers Explained* section.

Answers Explained

PART I

1. The y-intercept of the line is located at $(0, b)$. If b is increased by 4, the y-coordinate of the y-intercept will also be increased by 4. This is a shift of 4 units up.

   If, for example, b was originally 1, the line would have a slope of –2 and a y-intercept of $(0, 1)$. If the b-value was increased by 4, the new value of b would be 5. The new line would have a slope of –2 and a y-intercept of $(0, 5)$, which is a shift up by 4 units.

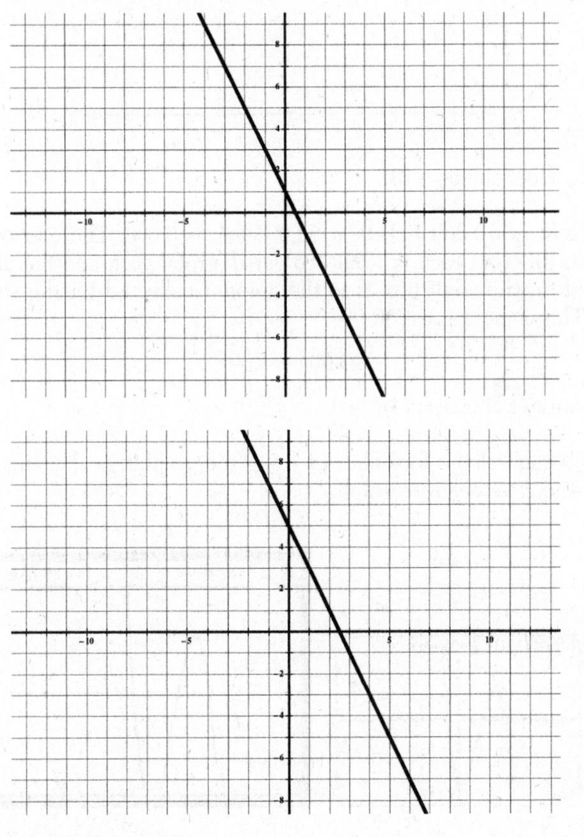

   The correct choice is **(2)**.

2. Rowan's graph can be shown by the equation $y = 50 + 5x$, while Jonah's graph can be shown by the equation $y = 10 + 15x$. Jonah's line has a slope of 15, while Rowan's line has a slope of 5. Since 15 is greater than 5, Jonah's line is steeper than Rowan's. Based on the graph, at first Rowan's line lies above Jonah's. After $x = 4$ weeks, Jonah's line lies above Rowan's.

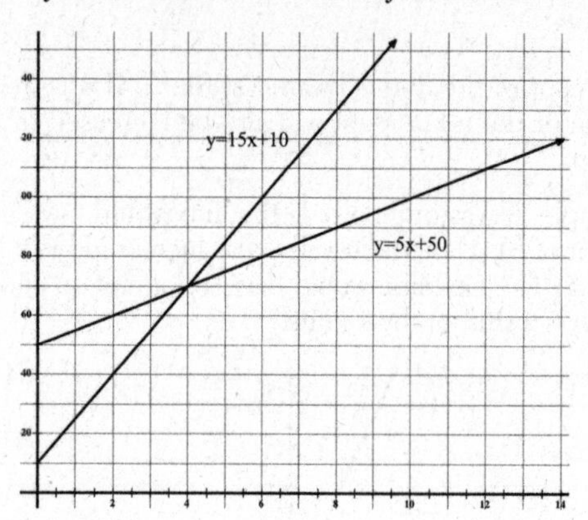

The correct choice is **(3)**.

3. The cost of a adult tickets at \$3.00 each is $3.00a$. The cost of s student tickets at \$1.50 each is $1.50s$. To get the total cost, add the amount collected for the adult tickets to the amount collected for the student tickets. This is the expression

$$3.00a + 1.50s.$$

The correct choice is **(4)**.

4. Graph each of the four choices using the graphing calculator, and compare them to the given graph.

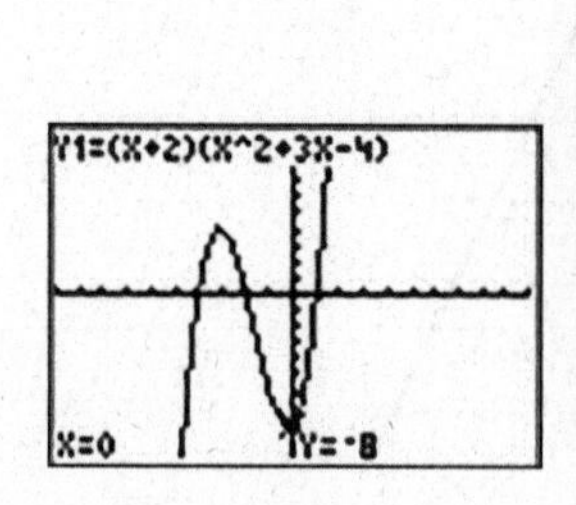

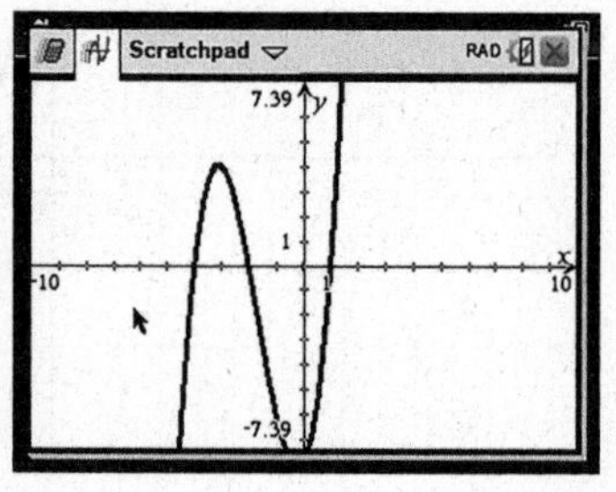

Without using the graphing calculator, the equation can be found by locating the x-intercepts of the graph. For the given graph, there are x-intercepts at $(-4, 0)$, $(-2, 0)$, and $(1, 0)$. So the function the graph is based on will have the factors $(x + 4)$, $(x + 2)$, and $(x - 1)$. In general, if there is an x-intercept at $(a, 0)$, there will be a factor of $(x - a)$.

A function that has these three factors is $f(x) = (x + 4)(x + 2)(x - 1)$. If the first and third factors are combined, the function becomes

$$f(x) = (x + 2)(x^2 + 3x - 4).$$

The correct choice is **(1)**.

5. The cost of the seven packs of chewing gum is $0.75 \cdot 7$. The cost of b bottles of juice is $1.25b$. Together the gum and the juice must cost less than or equal to \$22.00. So the inequality is $0.75(7) + 1.25b \leq 22$.

   The correct choice is **(4)**.

6. To graph the inequality $y \leq x + 3$, first graph $y = x + 3$ with a solid line (if the inequality contained a < sign instead of a $\leq$, it would require a dotted line). To decide which side of the line to shade, test to see if $(0, 0)$ makes the inequality true:

$$0 \leq 0 + 3$$
$$0 \leq 3$$

Since this is true, the side of the line containing $(0, 0)$ gets shaded.

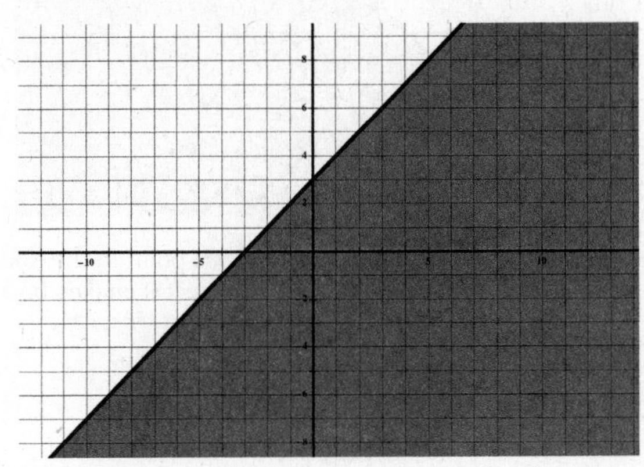

For the inequality $y \geq -2x - 2$, first graph $y = -2x - 2$ with a solid line. Test to see if $(0, 0)$ makes the inequality true:

$$0 \geq -2 \cdot 0 - 2$$
$$0 \geq -2$$

Since this is true, the side of the line containing (0, 0) gets shaded.

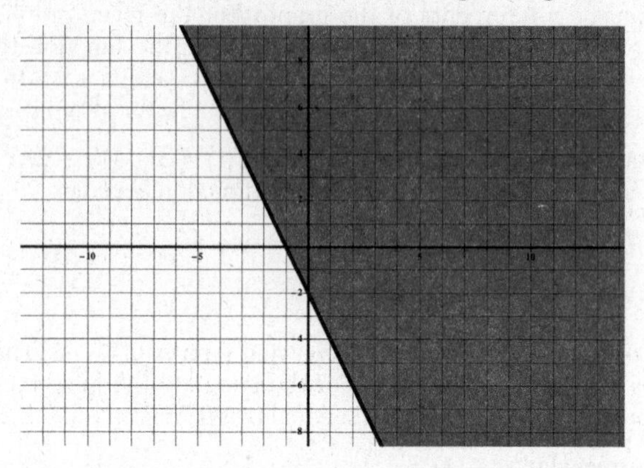

When both of these graphs are put onto the same set of axes, the graph looks like the following.

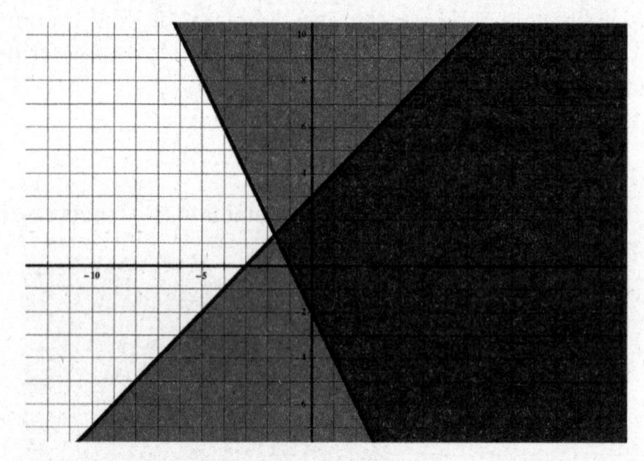

Another way to solve this question is to pick a point in the double-shaded region for each of the answer choices to see if it makes both equations true.

Testing choice (1):
(–3, 1) is a point in the double-shaded region.

$$1 \leq -3 + 3$$
$$1 \leq 0$$

This is not true, so choice (1) can be eliminated.

Testing choice (2):
(–2, –1) is a point in the double-shaded region.

$$-1 \leq -2 + 3$$
$$-1 \leq 1$$

This is true for the first inequality. Test the second inequality

$$-1 \geq -2 \cdot -2 - 2$$
$$-1 \geq 4 - 2$$
$$-1 \geq 2$$

This is not true, so choice (2) can be eliminated.

Testing choice (3):
(2, 1) is a point in the double-shaded region.

$$1 \leq 2 + 3$$
$$1 \leq 5$$

This is true for the first inequality. Test for the second inequality.

$$1 \geq -2 \cdot 2 - 2$$
$$1 \geq -4 - 2$$
$$1 \geq -6$$

This is also true, so choice (3) works.

Yet another way to answer this question is to produce the graph on a graphing calculator.

The TI-Nspire allows you to graph inequalities like this. When inputting the function, delete the = and replace with the appropriate sign.

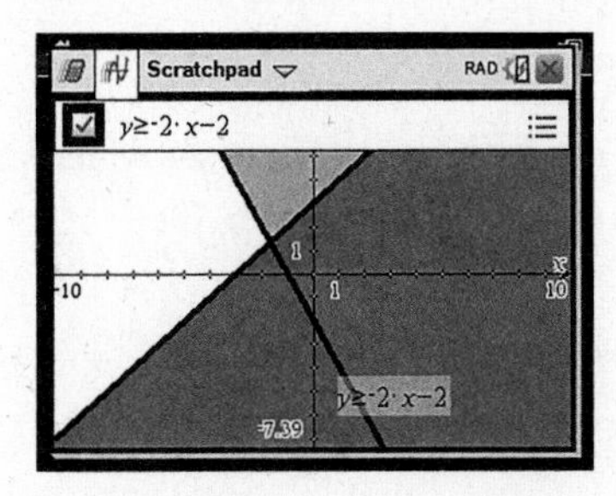

For the TI-84, you can graph the lines, but you have to determine which side of the lines to shade by testing a point like (0, 0). Then the shading can be set up in the Y= menu. Move the cursor to the \ symbol to the left of Y_1, and press [ENTER] three times to change the symbol to shade below the line. Move the cursor to the \ symbol to the left of Y2, and press [ENTER] two times to change the symbol to shade above the line.

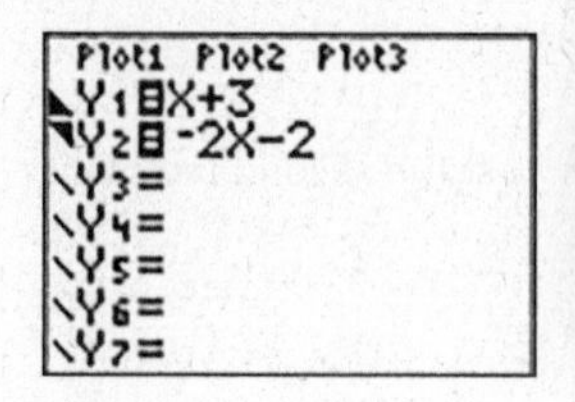

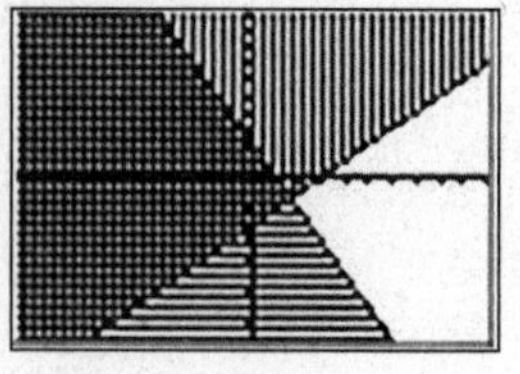

The correct choice is **(3)**.

7. An exponential function based on a real-world scenario often has the form $f(x) = a(1 + r)^x$, where a is the starting value and r is the growth rate. Note that r can be negative if the scenario is based on exponential decay. For this example, the a-value is 9.05 million which, in scientific notation, is 9.05×10^6. The r-value is 0.031 for 3.1%. (It is not 0.31 which would represent 31%.) Since the question asks for an expression for seven years from now, the value of x should be 7. The function is

$$f(t) = (9.05 \times 10^6)(1 + 0.031)^7.$$

The correct choice is **(3)**.

8. One way to solve this question is to calculate the cost for a specific number of gigabytes, for example 10. If 10 gigabytes are used, the cost will be \$62 plus \$30 · 8 or \$62 + \$240 = \$302. From the answer choices, choice (4) becomes $C = 62 + 30(10 - 2) = 62 + 30 \cdot 8$, which is the same calculation needed to find the cost for 10 gigabytes of data.

The correct choice is **(4)**.

9. Multiply each of the four expressions to see which simplify to

$4x^2 - 4x - 120$.

Testing I:
By the distributive property,

$2(2x^2 - 2x - 60) = 4x^2 - 4x - 120$

This works.

Testing II:
By the distributive property,

$4(x^2 - x - 30) = 4x^2 - 4x - 120$

This works.

Testing III:
Multiply the two binomials using the FOIL process (First, Outer, Inner, Last)

$$4(x + 6)(x - 5)$$

$$4(x \cdot x + x(-5) + 6(x) + 6(-5))$$
$$4(x^2 - 5x + 6x - 30)$$
$$4(x^2 + x - 30)$$

By the distributive property,

$4(x^2 + x - 30) = 4x^2 + 4x - 120$

This does not work because it has $+ 4x$ instead of $- 4x$.

Testing IV:
By the distributive property,

$$4x(x - 1) - 120$$
$$4x(x) - (4x)(1) - 120$$
$$4x^2 - 4x - 120$$

This works.

The correct choice is **(3)**.

10. Set up a system of equations. Let x be the number of small candles sold. Let y be the number of large candles sold.

Since 20 candles are sold, one equation is $x + y = 20$.

The cost of x small candles is $10.98x$. The cost of y large candles is $27.98y$. Since the total cost is known to be \$355.60, the other equation is $10.98x + 27.98y = 355.60$.

$$\begin{array}{rcrcl} x & + & y & = & 20 \\ 10.98x & + & 27.98y & = & 355.60 \end{array}$$

Multiply both sides of the top equation by -10.98.

$$\begin{array}{rcrcl} -10.98(x & + & y) & = & -10.98(20) \\ 10.98x & + & 27.98y & = & 355.60 \end{array}$$

$$\begin{array}{rrcrcl} & -10.98x & - & 10.98y & = & -219.60 \\ + & & & & & \\ & 10.98x & + & 27.98y & = & 355.60 \\ \hline & & & 17y & = & 136 \\ & & & \dfrac{17y}{17} & = & \dfrac{136}{17} \\ & & & y & = & 8 \end{array}$$

Since this is a multiple-choice question, there is an alternative way to solve it that does not involve creating and solving the system of equations. The correct answer can be found by testing each of the answer choices until you find one that agrees with the given information.

Testing choice (1):
If 6 large candles were sold, $20 - 6 = 14$ small candles were sold. The price of 6 large candles and 14 small candles is $6 \cdot 27.98 + 14 \cdot 10.98 = 321.60$, which is not 355.60.

Testing choice (2):
If 8 large candles were sold, $20 - 8 = 12$ small candles were sold. The price of 8 large candles and 12 small candles is $8 \cdot 27.98 + 12 \cdot 10.98 = 355.60$.

The correct choice is **(2)**.

11. For a set of ordered pairs to be a function, each x-coordinate must have one and only one y-coordinate.

Checking I:
There is a row on the chart with a 2 for the x and a 6 for the y and another row with a 2 for the x and a -6 for the y. These correspond to the ordered pairs $(2, 6)$ and $(2, -6)$. However, a function cannot have two ordered pairs with the same x-coordinate but with different y-coordinates.

This is not a function.

Checking II:
In this list of ordered pairs, each has a different x-coordinate. So this is a function. The ordered pairs $(1, 1)$ and $(2, 1)$ have the same y-coordinate, but that is permitted. Only if two ordered pairs have the same x-coordinate but different y-coordinates would it then not be a function.

This is a function.

Checking III:
This graph fails the vertical line test. Since it is possible to draw a vertical line that passes through two points of the graph, those two points have the same x-coordinate but different y-coordinates.

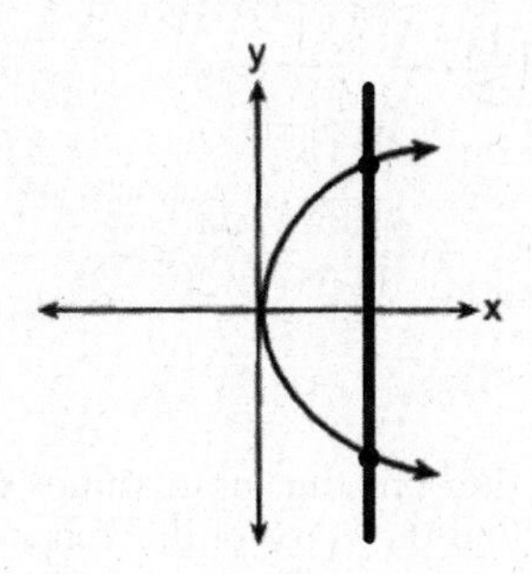

This is not a function.

Checking IV:
The graph of the solution set of the equation $y = 2x + 1$ is a line that passes the vertical line test. There is no vertical line that passes through more than one point on the line.

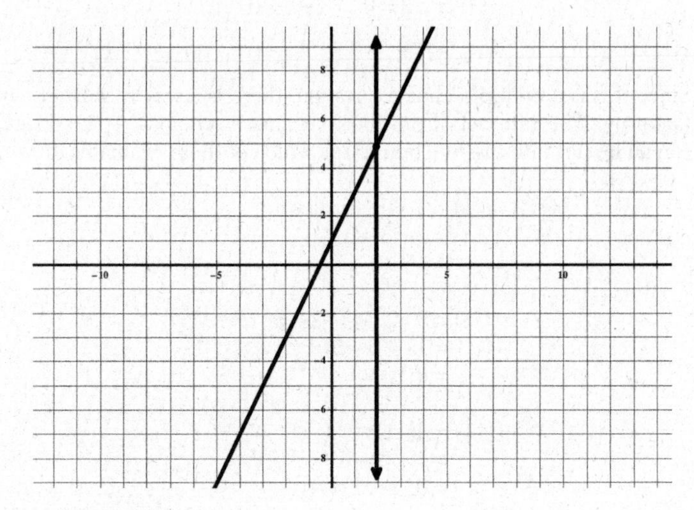

This is a function.

The correct choice is **(2)**.

12. To evaluate $f\left(\frac{1}{2}\right)$, substitute $\frac{1}{2}$ for both occurrences of x in the function definition.

$$f\left(\frac{1}{2}\right)=\frac{\sqrt{2\left(\frac{1}{2}\right)+3}}{6\left(\frac{1}{2}\right)-5}$$

$$=\frac{\sqrt{1+3}}{3-5}=\frac{\sqrt{4}}{-2}=\frac{2}{-2}=-1$$

The correct choice is **(3)**.

13. The zeros of a function are the input values that make the function evaluate to zero. To find the zeros of this function, find the solutions to the equation $0 = 3x^2 - 3x - 6$. This equation is most quickly solved by factoring.

$$0=3x^2-3x-6$$
$$0=3(x^2-x-2)$$
$$0=3(x-2)(x+1)$$
$$x-2=0 \text{ or } x+1=0$$
$$x=2 \text{ or } x=-1$$

Since this is a multiple-choice question, there is a way to solve it without factoring. The numbers from the four answer choices, 1, 2, –1, and –2, can all be put into the function to see which of them evaluates to zero.

$$\begin{aligned} f(1) &= 3\cdot 1^2 - 3\cdot 1 - 6 \\ &= 3 - 3 - 6 \\ &= -6 \end{aligned}$$

$$\begin{aligned} f(2) &= 3\cdot 2^2 - 3\cdot 2 - 6 \\ &= 12 - 6 - 6 \\ &= 0 \end{aligned}$$

$$\begin{aligned} f(-1) &= 3(-1)^2 - 3(-1) - 6 \\ &= 3 + 3 - 6 \\ &= 0 \end{aligned}$$

$$\begin{aligned} f(-2) &= 3(-2)^2 - 3(-2) - 6 \\ &= 12 + 6 - 6 \\ &= 12 \end{aligned}$$

Since $f(2) = 0$ and $f(-1) = 0$, the zeros of the function are 2 and -1.

The correct choice is **(4)**.

14. A sequence with a first term equal to 10 and a common difference of 4 has the numbers 10, 14, 18, 22, ... The recursively defined function must have $f(1) = 10$ and $f(2) = 14$. Only choices (1) and (3) have $f(1) = 10$. Calculate $f(2)$ for those two choices.

Testing choice (1):

$$\begin{aligned} f(2) &= f(2-1) + 4 \\ &= f(1) + 4 \\ &= 10 + 4 \\ &= 14 \end{aligned}$$

This works.

Testing choice (3):

$$\begin{aligned} f(2) &= 4f(2-1) \\ &= 4f(1) \\ &= 4(10) \\ &= 40 \end{aligned}$$

This does not work.

The correct choice is **(1)**.

15. The average rate of change is calculated by dividing the change in temperature by the change in time.

Testing choice (1):

$$\begin{aligned} &\frac{700-200}{1-0} \\ &= \frac{500}{1} \\ &= 500 \end{aligned}$$

Testing choice (2):

$$\begin{aligned} &\frac{900-700}{1.5-1} \\ &= \frac{200}{0.5} \\ &= 400 \end{aligned}$$

Testing choice (3):

$$\frac{1640-1300}{5-2.5}$$
$$=\frac{340}{2.5}$$
$$=136$$

Testing choice (4):

$$\frac{1800-1640}{8-5}$$
$$=\frac{160}{3}$$
$$\approx 53.33$$

This question can also be answered without doing any calculations since the average rate of change is related to the slope of the line segments joining the endpoints of the interval. Since the line segment connecting (0, 200) to (1, 700) is the steepest of the four intervals, the rate of change on that interval is the greatest.

The correct choice is **(1)**.

16. A piecewise function behaves differently for different input values. In this example, the graph should look like $y = |x|$ for x-values less than 1 and $y = \sqrt{x}$ for x-values greater than or equal to 1.

The graph of $y = |x|$ looks like this:

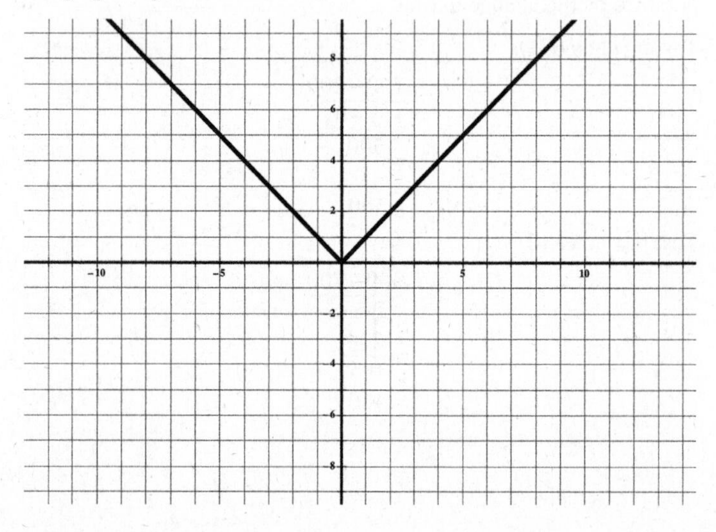

The graph of $y = \sqrt{x}$ looks like this:

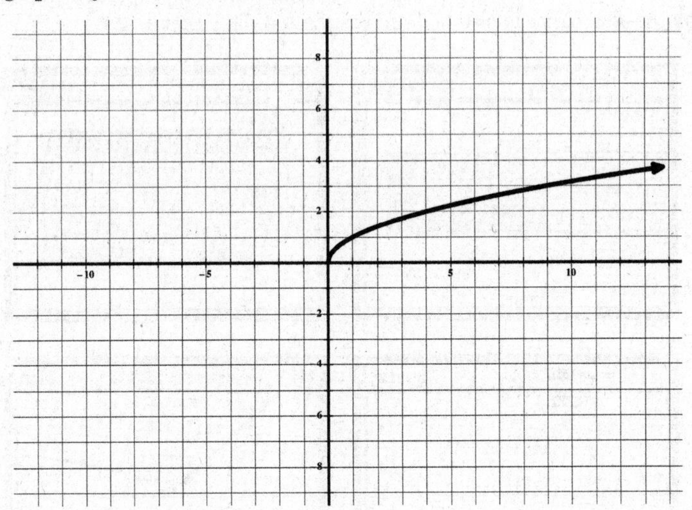

Taking the portion of $y = |x|$ to the left of $x = 1$ and the portion of $y = \sqrt{x}$ to the right of $x = 1$ looks like this:

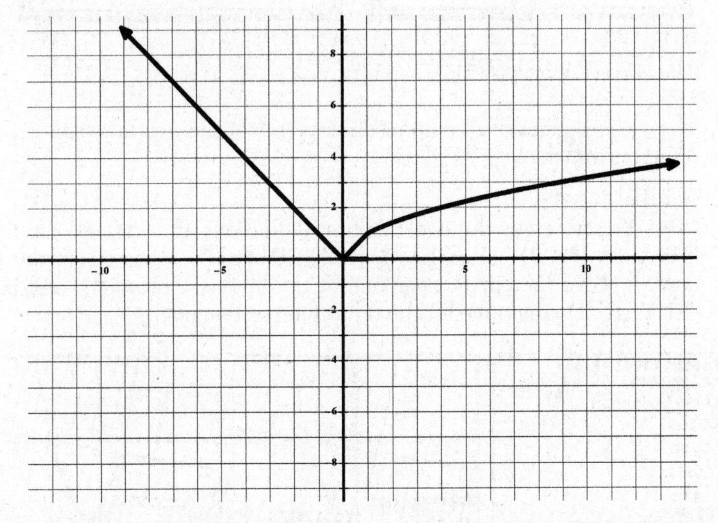

On the TI-Nspire calculator, piecewise functions can be graphed. From the Graph Scratchpad in the entry line, press the [TEMPLATE] key next to the [9]. Select the piecewise template on the top row, third from the right.

In the "Number of function pieces" field, put 2. To enter the function pieces with the conditions, press [ctrl] and [=] to select the appropriate symbol. The absolute value bars can also be found in the template menu.

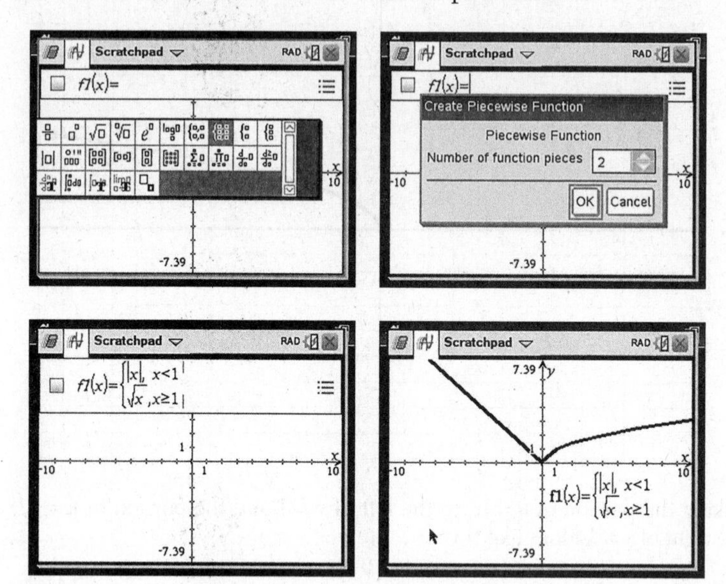

The correct choice is **(2)**.

17. The x-coordinates of the intersection points of the two functions are the numbers for which $f(x) = g(x)$.

For the TI-84:
Press Y= and enter the two function definitions after Y1 and Y2. Press [GRAPH] [2ND] [TRACE] [5] [ENTER] [ENTER] to select the two curves. Move the cursor near one of the intersection points, and press [ENTER]. Do the same for the other intersection points.

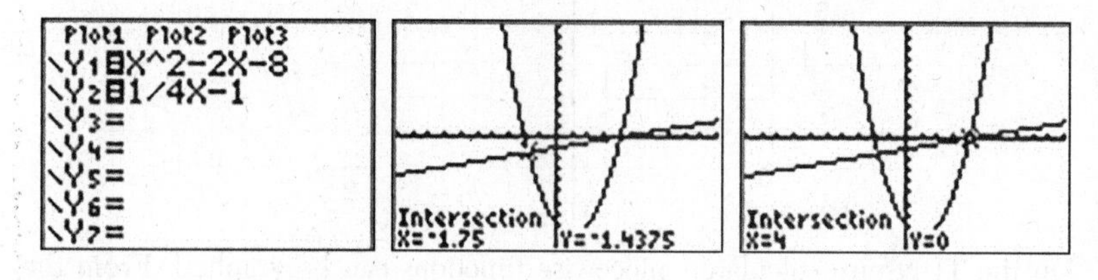

For the TI-Nspire:
From the Graph Scratchpad, enter the two functions into the entry line and press [ENTER]. Press [MENU] [6] [4]. Move the cursor to the left of an intersection point, and press [CLICK]. Move the cursor to the right of the same intersection point, and press [CLICK] again.

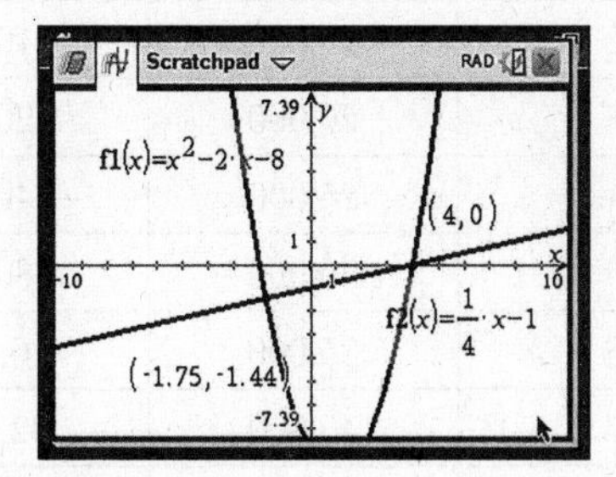

Since this is a multiple-choice question, it can also be solved by substituting the numbers from each choice into both functions to see if they evaluate to the same number.

Testing choice (2):

$$f(-1.75) = (-1.75)^2 - 2(-1.75) - 8$$
$$= -1.4375$$
$$g(-1.75) = \frac{1}{4}(-1.75) - 1$$
$$= -1.4375$$

$$f(4) = 4^2 - 2(4) - 8$$
$$= 0$$
$$g(4) = \frac{1}{4}(4) - 1$$
$$= 0$$

Only for choice (2) is it true that $f(x) = g(x)$ for both numbers.

The correct choice is **(2)**.

18. The quickest way to solve this question is to create a chart:

Month	Company A	Company B
1	10,000	500
2	15,000	1000
3	20,000	2000
4	25,000	4000
5	30,000	8000
6	35,000	16,000
7	40,000	32,000
8	45,000	64,000

Month 8 is the first time that company B's payment exceeds company A's payment.

The correct choice is **(3)**.

19. Use the graphing calculator. The mean for team *A* is 6, and the team's standard deviation is approximately 3.16. The mean for team *B* is 6.85, and the team's standard deviation is approximately 3.06. So team *A* has a mean less than the mean of team *B*. Team *A* has a standard deviation greater than the standard deviation of team *B*.

For the TI-84:
Press [STAT] [1] to enter the data. Go to the home page. Then press [STAT] [RIGHT] [1] to see the statistics.

L1 L2 L3 2
L2(1)=

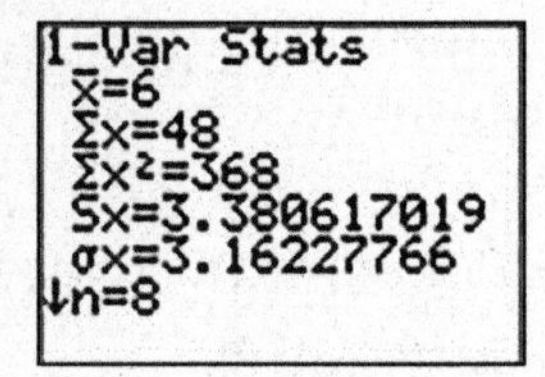

For the TI-Nspire:
From the home screen, select "Add Lists & Spreadsheet to: New Document." Enter the data into the A column starting at A1. Press [MENU] [4] [1] [1]. Enter 1 into the "Num of Lists" field. Enter a[] into the X1 List field. Press the "OK" button.

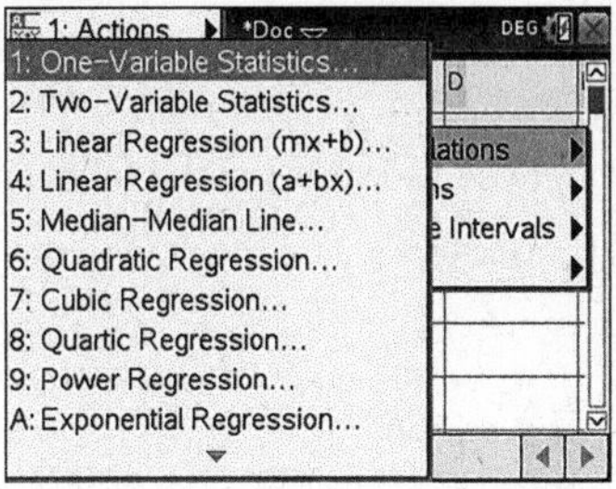

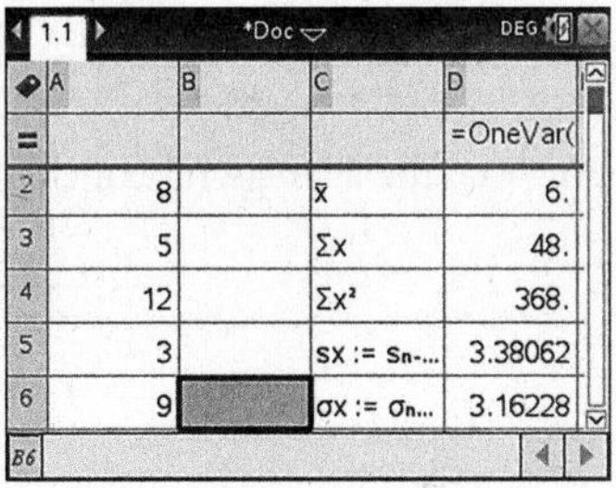

The correct choice is **(1)**.

20. To complete the square for quadratic function $f(x) = x^2 + bx + c$, first rewrite it as

$$f(x) = x^2 + bx + \left(\frac{b}{2}\right)^2 - \left(\frac{b}{2}\right)^2 + c$$

and combine the last two terms.

$$f(x) = x^2 - 12x + 6^2 - 6^2 + 7$$
$$f(x) = x^2 - 12x + 36 - 36 + 7$$
$$f(x) = x^2 - 12x + 36 - 29$$

Then factor the first three terms using the perfect square trinomial pattern.

$$f(x) = (x - 6)^2 - 29$$

The a-value for this equation is 6. Had the question said $f(x) = (x + a)^2 + b$, then the a-value would have been –6.

The correct choice is **(1)**.

21. Test each choice to determine the answer.

Testing choice (1):
For g(x), the average rate of change between –1 and 1 is the change in the function g(1) – g(–1) divided by the change in the x-value 1 – (–1).

$$\frac{g(1)-g(-1)}{1-(-1)}=\frac{(-1^2-1+6)-[-(-1)^2-(-1)+6]}{2}$$
$$=\frac{4-6}{2}$$
$$=\frac{-2}{2}$$
$$=-1$$

For $n(x)$, the average rate of change between –1 and 1 is the following:

$$\frac{n(1)-n(-1)}{1-(-1)}=\frac{9-5}{2}$$
$$=\frac{4}{2}$$
$$=2$$

The average rate of change for $n(x)$ is greater than the average rate of change for $g(x)$ between –1 and 1, so choice (1) is not correct.

Testing choice (2):
The y-intercept of $g(x)$ is $g(0) = -0^2 - 0 + 6 = 6$. The y-intercept of $n(x)$ is $n(0) = 8$. The y-intercept of $n(x)$ is greater than the y-intercept of $g(x)$, so choice (2) is not correct.

Testing choice (3):
The maximum value of $n(x)$ is 9 at $x = 1$. The maximum value of $g(x)$ can be found with a graphing calculator. Graph the function $f(x) = -x^2 - x + 6$. Then for the TI-84, press [2ND] [TRACE] [4] to find the maximum point. For the TI-Nspire press [MENU] [6] [3] to find the maximum point.

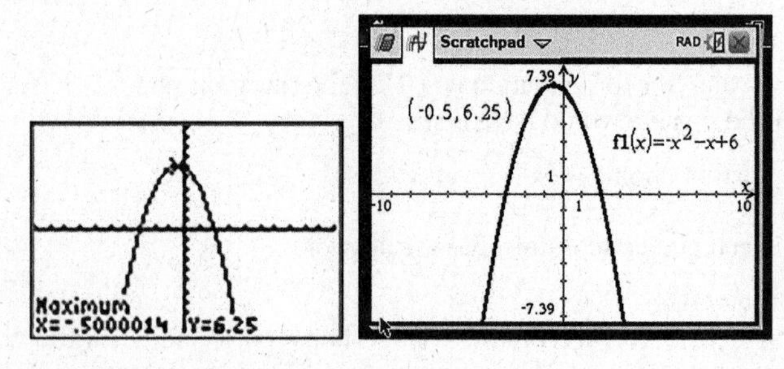

Without the graphing calculator, the maximum point of a quadratic can be calculated. The x-coordinate is $x = \frac{-b}{2a} = \frac{-(-1)}{2(-1)} = -0.5$. The y-coordinate is $g(-0.5) = 6.25$.

The maximum point is at $(-\frac{1}{2}, 6.25)$. So the maximum value is 6.25, which is less than 9. The maximum value of $n(x)$ is greater than the maximum value of $g(x)$, so choice (3) is not correct.

Testing choice (4):
The roots of a function are the input values that make that function evaluate to zero. For $n(x)$ since both $n(-2)$ and $n(4)$ equal 0, the roots of $n(x)$ are -2 and 4. For $g(x)$, the roots can be found by solving the equation $0 = -x^2 - x + 6$.

$$
\begin{aligned}
0 &= -x^2 - x + 6 \\
0 &= -1(x^2 + x - 6) \\
0 &= -1(x + 3)(x - 2) \\
x + 3 = 0 &\text{ or } x - 2 = 0 \\
x = -3 &\text{ or } x = 2
\end{aligned}
$$

The roots of $g(x)$ are -3 and 2.

The sum of the roots of $n(x)$ is $-2 + 4 = 2$. The sum of the roots of $g(x)$ is $-3 + 2 = -1$. So the sum of the roots of $n(x)$ is greater than the sum of the roots of $g(x)$.

Another way to test the choices is to make the graphs of $g(x)$ and $n(x)$ on the same set of axes.

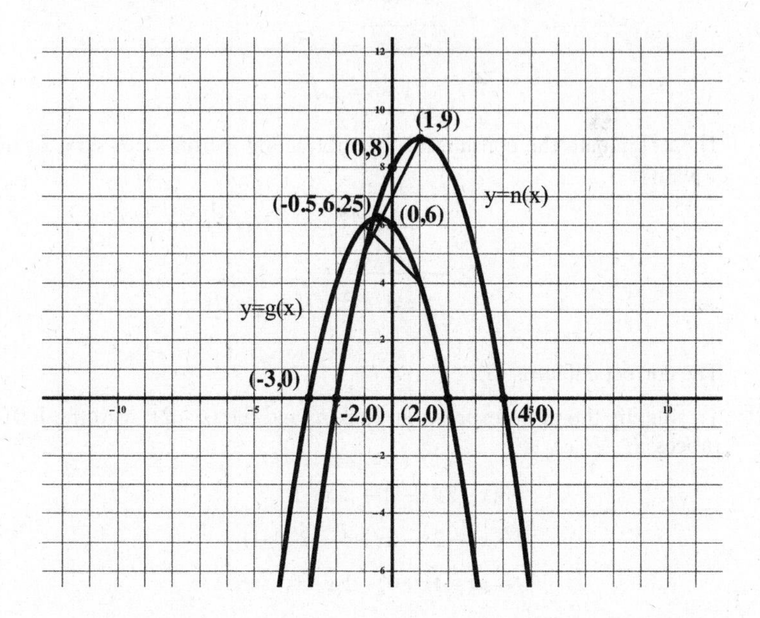

The y-intercept of $g(x)$ is less than the y-intercept of $n(x)$. The maximum value of $g(x)$ is less than the maximum value of $n(x)$. The roots correspond to the x-intercepts. The rate of change between $x = -1$ and $x = 1$ can be compared by looking at the slopes of the line segments between $x = -1$ and $x = 1$ for both parabolas. Since the line segment between $x = -1$ and $x = 1$ for $n(x)$ goes up from left to right, the slope or rate of change is positive. Since the line segment between $x = -1$ and $x = 1$ for $g(x)$ goes down from left to right, the slope or rate of change is negative. A positive rate of change is greater than any negative rate of change.

The correct choice is **(4)**.

22. The sum of two rational numbers is always rational. Since 4 and 9 are both perfect squares, $\frac{1}{\sqrt{4}} = \frac{1}{2}$ and $\frac{1}{\sqrt{9}} = \frac{1}{3}$, which are both rational numbers. So the sum $\frac{1}{2} + \frac{1}{3} = \frac{5}{6}$ is also rational.

    The correct choice is **(2)**.

23. To solve an equation in this form, first eliminate the exponent by taking the square root of both sides of the equation:

$$(x+3)^2 = 7$$
$$\sqrt{(x+3)^2} = \pm\sqrt{7}$$
$$x+3 = \pm\sqrt{7}$$

Then eliminate the constant, 3, by subtracting it from both sides of the equation:

$$\begin{array}{r} x+3=\pm\sqrt{7} \\ -3=-3 \\ \hline x=-3\pm\sqrt{7} \end{array}$$

The correct choice is **(3)**.

24. To simplify the given expression, first expand the $(x - 2)^2$ with the FOIL process:

$$3(x-2)(x-2)-2(x-1)$$
$$3(x^2-2x-2x+4)-2(x-1)$$
$$3(x^2-4x+4)-2(x-1)$$

Now, distribute the 3 through the parentheses on the left:

$$3x^2 - 12x + 12 - 2(x - 1)$$

Now, distribute the –2 through the parentheses on the right:

$$3x^2 - 12x + 12 - 2x + 2$$

Notice that the last term is +2, not –2, because (–2)(–1) = + 2.

Finally, combine like terms:

$$3x^2 - 12x - 2x + 12 + 2$$

$$3x^2 - 14x + 14$$

An alternative way to compare the given expression to each of the answer choices is to compare the graph of $y = 3(x - 2)^2 - 2(x - 1)$ to the graphs of $y = 3x^2 - 2x - 10$, $y = 3x^2 - 2x - 14$, $y = 3x^2 - 14x + 10$, and $y = 3x^2 - 14x + 14$.

Since the graph of $y = 3(x - 2)^2 - 2(x - 1)$ is the same as the graph of $y = 3x^2 - 14x + 14$, those expressions are equivalent.

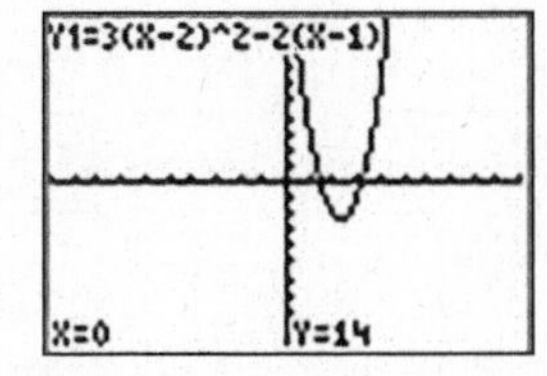

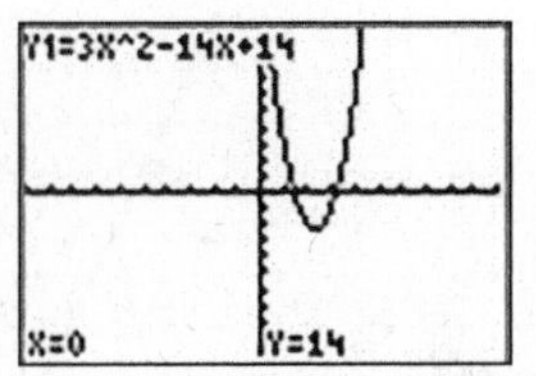

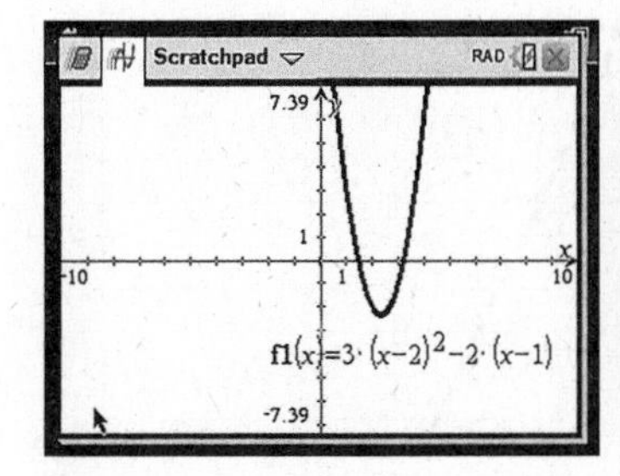

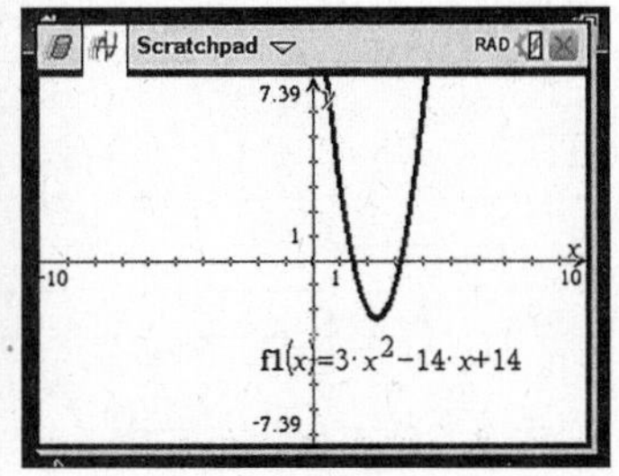

The correct choice is **(4)**.

PART II

25. Since this is an arithmetic sequence, the formula for arithmetic sequences given in the reference sheet can be used: $a_n = a_1 + (n - 1)d$, where d is the difference between two consecutive terms. This example uses function notation. So the formula is $h(n) = h(1) + (n - 1)d$, where $h(1) = 3.0$ and $d = 1.5$.

    One valid solution is $h(n) = 3.0 + (n - 1)1.5$.

26. The graph of a linear inequality is the shading of all the points on one side of a boundary line. To graph the boundary line, change the > into an = and rewrite in slope-intercept form, $y = mx + b$:

$$\begin{array}{r} 2x + y = 1 \\ -2x = -2x \\ \hline y = -2x + 1 \end{array}$$

Graph this boundary line as a dotted line since the points on the line are not part of the solution. If the inequality symbol had been a ≥ symbol, then the boundary line would need to be graphed as a solid line.

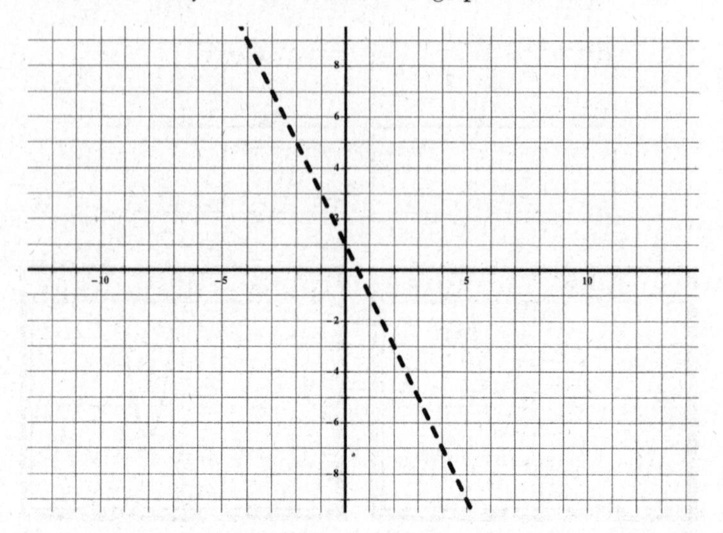

To determine which side of the line to shade, test to see whether or not (0, 0) is part of the solution set. Do this by substituting 0 for both x and y into the original inequality and see if this point makes the inequality true.

$$2(0) + 0 > 1$$
$$0 > 1$$

Since this is not true, the side of the line that does not contain (0, 0) should be shaded.

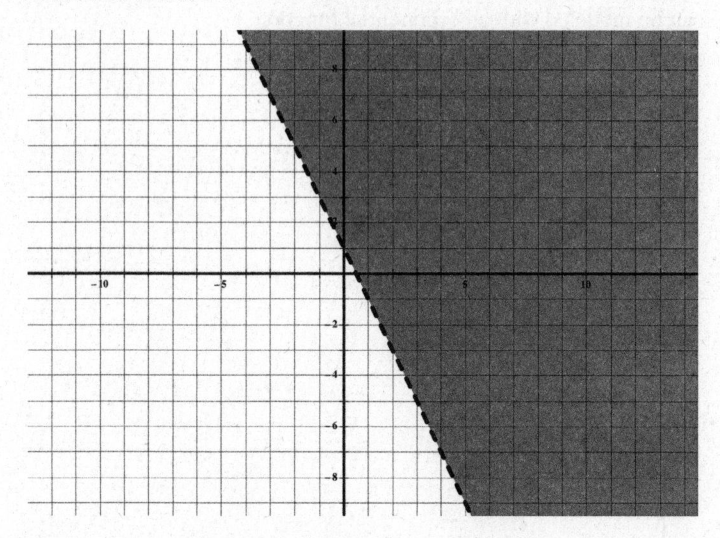

On the TI-Nspire, the inequality can be graphed by rewriting it into slope-intercept form and replacing the = with a >.

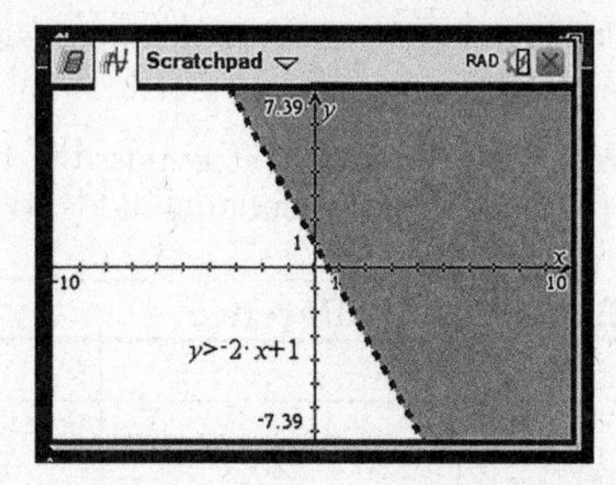

On the TI-84, you have to decide which side to shade by testing (0, 0) and then setting the appropriate shading on the Y= menu by moving the cursor to the left of the Y1 on the \ and pressing [ENTER] twice. The TI-84 will not graph the boundary line as a dotted line.

```
Plot1 Plot2 Plot3
Y1 = -2X+1
\Y2=
\Y3=
\Y4=
\Y5=
\Y6=
\Y7=
```

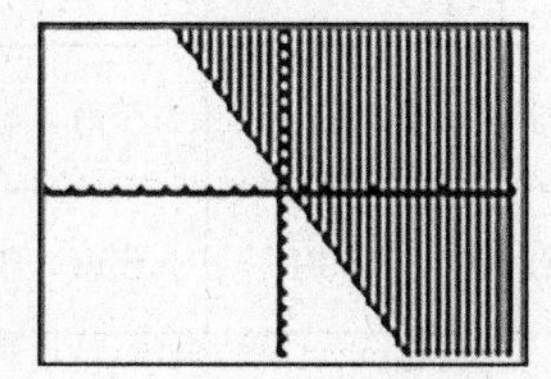

27. If the shape of the graph of $B(x)$ resembles a straight line, it can be modeled with a linear function. If the shape of the graph of $B(x)$ is a curve, it can be modeled with an exponential function.

The graph of the 10 given points looks like this:

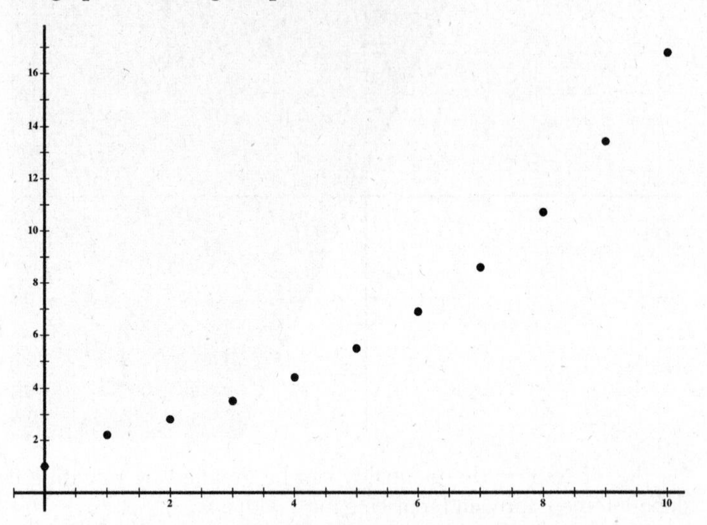

Since the graph looks more like a curve than a straight line, use an exponential model.

Even without producing the graph, it is possible to use a chart to see whether a linear function or an exponential model is more appropriate.

x	$B(x)$	Difference	Ratio
1	220		
2	280	$280 - 220 = 60$	$\frac{280}{220} \approx 1.3$
3	350	$350 - 280 = 70$	$\frac{350}{280} \approx 1.3$
4	440	$440 - 350 = 90$	$\frac{440}{350} \approx 1.3$
5	550	$550 - 440 = 110$	$\frac{550}{440} \approx 1.3$
6	690	$690 - 550 = 140$	$\frac{690}{550} \approx 1.3$

x	B(x)	Difference	Ratio
7	860	$860 - 690 = 170$	$\frac{860}{690} \approx 1.2$
8	1070	$1070 - 860 = 210$	$\frac{1070}{860} \approx 1.2$
9	1340	$1340 - 1070 = 270$	$\frac{1340}{1070} \approx 1.3$
10	1680	$1680 - 1340 = 340$	$\frac{1680}{1340} \approx 1.3$

If the differences in column 3 were all approximately the same number, a linear model would be more accurate. Since the ratios in column 4 are all approximately the same number, an exponential model is more accurate.

28. At the beginning of the trip, the car has not traveled any distance. This corresponds to the point (0, 0). After 2 hours, the car has traveled $2 \cdot 60 = 120$ miles. This corresponds to the point (2, 120). Since the car does not move for the next 30 minutes, it is still at 120 miles from the starting point 2.5 hours after starting. This corresponds to the point (2.5, 120). The car then travels 30 miles per hour for the next hour for a total of 30 more miles. The point corresponding to the end of the trip, then, is (3.5, 150). When these four points are joined with line segments, the graph looks like this:

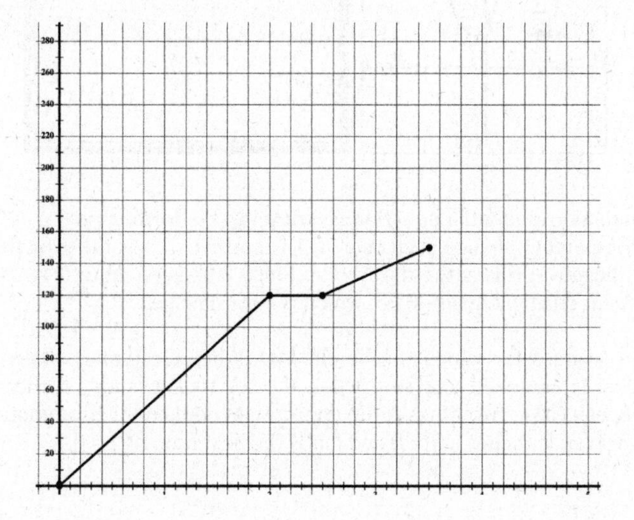

29. The solutions to a quadratic equation can be calculated with the quadratic formula found on the reference sheet, $x = \frac{-b \pm \sqrt{b^2 - 4ac}}{2a}$, where a, b, and c are the coefficients when the equation is in the form $ax^2 + bx + c = 0$. For this question, $a = 1$, $b = -2$, and $c = 5$.

    Use the quadratic formula,

$$x = \frac{-(-2) \pm \sqrt{(-2)^2 - 4(1)(5)}}{2(1)}$$
$$= \frac{2 \pm \sqrt{4 - 20}}{2}$$
$$= \frac{2 \pm \sqrt{-16}}{2}$$

    Since there is a negative number inside the square root sign, there are no real solutions to the equation.

    Another way to solve this is to make a graph of $y = x^2 - 2x + 5$ and see if it has any x-intercepts. Since there are no x-intercepts, the equation $x^2 - 2x + 5 = 0$ has no real solutions.

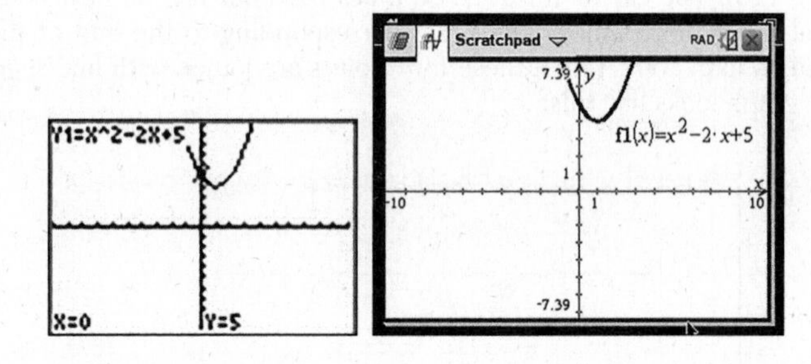

30. When an exponential equation is written in the form $y = a(1 + r)^x$, then r is the percent change each year. If r is positive, this is the growth rate. If r is negative, this is the decay rate. Since 0.95 is equal to $(1 - 0.05)$, the percent change equals -0.05, which is a 5% change.

31. The t-values for which the height $h(t)$ is measured start when the toy rocket is launched and stop when the toy rocket lands on the ground. Between those two times is the appropriate domain for this function. The toy rocket launches at $t = 0$ and lands the next time $h(t) = 0$.

Solve the equation $0 = -16t^2 + 64t$ to find when the toy rocket lands:

$$0 = -16t^2 + 64t$$
$$0 = -16t(t - 4)$$

$$-16t = 0 \text{ or } t - 4 = 0$$
$$t = 0 \text{ or } t = 4$$

So the toy rocket lands after 4 seconds, making the domain $0 \le t \le 4$.

Graphing the function on the graphing calculator also reveals that the toy rocket is launched at $t = 0$ and lands at $t = 4$.

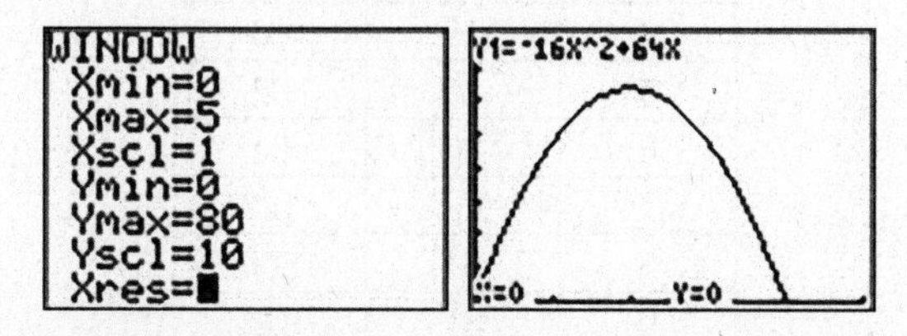

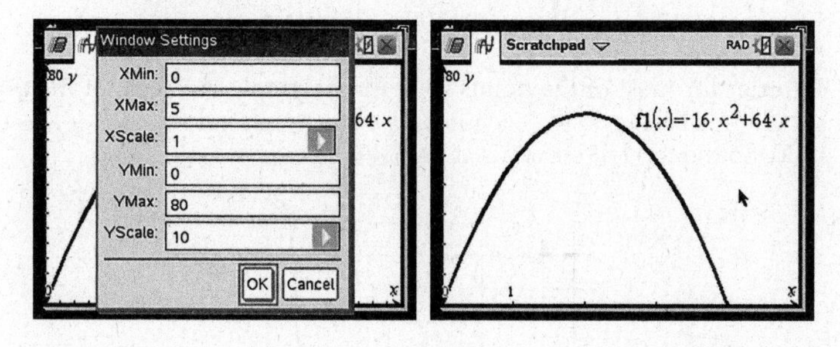

32. The number of minutes Jackson spends on the treadmill can be written as an arithmetic sequence 30, 32, 34, 36, 38, . . . where the first term is 30 and the common difference is 2.

    For the equation, the formula for the *n*th term of an arithmetic sequence provided on the reference sheet can be used, $a_n = a_1 + (n - 1)d$, where a_1 is the first term and d is the common difference.

    For this question, the function uses d instead of n for the term number.

    The function is $T(d) = 30 + 2(d - 1)$, which can be simplified to $T(d) = 2d + 28$.

    For the second part, substitute 6 for d into the expression:

$$T(6) = 30 + 2(6 - 1)$$
$$= 30 + 2 \cdot 5$$
$$= 40$$

PART III

33. For $x = 0$ to $x = 10$, make a chart for $f(x)$ and $g(x)$.

x	$f(x) = x^2$	$g(x) = 2^x$
0	0	1
1	1	2
2	4	4
3	9	8
4	16	16
5	25	32
6	36	64
7	49	128
8	64	256
9	81	512
10	100	1024

In order for most of the points to fit on the graph, the scale of the y-axis should be at least 30 for each box. For the graph below, each y-axis unit is 30. So the top left corner is at (0, 600).

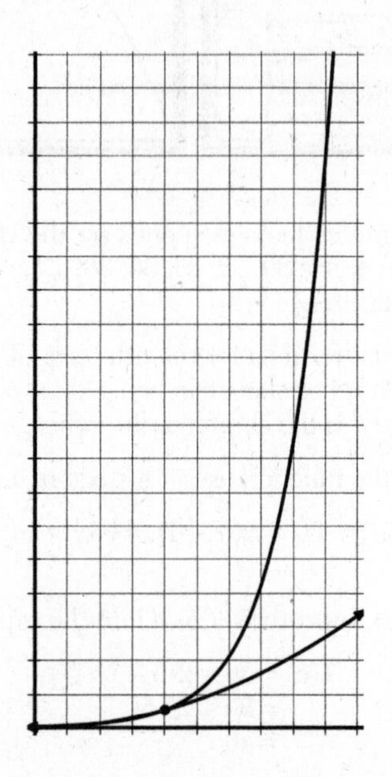

According to the graph it seems that for any x-value greater than 4 the graph of $g(x)$ is above the graph of $f(x)$. So for $x = 20$, $g(x) > f(x)$.

To be sure, you can calculate $f(20) = 20^2 = 400$ and $g(20) = 2^{20} = 1{,}048{,}576$. So $g(20)$ is much greater than $f(20)$.

34. To solve the inequality, first simplify both sides:

$$7x - 3(4x - 8) \leq 6x + 12 - 9x$$
$$7x - 12x + 24 \leq -3x + 12$$
$$-5x + 24 \leq -3x + 12$$

Be careful in the second step to write +24 since $-3(-8) = +24$.

Add $3x$ to both sides of the inequality and subtract 24 from both sides of the inequality:

$$\begin{array}{r} -5x + 24 \leq -3x + 12 \\ \underline{+3x = +3x} \\ -2x + 24 \leq 12 \\ \underline{-24 = -24} \\ -2x \leq -12 \end{array}$$

Divide both sides of the inequality by -2. When both sides of an inequality are multiplied by or divided by a negative, the direction of the inequality sign must be reversed to keep the inequality true:

$$\frac{-2x}{-2} \geq \frac{-12}{-2}$$
$$x \geq 6$$

The solution set is $x \geq 6$.

The numbers in this solution set that are also integers between 4 and 8 are just 6, 7, and 8.

35. To isolate the r-term in the equation $V = \pi r^2 h$, first treat the other variables, V and h, as if they were constants:

$$V = \pi r^2 h$$
$$\frac{V}{\pi h} = \frac{\pi r^2 h}{\pi h}$$
$$\frac{V}{\pi h} = r^2$$

Finish by taking the square root of both sides:

$$\sqrt{\frac{V}{\pi h}} = \sqrt{r^2}$$
$$\sqrt{\frac{V}{\pi h}} = r$$

Since the radius must be a positive number, the ± symbol is not necessary.

The formula for r can be used to calculate the radius of the can. Since the diameter is double the radius, multiply r by 2:

$$\begin{aligned} r &= \sqrt{\frac{V}{\pi h}} \\ &= \sqrt{\frac{66}{\pi \cdot 3.3}} \\ &\approx 2.52 \end{aligned}$$

$$\begin{aligned} d &= 2r \\ &= 2 \cdot 2.52 \\ &= 5 \text{ inches} \end{aligned}$$

36. The line of best fit can be found with a graphing calculator.

    For the TI-84:
    Turn "diagnostics" on by pressing [2ND] [0] and scroll down to DiagnosticOn and press [ENTER] [ENTER]. Press [STAT] [1]. Clear the contents of L1 and L2 by moving the cursor to the top row and pressing [CLEAR] [ENTER] for the first two columns. Input the numbers 0, 1, 2, 4, and 6 into L1. Enter 8.3, 8.5, 8.5, 8.8, and 9.3 into L2. Press [STAT] [RIGHT]. Press [4] for LinReg($ax + b$), and then press [ENTER].

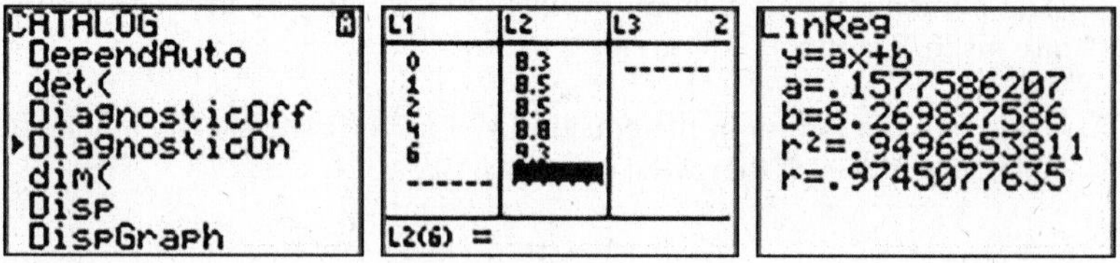

CATALOG
DependAuto
det(
DiagnosticOff
▸DiagnosticOn
dim(
Disp
DispGraph

L1	L2	L3
0	8.3	------
1	8.5	
2	8.5	
4	8.8	
6	9.3	

L2(6) =

LinReg
y=ax+b
a=.1577586207
b=8.269827586
r²=.9496653811
r=.9745077635

For the TI-Nspire:
From the home screen go to “Add Lists & Spreadsheet to: New Document.” In cells A1 to A5, input the numbers 0, 1, 2, 4, and 6. In cells B1 to B5, enter the numbers 8.3, 8.5, 8.5, 8.8, and 9.3. Press [menu] [4] [1] [3] for Linear Regression ($mx + b$). In the X List field, enter “a[ ]” (use [ctrl] [(] to create the square brackets). In the Y List field enter “b[ ].” Press the OK button.

The equation with values rounded to the nearest hundredth is

$$y = 0.16x + 8.27.$$

The correlation coefficient, r, rounded to the nearest hundredth is 0.97. Since this is very close to 1, there is a strong association.

PART IV

37. The diagram below shows the picture with the frame. The dimensions of the picture are 6 by 8, while the width of the frame is the unknown, denoted by x.

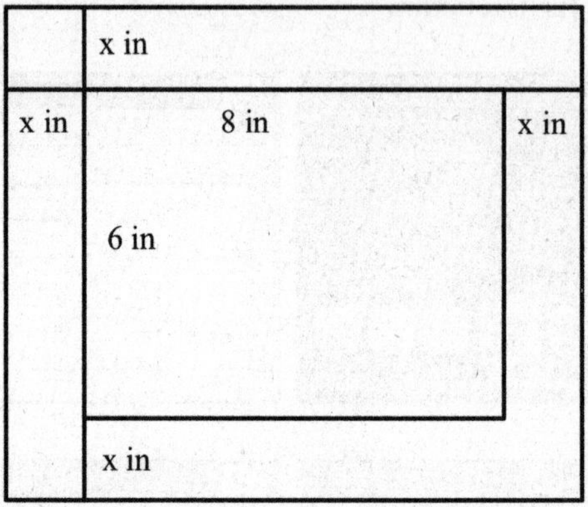

x in

x in 8 in x in

6 in

x in

The width of the large rectangle is $2x + 6$, and the length of the large rectangle is $2x + 8$. Since the area of the large rectangle, which includes the frame, must be less than or equal to 100 and the area of the large rectangle is $(2x + 6)(2x + 8)$, the inequality is $(2x + 6)(2x + 8) \leq 100$.

The maximum width for the frame occurs when the area is exactly 100. So the equation $(2x + 6)(2x + 8) = 100$ can be used to find the maximum width:

$$\begin{aligned}
(2x \cdot 2x) + (2x \cdot 8) + (6 \cdot 2x) + (6 \cdot 8) &= 100 \\
4x^2 + 16x + 12x + 48 &= 100 \\
4x^2 + 28x + 48 &= 100 \\
-100 &= -100 \\
\hline
4x^2 + 28x - 52 &= 0 \\
4(x^2 + 7x - 13) &= 0
\end{aligned}$$

The quadratic equation $x^2 + 7x - 13 = 0$ can be solved with the quadratic formula where $a = 1$, $b = 7$, and $c = -13$:

$$x = \frac{-7 \pm \sqrt{7^2 - 4(1)(-13)}}{2(1)}$$
$$= \frac{-7 \pm \sqrt{49+52}}{2}$$
$$= \frac{-7 \pm \sqrt{101}}{2}$$
$$= 1.5 \text{ inches}$$

The other solution, $\frac{-7-\sqrt{101}}{2} \approx -8.5$, should be rejected since the width of the frame must be a positive number of inches.

To check that the answer is correct, see if the area of the large rectangle is close to 100. If the frame has a width of 1.5, the width of the rectangle is $2 \cdot 1.5 + 6 = 9$. Then the length of the rectangle is $2 \cdot 1.5 + 8 = 11$. The area is $9 \cdot 11 = 99$ square inches, which is close to 100 square inches.

Topic	Question Numbers	Number of Points	Your Points	Your Percentage
1. Polynomials	9, 24	2 + 2 = 4		
2. Properties of Algebra	35	4		
3. Functions	4, 11, 12	2 + 2 + 2 = 6		
4. Creating and Interpreting Equations	2, 3, 8, 15	2 + 2 + 2 +2 = 8		
5. Inequalities	5, 6, 26, 34	2 + 2 + 2 + 4 = 10		
6. Sequences and Series	14, 18, 25, 32	2 + 2 + 2 + 2 = 8		
7. Systems of Equations	10	2		
8. Quadratic Equations and Factoring	13, 17, 20, 23, 29, 31, 37	2 + 2 + 2 + 2 + 2 + 2 + 6 = 18		
9. Regression	36	4		
10. Exponential Equations	7, 27, 30, 33	2 + 2 + 2 + 4 = 10		
11. Graphing	1, 16, 21, 28	2 + 2 + 2 + 2 = 8		
12. Statistics	19	2		
13. Number Properties	22	2		

HOW TO CONVERT YOUR RAW SCORE TO YOUR ALGEBRA I REGENTS EXAMINATION SCORE

The accompanying conversion chart must be used to determine your final score on the August 2015 Regents Examination in Algebra I. To find your final exam score, locate in the column labeled "Raw Score" the total number of points you scored out of a possible 86 points. Since partial credit is allowed in Parts II, III, and IV of the test, you may need to approximate the credit you would receive for a solution that is not completely correct. Then locate in the adjacent column to the right the scale score that corresponds to your raw score. The scale score is your final Algebra I Regents Examination score.

Regents Examination in Algebra I—August 2015 Chart for Converting Total Test Raw Scores to Final Examination Scores (Scaled Scores)

Raw Score	Scale Score	Performance Level	Raw Score	Scale Score	Performance Level	Raw Score	Scale Score	Performance Level
86	100	5	57	75	4	28	64	2
85	99	5	56	75	4	27	63	2
84	97	5	55	75	4	26	62	2
83	96	5	54	74	4	25	61	2
82	96	5	53	73	3	24	60	2
81	93	5	52	73	3	23	59	2
80	92	5	51	73	3	22	58	2
79	91	5	50	72	3	21	57	2
78	90	5	49	72	3	20	55	2
77	89	5	48	72	3	19	54	1
76	87	5	47	72	3	18	52	1
75	86	5	46	71	3	17	51	1
74	85	5	45	71	3	16	49	1
73	84	4	44	71	3	15	47	1
72	83	4	43	71	3	14	45	1
71	83	4	42	70	3	13	43	1
70	82	4	41	70	3	12	41	1
69	81	4	40	70	3	11	38	1
68	80	4	39	69	3	10	36	1
67	80	4	38	69	3	9	33	1
66	79	4	37	69	3	8	30	1
65	78	4	36	68	3	7	27	1
64	78	4	35	68	3	6	24	1
63	77	4	34	67	3	5	21	1
62	77	4	33	67	3	4	17	1
61	76	4	32	66	3	3	13	1
60	76	4	31	66	3	2	9	1
59	76	4	30	65	3	1	5	1
58	75	4	29	64	2	0	0	1

NOTES

NOTES